Klaus Denecke

Algebra and Discrete Mathematics for Economics

Also of Interest

Linear Algebra
Vector and Inner Product Spaces
Saurabh Chandra Maury, 2025
ISBN 978-3-11-151570-0, e-ISBN (PDF) 978-3-11-151603-5

Discrete Mathematics
Combinatorics, Counting, Proofs, Recurrences, Solutions
Walter Hower, 2025
ISBN 978-3-11-120643-1, e-ISBN (PDF) 978-3-11-120689-9

Algebras of Unbounded Operators
Algebraic and Topological Aspects of Murray–von Neumann Algebras
Aleksey Ber, Vladimir Chilin, Galina Levitina, Fedor Sukochev, Dmitriy Zanin, 2025
ISBN 978-3-11-159791-1, e-ISBN (PDF) 978-3-11-159968-7

Abstract Algebra
With Applications to Galois Theory, Algebraic Geometry, Representation Theory and Cryptography
Gerhard Rosenberger, Annika Schürenberg, Leonard Wienke, 2024
ISBN 978-3-11-113951-7, e-ISBN (PDF) 978-3-11-114252-4

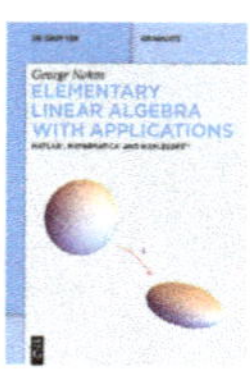

Elementary Linear Algebra with Applications
MATLAB®, Mathematica® and Maplesoft™
George Nakos, 2024
ISBN 978-3-11-133179-9, e-ISBN (PDF) 978-3-11-133185-0

Elements of Discrete Mathematics
Numbers and Counting, Groups, Graphs, Orders and Lattices
Volker Diekert, Manfred Kufleitner, Gerhard Rosenberger, Ulrich Hertrampf, 2024
ISBN 978-3-11-106069-9, e-ISBN (PDF) 978-3-11-106255-6

Klaus Denecke

Algebra and Discrete Mathematics for Economics

Numbers, Graphs and Matrices in Input-Output Economics

DE GRUYTER

Mathematics Subject Classification 2020
Primary: 08A99, 05C90, 15A30; Secondary: 91B02, 91B42, 91B38, 01A73

Author
Klaus Denecke, Professor emeritus
Institut für Mathematik
Universität Potsdam
Karl-Liebknecht Str. 24-25
D-14476 Potsdam
Deutschland

ISBN 978-3-11-223442-6
ISBN 978-3-11-223443-3 (PDF)
ISBN 978-3-11-223444-0 (E-PUB)
DOI https://doi.org/10.1515/9783112234433

Library of Congress Control Number: 2026933096

Bibliographic information published by the Deutsche Nationalbibliothek
The Deutsche Nationalbibliothek lists this publication in the Deutsche Nationalbibliografie; detailed bibliographic data are available on the Internet at https://dnb.dnb.de.

De Gruyter and Walter de Gruyter GmbH are part of De Gruyter Brill.
www.degruyterbrill.com

Questions about General Product Safety Regulation: productsafety@degruyterbrill.com

Cover illustration: Klaus Denecke

There is no royal road to science, and only those who do not dread the fatiguing climb of its steep paths have a chance of gaining its luminous summits

(Karl Marx)

IN MEMORIAM HANS-JÜRGEN HOEHNKE (1925–2007)

Contents

Introduction — XI

1 General basic concepts — 1
1.1 A survey on number systems — **2**
1.2 Basic concepts of logic — **6**
1.3 Basic concepts of set theory — **12**
1.4 Equivalence relations — **17**
1.5 Partial order relations — **20**
1.6 Partial order relations and efficiency of economic activities — **23**
1.7 Functions — **26**
1.8 Historical remarks — **33**

2 Commodities and money – a relational model — 35
2.1 Use and exchange values of a commodity — **36**
2.2 Elementary and expanded form of the value of a commodity — **36**
2.3 The main theorem for the exchange of commodities — **37**
2.4 Historical remarks — **42**

3 Numbers — 43
3.1 Natural numbers — **44**
3.2 Representation of natural numbers — **51**
3.3 Nonnegative rational numbers — **52**
3.4 Representation of nonnegative rational numbers — **60**
3.5 Integers — **65**
3.6 Elementary number theory — **70**
3.7 Rational numbers — **86**
3.8 Historical remarks — **88**

4 Linear diophantine equations and consumer theory — 91
4.1 The consumer problem — **92**
4.2 The budget equation — **92**
4.3 The cases $n = 2$ and $n = 3$ — **93**
4.4 Consumer's preferences — **97**
4.5 Perfect substitutes and perfect complements — **98**
4.6 Preferences and utility functions — **98**
4.7 Historical remarks — **100**

5 Graphs — 103
5.1 Basic concepts — **104**
5.2 Eulerian and Hamiltonian graphs — **109**

5.3 Trees — 111
5.4 Bipartite graphs — 113
5.5 Planar graphs — 115
5.6 Coloring of graphs — 118
5.7 Historical remarks — 120

6 Matrices — 121
6.1 Calculation with matrices — 122
6.2 Vector spaces — 125
6.3 Regular matrices and determinants — 136
6.4 Systems of linear equations — 147
6.5 Matrix equations — 151
6.6 Historical remarks — 157

7 Universal algebra — 159
7.1 The concept of an algebra — 160
7.2 Examples — 161
7.3 Subalgebras — 164
7.4 Homomorphisms — 167
7.5 Direct and subdirect products — 174
7.6 Terms, trees, and term operations — 179
7.7 Identities and models — 182
7.8 Fully invariant congruence relations — 185
7.9 Relatively free algebras — 186
7.10 Varieties — 187
7.11 Generalizations — 190
7.12 Historical remarks — 198

8 An algebraic approach to production systems — 201
8.1 Introduction — 201
8.2 Input–output graphs and matrices — 202
8.3 Activities and techniques — 209
8.4 The area of restrictions — 218
8.5 One-based, one-step techniques — 223
8.6 Multi-stage techniques — 228
8.7 Cyclic techniques and their matrices — 237
8.8 Historical remarks — 241

9 Graphs and matrices in material management — 245
9.1 Operations with matrices — 246
9.2 Material requirement planning (MRP) — 246
9.3 The gozintograph – an example — 247

9.4 More than two levels — **253**
9.5 Across levels — **255**
9.6 Calculation for fixed amounts of raw materials — **259**
9.7 Historical remarks — **260**

10 The use of matrices for internal service billings — 261
10.1 Methods of internal service billings – an example — **262**
10.2 General approach — **264**
10.3 A special case — **265**
10.4 Historical remarks — **267**

11 Matrices in economic systems — 269
11.1 Leontief's theorem — **270**
11.2 Algebraic properties of $\mathcal{M}_{(n,n)}(\mathbb{Q}^+)$ — **277**
11.3 Historical remarks — **279**

12 Formal concept analysis in economics — 281
12.1 Complete lattices and Galois connections — **282**
12.2 Concept lattices — **283**
12.3 The concept lattice of the main legal forms of a company — **284**
12.4 Historical remarks — **286**

Bibliography — 289

Index — 291

Introduction

Different from data in natural sciences, data in economical sciences are results of human activities. Therefore, sets of data which are considered in economics are finite sets of rational numbers, mostly positive. Functions which are considered in economics are discrete. So, one should think that methods from algebra and discrete mathematics are important in economics. But usually in textbooks on mathematics for economics, methods from calculus, differential equations, and topology dominate those from algebra and discrete mathematics.

At first, we will present some examples which show that even for basic economics concepts, methods from algebra and discrete mathematics can be important. Commodities are the fundamentals of the capitalist mode of production. At the beginning of the historical development of the "commodities-exchange-societies," x units of commodity A were exchanged for y units of commodity B. In an expanded form,

$$x \text{ units of } A \sim \begin{cases} y_1 & \text{units of } B_1, \\ \text{or } y_2 & \text{units of } B_2, \\ \vdots & \\ \text{or } y_n & \text{units of } B_n. \end{cases}$$

Step by step, money became the universal exchange equivalent for any commodity. Commodities are exchanged on a ware market since they have some value. The value of a commodity is twofold. Commodities have use and exchange value. The use value expresses the value of concrete labor which is embodied in a commodity. But for the exchange of commodities on a market, the exchange value is important. This value is determined by the socially necessary labor embodied in the commodity. This idea, going back to K. Marx' book "Capital," is used to define a function lab and to apply an equivalence relation, the kernel of function lab, to give a mathematical definition of the concept "value of a commodity." Then the well-known diagram lemma is used for a characterization of the exchange relation.

How will a rational consumer make consumption decisions? This is the main question in consumer theory. Being a rational consumer means respecting the budget restriction and asking for the usefulness of the bundle $(x_1, \dots, x_n)$ of commodities the consumer wants to buy. We assume that in an equilibrium between supply and demand the consumer has to pay the prices $p_1, \dots, p_n$ for the commodities $x_1, \dots, x_n$. If the budget is given by m then the linear equation

$$p_1x_1 + \cdots + p_nx_n = m$$

is called the *budget equation*. Here $p_1, \dots, p_n, m$ are nonnegative rational numbers, i. e., fractions $p_i = \frac{a_i}{b_i}$, $m = \frac{m_1}{m_2}$, $m_1, m_2, a_i, b_i \in \mathbb{N}, m_2, b_i \neq 0$. Multiplying by the least com-

https://doi.org/10.1515/9783112234433-203

mon multiple $\operatorname{lcm}(b_1, \ldots, b_n, m_2)$, one gets a linear equation with natural numbers as coefficients,

$$p'_1 x_1 + \cdots + p'_n x_n = m',$$

i. e., a linear diophantine equation which is solvable with n-tuples $(x_1, \ldots, x_n)$ of integers if and only if the greatest common divisor $\gcd(p'_1, \ldots, p'_n)$ divides m'. The solutions are easy to calculate by repeated use of the Euclidean algorithm. Consumer preferences can be expressed by a utility function u, and if the consumer is indifferent between two bundles $(x_1, \ldots, x_n)$ and $(y_1, \ldots, y_n)$ then $u(x_1, \ldots, x_n) = u(y_1, \ldots, y_n)$. This means that the pair of bundles belongs to the kernel of function u. With such an equivalence relation, the kernel of function u defines a partition of the set of all n-tuples of integers. In economics textbooks for students, usually only the case $n = 2$ is considered, but for arbitrary real coefficients p_1, p_2 and real solutions x_1, x_2. The set of indifferences will be considered as a curve in the two-dimensional Euclidean space (see Chapter 1, Example 1.2.5). Students have to prove that each point of the two-dimensional space is on such a curve and that two different curves of indifferences have an empty intersection. But this is a property of a partition.

Nowadays globalization connects the economies of different countries. Goods and services are exchanged worldwide. Input–output analysis uses the advantage of the relatively stable flow of goods and services not only between different national economies, but also between different branches of the economy of one country or between different departments of one factory. The essential principles of input–output analysis can easily be described by considering an input–output table. In his book "Input–Output Economics" from 1986 [45], the "father" of such input–output tables, the American economist Wassily W. Leontief develops the resulting theory behind. His work was honored by the Nobel Price for Economics in 1973. To the input–output table there corresponds a system of linear equations. Any such system of linear equations can be written as only one equation consisting of matrices whose entries are nonnegative rational numbers. The flow itself can be described by a directed edge-valued graph where the sets of values consist of positive rational numbers. Therefore numbers, especially rational and in particular positive rational, graphs, and matrices play an important role in the input–output analysis. Knowledge from algebra and discrete mathematics, especially of numbers, algebraic structures, graphs, and linear algebra can give a better understanding of the input–output economics. This knowledge will be applied to several situations in economics.

The input–output model is traditionally used to describe production processes. Inputs can be hours of human labor, machine hours, capital, raw materials, energy, water, semifinished products, etc. They will be transformed into outputs. The main outputs are the products which shall be produced. These processes can be described by a particular class of graphs which we call input–output graphs and which are in the simplest cases bipartite graphs. To each input–output graph there corresponds a particular matrix which we call the input–output matrix. There is a one-to-one correspondence between the set of input–output graphs and the set of all input–output matrices. On each set, we define a binary partial operation. Both partial operations satisfy the associative law as a weak identity. So, we obtain two partial semigroups and show that they are weakly isomorphic. This result is important to model multiple-step production processes.

The second economics problem to be considered is material requirement planning. Material means raw substances, subassemblies, parts, other supplies, and any industrial services required for product manufacturing. Material requirement planning helps ensure that the right inventory is available for the production process exactly when it is needed in quantities which are needed and at the lowest possible cost. We show how directed edge-valued graphs can help us understand complex production procedures. Addition and multiplication of matrices allow calculating the quantities of various raw materials which are needed to satisfy a certain demand of final products. We show also that alternatively one can use input–output tables to start the calculation.

The third area to be considered is the internal cost allocation of a business. Also in this case, systems of linear and matrix equations help solve economics problems.

In all these situations, we can set up input–output tables, from which systems of linear equations can be derived, corresponding to matrix equations of the form $A \cdot X = b$ where A is a matrix with nonnegative rational numbers as entries. Then we ask: Is there a necessary and sufficient condition for the matrix A which leads for every demand b to an economy equilibrium? The answer is given by Leontief's theorem. We give a proof of this theorem which generalizes the three economics situations described before. So, the mathematical problem is under which conditions for any A and b with nonnegative rational entries there exists an X with nonnegative rational entries such that $A \cdot X = b$ is satisfied.

The last economics applications form the contents of Chapter 11. Meanwhile, in Chapter 1 we introduce the basic mathematical concepts which are used in the whole text: numbers, sets, functions, and relations, especially equivalence and partial order relations. In this chapter we formulate and prove the main theorem for equivalence relations and the diagram lemma for functions. These topics will be applied in Chapter 2 to define the value of a commodity and to prove the main theorem for the exchange of commodities. In Chapter 3 we define natural numbers as cardinalities of finite sets and as Peano structures, integers as equivalence classes of differences of natural numbers, rational numbers as equivalence classes of quotients of integers, as well as positive rational numbers as equivalence classes of quotients of natural numbers. In connection with the properties of integers, we introduce the most important concepts and results of elementary number theory, especially calculation with remainder classes and the solution of linear diophantine equations. The latter topics will be applied in Chapter 4 to solve the consumer problem. The basic arithmetic laws satisfied by the different sets of numbers motivate us to consider algebraic structures, especially groups, rings, and fields, but also semigroups, semirings, and semifields. Later we define lattices and take a look at the generalization of special algebraic structures, namely universal algebras, partial algebras, and many-sorted algebras. In Chapter 5 we define graphs, directed graphs, edge-valued graphs, bipartite and multiple-partite graphs. As input–output graphs we will use the latter classes of graphs to describe important situations in economics. In Chapter 7 on linear algebra, vector spaces are introduced as two-sorted algebras. We prove the main theorems on the solution of systems of linear equations using matrices and determinants. We apply graph theory and linear algebra to production systems in Chapter 8, to material management problems in Chapter 9, and to internal service billings in Chapter 10. In Chapter 11 we formulate and prove Leontief's theorem. In Chapter 12 we come back to Chapter 2 where we used equivalence relations to define the concept a "value of a commodity." We give a short introduction to formal concept analysis and apply these ideas to build the concept lattice called "the main legal forms of a company."

There are many good textbooks on mathematical methods in economics (see e. g. [56] or [60]). Our goal is to present basic knowledge from algebra and discrete mathematics with applications in economics. The text should be interesting for economists and for mathematicians who are interested in applications. Each chapter ends with "historical remarks" to express that the logic and history are intertwined and that mathematics is much more than a collection of problems, "hard to understand and difficult to solve by the students."

The text is based on lectures which the author gave at the University of Potsdam.

I would like to thank editor and staff of De Gruyter Brill publishing house, especially Ute Skambraks and Kristina Grigorjeva for their kindness and help.

1 General basic concepts

In this chapter we will take the first steps into the fascinating world of mathematics. The earliest mathematical texts available are from Mesopotamia and Egypt and are more than 4000 years old. Mathematics started with counting, which needs numbers.

At first, we will give a survey on number systems, operations with numbers, and the properties of these operations. We will speak about sets and operations defined on sets. Then we will introduce the important concepts of a function and a relation. To understand the way in which mathematicians create, formulate, and prove new results, we need some understanding of logic.

https://doi.org/10.1515/9783112234433-001

1.1 A survey on number systems

For the systems of the natural, integer, rational, nonnegative rational, real, and complex numbers, we use the symbols $\mathbb{N}, \mathbb{Z}, \mathbb{Q}, \mathbb{Q}^+, \mathbb{R}$, and $\mathbb{C}$, respectively. Everybody learnt in school how to calculate with numbers. Our aim is to clear up the mathematical, especially the algebraic, background of calculations with numbers. These are the number systems in which we are interested:

Natural numbers

The system $\mathbb{N}$ consists of the numbers $0, 1, 2, \ldots$ Using the concept of a set and considering concrete sets of objects in the real world, small children learn the first numbers $1, 2, 3, 4, 5, 6$ by a process of abstraction. For addition, they form the union of sets, and to understand which number is greater or smaller than another one, they compare cardinalities of concrete sets. Greater natural numbers and calculations with them are understood by continued counting. In Chapter 2 we will introduce mathematical models for both ways.

Integer numbers

The system $\mathbb{Z}$ consists of $\ldots, -3, -2, -1, 0, 1, 2, 3, \ldots$; $\mathbb{Z}$ will be introduced as the union of $\mathbb{N}$ and the set of all numbers opposite to those in $\mathbb{N}$. Another way to come to the integers goes via the formation of so-called *differences*.

Nonnegative rational numbers

The nonnegative rational numbers $\mathbb{Q}^+$ are the *fractions*

$$\frac{p}{q}, \quad \text{where } p, q \in \mathbb{N}, q \neq 0.$$

Any natural number n can be written as a fraction $\frac{n}{1}$.

Rational numbers

The rational numbers are the fractions of the form

$$\frac{p}{q}, \quad \text{where } p, q \in \mathbb{Z}, q \neq 0.$$

Rational numbers can be also written as *decimal fractions* of the form

$$a = a_0.a_1a_2 \ldots a_l\overline{b_1b_2 \ldots b_k} \quad \text{with } a_0 \in \mathbb{Z}, a_i, b_j, \in \mathbb{N}, 0 \leq a_i, b_j \leq 9,$$
$$i = 1, \ldots, l, j = 1, \ldots, k.$$

Such decimal fractions are said to be *periodic* and the sequence $b_1, b_2, \ldots, b_k$ of digits is said to be a *period*. Finite decimal fractions can be written as periodic ones with period 0. Later on we will see that, for reasons of uniqueness, we have to exclude periodic decimal fractions with period 9.

Real numbers

The system $\mathbb{R}$ is the set of all infinite decimal fractions

$$a = a_0.a_1a_2\ldots.$$

Every real number corresponds uniquely to a point of the number line and, conversely, every point of the number line corresponds uniquely to a real number. The rational numbers are not enough to fill in the number line. The Pythagoreans knew already nearly 2000 years ago that there exist points of the number line which do not correspond to rational numbers, for instance, the point in between the rational numbers 1.4 and 1.5 which corresponds to $\sqrt{2}$. The set of real numbers consists of the subset of rational numbers and the subset of irrational numbers, which cannot be written as fractions $\frac{p}{q}$, with $p, q \in \mathbb{Z}$, $q \neq 0$, or equivalently, which can be written as nonperiodic infinite decimal fractions.

Complex numbers

There are equations which have no solution in $\mathbb{R}$. A simple example is the equation $x^2 = -1$. Since the square of a real number is always greater than or equal to 0, there is no $x \in \mathbb{R}$ which solves this equation. Complex numbers can be written as $x = a + b \cdot i$, where $a, b \in \mathbb{R}$ and $i^2 = -1$, with $i \notin \mathbb{R}$. Complex numbers $a + b \cdot i$ correspond uniquely to the pairs (a, b) of real numbers which can be represented as points in a plane containing a coordinate system.

Clearly, $\mathbb{N}$ is contained in $\mathbb{Z}$ and also in $\mathbb{Q}^+$, which is contained in $\mathbb{Q}$. Each rational number is also a real number, and each real number is a complex number, since $x \in \mathbb{R}$ can be written as complex number $x = a + 0 \cdot i$. These relationships are pictured in Figure 1.1.

A number system is not only a set, it is also characterized by the operations which can be defined on the set of numbers chosen. If the operations of addition and multiplication are defined on a set of numbers, then this means that to arbitrary pairs (a, b) of elements from the set there is a uniquely assigned sum $a + b$, respectively a product $a \cdot b$. Subtraction and division are the inverse operations to addition and multiplication, respectively. If one of the summands (factors) and the sum (product) are given, we ask for the second summand (factor). We also want to compare two numbers and define an order relation on a number system. If addition and multiplication are compatible with the order relation, i. e., if addition and multiplication preserve the order relation, we obtain the property of *monotony*.

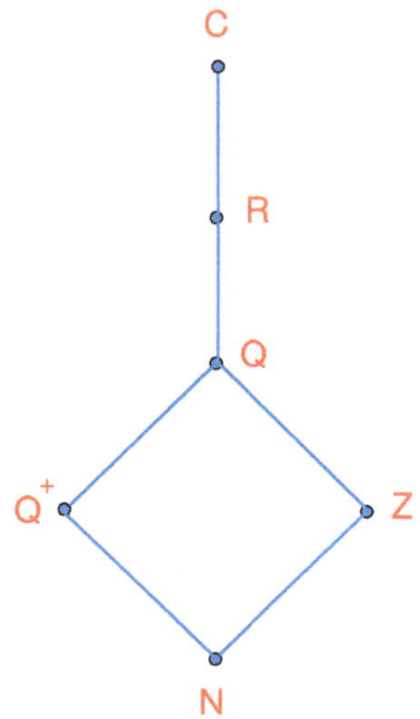

Figure 1.1: Number systems.

In $\mathbb{N}$, $\mathbb{Z}$, $\mathbb{Q}^+$, $\mathbb{Q}$, $\mathbb{R}$, and $\mathbb{C}$, the following operations are always defined:

- in $\mathbb{N}$, addition and multiplication;
- in $\mathbb{Z}$, addition, multiplication, and subtraction;
- in $\mathbb{Q}^+$, addition, multiplication, and division, except for division by 0;
- in $\mathbb{Q}$, addition, multiplication, subtraction, and division, except for division by 0;
- in $\mathbb{R}$, addition, multiplication, subtraction, and division, except for division by 0; moreover, each equation of the form $x^n - a = 0$, $a \in \mathbb{R}$, $a \geq 0$, $n \in \mathbb{N} \setminus \{0\}$ is solvable in $\mathbb{R}$, i. e., has a solution $x \in \mathbb{R}$;
- in $\mathbb{C}$, addition, multiplication, subtraction, and division, except for division by 0; moreover, each equation of the form $x^n - a = 0$, $a \in \mathbb{R}$, $n \in \mathbb{N} \setminus \{0\}$ is solvable in $\mathbb{C}$, i. e., has a solution $x \in \mathbb{C}$.

Finally, a number system is not only characterized by its elements, operations, and order relations, but also by the laws which are valid in this number system.

For addition and multiplication, the following laws are satisfied:

Let $A \in \{\mathbb{N}, \mathbb{Z}, \mathbb{Q}^+, \mathbb{Q}, \mathbb{R}, \mathbb{C}\}$ be a number system. For all elements $a, b, c \in A$, there hold:

- the *additive associative law*

$$a + (b + c) = (a + b) + c;$$

- the *multiplicative associative law*

$$a \cdot (b \cdot c) = (a \cdot b) \cdot c;$$

- the *commutative laws* of the addition

$$a + b = b + a$$

 and multiplication

$$a \cdot b = b \cdot a;$$

- the *distributive laws*

$$(a + b) \cdot c = a \cdot c + b \cdot c \quad \text{and} \quad a \cdot (b + c) = a \cdot b + a \cdot c.$$

As usual, instead of $a \cdot b$ one also writes ab.

The *invertibility* of addition and multiplication is the problem to solve equations of the form

$$a + x = b \quad \text{and} \quad ax = b, \text{ respectively,}$$

and this is the question of whether for all $a, b \in A$ there exist elements $x \in A$ which solve the corresponding equation.

The real numbers (and therefore also the natural, nonnegative rational, rational, and integer numbers) can be ordered by the relations "*a is less than b*", denoted $a < b$, or "*a is less than or equal to b*", denoted $a \leq b$.

For all elements $a, b \in A$, the following properties are satisfied:

(1) $a \not< a$;
(2) $a < b$ and $b < c$ imply $a < c$;
(3) $a < b$ implies not $(b < a)$;
(4) $a \not< b$ and $b \not< a$ imply $a = b$.

For the relation "*a* is less than or equal to *b*" ($a \leq b$), there hold:

(1) $a \leq b$ and $b \leq a$ imply $a = b$;
(2) For all $a, b \in A$, one has $a \leq b$ or $b \leq a$.

These relations are compatible with addition and multiplication:

(1) For all $a, b, c \in A$, from $a < b$ follows $a + c < b + c$;
(2) From $a < b$ and $c \lesseqgtr 0$ follows $ac \gtreqless bc$. Here $\lesseqgtr$ means either $<$ or $>$, respectively and the meaning of $\gtreqless$ is clear.

All these laws are also valid for the relation $\leq$.

Numbers in economics are used to measure raw materials and other inputs, or results of production processes, industries of a country, or profit given in a specific currency. All these numbers are natural numbers, fractions, or finite decimal fractions, i. e., nonnegative rational numbers. Sometimes, for instance, if results are decreasing, one uses negative rational numbers. This means that in economics the rational number system is of main interest. This is quite different from the use of numbers in natural sciences which describe processes in nature and can be not only discrete, but also continuous and even infinite.

1.2 Basic concepts of logic

Statements describe the facts of reality or of our thinking. Statements can be written or spoken sentences of a natural language, formulas, pictures, or can occur in other forms. Forgetting about the concrete contents of a statement, the only property of a statement which is of interest in logic is its truth: Is statement p true (T) or false (F)? The concept of a statement is a basis concept of logic, which can simply be described as follows:

Every statement is either true or false.

This means that there is no other truth value except for "true" or "false." Especially, there is no "half-true", there is no "perhaps", there is no "probably." The basic property of a statement expresses even more, and this is also contained in the formulation "either ...or," namely, there is no statement which is true and false (at this moment, at the same time). Sometimes, the first property of a statement is called the "excluded third," and the second one is called the "excluded contradiction." The basic property is used to decide whether a certain sentence is a statement or not. In principle one must be able to say whether something is true or false. Sometimes, we do not know this today, it is an open problem whether some statement p is true or false, but we are hopeful that one day we can solve the problem.

Using certain connecting words, given statements can be compounded into further statements. Such connecting words are: "and," "or," "if ...then," "if and only if," "not." If we want to express a property of any statement, we use statement variables: A, B, C, p, q, r, etc. In the following table, we list the most common modes to compose statements.

compound statement	**notation**	**symbol**
non A	negation	$\neg A$ (or $\sim A, -A, \bar{A}$)
A and B	conjunction	$A \wedge B$
A or B	alternative (or disjunction)	$A \vee B$
if A then B	implication	$A \Rightarrow B$
A if and only if B	equivalence	$A \Leftrightarrow B$

Remark 1.2.1.

(1) In the implication $A \Rightarrow B, A$ is said to be *the premise* (or presumption) and B *conclusion* (or assertion, also claim). We say if A then B, or A implies B, or A is sufficient for B, or B is necessary for A.
(2) If $A \Leftrightarrow B$, then A if and only if B, or A is sufficient and necessary for B.
(3) The symbols $\neg, \wedge, \vee, \Rightarrow, \Leftrightarrow$ are said to be the propositional functors.

A mode of statement composition will be said to be a *truth-functional* if the truth value of the compound statement is completely determined by the truth values of the components and not by their contents. This principle of abstraction is called *extensionality* and is the basis of the model of natural thinking going back to *Aristotle*. Each such mode of composition is exhaustively described by a *truth table* which gives the truth value of the compound statement for each assignment of truth values to the components:

A	B	$A \wedge B$	$A \vee B$	$A \Rightarrow B$	$A \Leftrightarrow B$
T	T	T	T	T	T
T	F	F	T	F	F
F	T	F	T	T	F
F	F	F	F	T	T

A	$\neg A$
W	F
F	W

The table tells us, for instance, that the conjunction is true if and only if both components are true, and the disjunction is false if and only if both components are false. The implication is false if and only if the first component is true and the second one is false, and the equivalence is the conjunction of $A \Rightarrow B$ and $B \Rightarrow A$. The negation is a unary truth function which is true if the only input is false and false if the input is true. Conjunction, disjunction, implication, and equivalence are binary truth functions since each has two inputs. When a truth-functional composition is applied iteratively to form a complex statement, we can always calculate the output for each possible input by a truth table. Sequences like $(A \Rightarrow B) \wedge (B \Rightarrow A)$ consisting of statement variables A, B, C, etc., and propositional functors $\wedge, \vee, \Rightarrow, \Leftrightarrow, \neg$ are said to be formulas of propositional logic if they are defined in the following inductive way:

(1) Every statement variable is a formula of propositional logic.
(2) If F, F_1, F_2 are formulas, then $\neg F, (F_1 \vee F_2), (F_1 \wedge F_2), (F_1 \Rightarrow F_2)$, and $(F_1 \Leftrightarrow F_2)$ are formulas of propositional logic.

If we set $T = 1, F = 0$, then truth functions are functions defined on the two-element set $\{0, 1\}$. To each formula of propositional logic there corresponds a uniquely determined truth-function, but to different formulas the same truth function can correspond, as the example $A \Rightarrow B$, $\neg B \Rightarrow \neg A$ shows. This observation leads to the following concept.

Definition 1.2.2. Let P, P' be two formulas of propositional logic. Then P is said to be proposition logical equivalent to P', written as $P \sim P'$, if the truth-functions corresponding to P and P' are equal.

We noticed already that

$$A \Rightarrow B \sim \neg B \Rightarrow \neg A.$$

For the proof that two formulas of propositional logic are equivalent, we simply compare the truth tables of the corresponding truth functions.

We remark that $\neg B \Rightarrow \neg A$ is said to be the *contraposition* of $A \Rightarrow B$.

If a formula is true (T) for every assignment of the occurring variables with T or F, it is called a *tautology*, and if it is always false (F), we have a contradiction.

Example 1.2.3. The statement $A \wedge \neg A$ is a contradiction, and $A \vee \neg A$ is a tautology.

It is easy to establish

Corollary 1.2.4. *The proposition logical formulas P and P' are proposition logical equivalent if and only if $P \Leftrightarrow P'$ is a tautology.*

The following formulas are important tautologies:

$$\begin{aligned} ((A \Rightarrow B) \wedge (B \Rightarrow A)) &\Leftrightarrow (A \Leftrightarrow B), \\ ((A \Rightarrow B) \wedge (B \Rightarrow C)) &\Rightarrow (A \Rightarrow C), \\ (A \Rightarrow B) &\Leftrightarrow (\neg B \Rightarrow \neg A), \\ \neg(A \wedge B) &\Leftrightarrow \neg A \vee \neg B, \\ \neg(A \vee B) &\Leftrightarrow \neg A \wedge \neg B. \end{aligned}$$

Tautologies and proposition logical equivalences are related with proof methods.

The truth table

A	B	$A \Rightarrow B$	$A \wedge (A \Rightarrow B)$	$(A \wedge (A \Rightarrow B)) \Rightarrow B$
T	T	T	T	T
T	F	F	F	T
F	T	T	F	T
F	F	T	F	T

shows that

$$((A \wedge (A \Rightarrow B)) \Rightarrow B)$$

is a tautology. If A is true and if the implication $A \Rightarrow B$ is also true, then B is true. This *rule of consequences* (also *rule of inferences*) is called *modus ponens* and is usually written in the form

$$\frac{A, A \Rightarrow B}{B}.$$

Another rule of consequences is

$$\frac{B, \neg A \Rightarrow \neg B}{A},$$

which is known as an *indirect proof*.

Finally, we mention another method to obtain proposition logical equivalences.

The *statement algebra* is based on the following equivalences which can easily be checked by the truth-table method:

$$\begin{aligned} A \vee B &\sim B \vee A, & A \wedge B &\sim B \wedge A, \\ A \vee (B \vee C) &\sim (A \vee B) \vee C, & A \wedge (B \wedge C) &\sim (A \wedge B) \wedge C, \\ A \vee A &\sim A, & A \wedge A &\sim A, \\ A \vee (A \wedge B) &\sim A, & A \wedge (A \vee B) &\sim A, \\ A \vee (B \wedge C) &\sim (A \vee B) \wedge (A \vee C), \\ A \wedge (B \vee C) &\sim (A \wedge B) \vee (A \wedge C). \end{aligned}$$

To get more proposition logical equivalences, we apply the usual methods to obtain new equations from given ones on the given equivalences.

We will give an example of an indirect proof from the theory of consumers.

Example 1.2.5. An indifference curve shows various combinations of two goods in various quantities, denoted by (x_1, x_2) where $x_1, x_2 \in \mathbb{Q}^+$ are the quantities of good 1 and good 2, respectively, which the consumer can choose. The pairs (x_1, x_2) correspond to points in a plane, and $x := (x_1, x_2)$, $y := (y_1, y_2)$ belong to the indifference curve I_1 if the consumer is indifferent with respect to the pairs x and y. We want to prove that the following proposition:

If the indifference curves I_1, I_2 are different (i. e., $I_1 \neq I_2$), then I_1 and I_2 do not intersect, i. e., I_1 and I_2 have no point in common.

Let statements A and B be as follows:

A: I_1 and I_2 do not intersect.

B: $I_1 \neq I_2$.

We want to prove that the implication $B \Rightarrow A$ is true and may assume that B is true, i. e., $I_1 \neq I_2$. If I_1 and I_2 do intersect, then there is a pair x which belongs to I_1 and also to I_2. Let y be an arbitrary point on I_2 and let z be any point on I_1. Then y and x have equal preference and x and z have equal preference. But then also y and z have equal preference and correspond to points which belong to I_1 and also to I_2. Since y and z were chosen as arbitrary points on I_2 and I_1, we get $I_1 = I_2$, which is the negation of B and so $\neg B \Rightarrow \neg A$ is true. Now applying the rule of indirect proof, we conclude that A is true, i. e., the different curves I_1, I_2 of preferences do not intersect (have no point in common).

The expressions $x < 3$, $x + y = 3$ are not statements. The basic property of a statement is not fulfilled because we cannot decide whether $x < 3$ or $x + y = 3$ are true or not. But if we substitute concrete numbers for x and y, then we obtain true or false statements. Therefore, it makes sense to say "there is an element $x \in \mathbb{N}$ such that after substitution we have a true or a false statement." For instance, for $x = 2, y = 1$, we get two statements which are true.

Definition 1.2.6. Let E be a set of objects, called a "basic area." The symbol x is said to be a *variable* on E if arbitrary elements from E can be substituted for x.

Definition 1.2.7. We say that H is a *statement form* on E if H satisfies the following properties:
1. H contains at least one variable on E (a free variable);
2. Substituting objects from E into H for all variables occurring in H gives statements.

The phrases "for all" and "there is" are said to be *quantifiers*. We will use the following symbols:

$$\begin{aligned} &\forall x \in E \quad \text{(universal quantifier)}, \\ &\exists x \in E \quad \text{(existential quantifier)}. \end{aligned}$$

Using quantifiers, the statement form $H(x)$ becomes one of the statements:

$$\begin{aligned} &\forall x \in E\ (H(x)) \quad \text{"for all } x \text{ from } E \text{ the statement } H(x) \text{ holds,"} \\ &\exists x \in E\ (H(x)) \quad \text{"there is an } x \text{ from } E \text{ for which } H(x) \text{ is true."} \end{aligned}$$

We will also use

$$\begin{aligned} &\exists!\, x \in E\ (H(x)) \quad \text{"there is at most one } x \text{ from } E \text{ for which } H(x) \text{ is true,"} \\ &\exists!!x \in E\ (H(x)) \quad \text{"there is exactly one } x \text{ from } E \text{ for which } H(x) \text{ is true."} \end{aligned}$$

The sentence "there is (at least) one x having property $H(x)$" means that there exists such an x, but possibly there are more. Thus, "there is an x" means more precisely that there is at least one x with the property $H(x)$. If there is at most one x having the property under consideration then the existence of the element x is not clear. But if an x exists having the corresponding property, then it is unique. "There exists exactly one x for which $H(x)$ is true" means there is at least one and at most one element x with $H(x)$ being true.

The main problem for quantified statements is to recognize logically equivalent statements. We speak of *predicate logical equivalences*.

At first, we investigate exchanging quantifiers.

Proposition 1.2.8. *Let E be a basic area for the variables $x, y, \ldots$ Then the following equivalences hold:*

(i) $\forall x\, \forall y\, (H(x,y)) \Leftrightarrow \forall y\, \forall x\, (H(x,y))$,
(ii) $\exists x\, \exists y\, (H(x,y)) \Leftrightarrow \exists y\, \exists x\, (H(x,y))$,
(iii) $\exists x\, \forall y\, (H(x,y)) \overset{\not\Leftarrow}{\Rightarrow} \forall y\, \exists x\, (H(x,y))$.

Proof. Claims (i) and (ii) are clear. For the proof of (iii), if there is a (fixed) x such that for all y the predicate form $H(x,y)$ is satisfied, then for all y (for every y) there is an x with $H(x,y)$ being true. But if for every y there exists an x (belonging to E) with $H(x,y)$ being true, then there is not necessarily an x, which is the same for all y, such that the statement $H(x,y)$ is satisfied. Thus the converse of (iii) is false. □

We consider the following example to (iii)

Example 1.2.9. There is an integer x (namely $x = 0$), which satisfies for every integer y the equation $y + x = y$. Then each integer y has (its own) integer x with $y + x = y$ (this is the integer 0). On the other hand, to each integer y there is an integer x ($x = -y$) with $y + x = 0$. This does not mean that there is an integer x, which is the opposite for all y.

One more problem is how to quantify conjunctions and disjunctions.

Proposition 1.2.10. *Let E be a basic area for variables $x, y, \ldots$ Then:*
(i) $\forall x\, (H_1(x) \wedge H_2(x)) \Leftrightarrow \forall x\, (H_1(x)) \wedge \forall x\, (H_2(x))$ *(since both quantifications on the right-hand side are independent, we can write* $\forall x\, (H_1(x)) \wedge \forall y\, (H_2(y))$*),*
(ii) $\exists x\, (H_1(x) \vee H_2(x)) \Leftrightarrow \exists x\, (H_1(x)) \vee \exists x\, (H_2(x))$,
(iii) $\forall x\, (H_1(x) \vee H_2(x)) \overset{\not\Rightarrow}{\Leftarrow} \forall x\, (H_1(x)) \vee \forall x\, (H_2(x))$.

Proof. Claims (i) and (ii) are clear. For the proof of (iii), we notice the following: if $H_1(x)$ is satisfied for all x or $H_2(x)$ is satisfied for all x, then $H_1(x) \vee H_2(x)$ is satisfied for all x. The converse is not true since $\forall x(H_1(x) \vee H_2(x))$ means that $H_1(x)$ is satisfied for some x, while $H_2(x)$ is satisfied for some others, but not that $H_1(x)$ holds for all x or $H_2(x)$ holds for all x. □

For a better understanding of (iii), we consider the following example:

Example 1.2.11. The following is true:

$$\forall x \in \mathbb{R}\left(x^2 \geq 2 \vee x^2 < 2\right),$$

but the following is not:

$$\forall x \in \mathbb{R}\left(x^2 \geq 2\right) \vee \forall x \in \mathbb{R}\left(x^2 < 2\right).$$

It does not matter whether we say that there is an x such that $H(x)$ is not satisfied or that is $H(x)$ satisfied for not all x. Both formulations are predicate logical equivalent. Here the negation of quantifiers occurs. Without proof, we state the following predicate logical equivalences.

Proposition 1.2.12.

(i) $\neg\forall x\,(H(x)) \Longleftrightarrow \exists x\,(\neg H(x))$,

(ii) $\neg\exists x\,(H(x)) \Longleftrightarrow \forall x\,(\neg H(x))$.

Predicate logical statements which are always true are called tautologies. The following example shows how to prove predicate logical tautologies.

Example 1.2.13. Show that the following predicate logical statement is a tautology:

$$\forall x\,\big(H_1(x) \Rightarrow H_2(x)\big) \Rightarrow \big(\forall x\,\big(H_1(x)\big) \Rightarrow \forall y\,\big(H_2(y)\big)\big). \qquad (*)$$

Step 1. For the proof, we use the truth table method to establish $A \Rightarrow B \sim \neg A \vee B$:

A	B	$A \Rightarrow B$	$\neg A \vee B$
T	T	T	T
T	F	F	F
F	T	T	T
F	F	T	T

Step 2. Using this equivalence, we can write (*) as follows:

$$\forall x\,\big(\neg H_1(x) \vee H_2(x)\big) \Rightarrow \big(\neg\forall x\,\big(H_1(x)\big) \vee \forall y\,\big(H_2(y)\big)\big)$$

and, moreover,

$$\forall x\,\big(\neg H_1(x) \vee H_2(x)\big) \Rightarrow \big(\exists x\,\big(\neg H_1(x)\big) \vee \forall y\,\big(H_2(y)\big)\big).$$

This implication is true, if the premise is false. Thus, we check all cases in which the premise is true. For these cases, we have to show that the conclusion is also true.

Step 2.1. If there is no x for which $\neg H_1(x)$ is true, then $H_2(x)$ has to be true for all x to obtain a true premise. Then the conclusion is also true and the implication is true.

Step 2.2. If there is an x for which $\neg H_1(x)$ is true, then the conclusion is true and so is the implication.

1.3 Basic concepts of set theory

Finite sets can be given by the finitely many elements which belong to the set, if these are not too many. Another way is a verbal description of the set using a natural language. To do that, one needs a defining property which exactly all elements of the set have in common and no other element of a given basic area has this property.

Example 1.3.1. The set of all real numbers whose square is smaller than 3 can be given by

$$\{x \in \mathbb{R} \mid x^2 < 3\}.$$

Here $H(x) =$ "$x^2 < 3$" is said to be *a generating propositional form*.

If $H(x)$ generates the set M, we will write

$$M = \{x \mid x \in E \text{ and } H(x)\}.$$

Here it is essential to give the basic area E for the variable x.

Axioms of set theory are propositions which are assumed to be true and from which other true statements of set theory can be derived using certain rules of inference. As examples we mention the following important axioms.

Axiom of comprehension. For any property (statement form) $H(x)$, where x is a variable from the basic area E, there exists exactly one set $\{x|H(x)\}$ consisting of all objects x from E, for which $H(x)$ is true.

Extensionality axiom. Any sets X and Y are equal if and only if X and Y contain the same elements.

Infinity axiom. There is at least one infinite set.

Axiom of choice. Let $\mathcal{M}$ be a system of pairwise disjoint nonempty sets. Then there is a set M which contains exactly one element from each set of the system.

The question of how to generate new sets from given ones leads to *set operations*.

Definition 1.3.2. The *union* of the sets A and B is the set of all elements which belong to A or B,

$$A \cup B := \{x \mid x \in A \vee x \in B\}.$$

If there are elements in A or B, then the union exists and is uniquely determined (by the axiom of comprehension). Therefore, formation of the union is an operation which can be used to "calculate" with sets. If A and B contain no elements, then the union does not exist. To avoid this situation, we define the *empty set*

$$\emptyset := \{x \mid x \in E \wedge x \neq x\}.$$

Definition 1.3.3. The *intersection* of the sets A and B contains exactly those elements of the basic area which belong to A and B,

$$A \cap B := \{x \mid x \in A \wedge x \in B\}.$$

Union and intersection satisfy the following laws:

Proposition 1.3.4. *Let A, B, C be arbitrary sets. Then:*

$$
\begin{aligned}
A \cup B &= B \cup A, & A \cap B &= B \cap A,\\
A \cup (B \cup C) &= (A \cup B) \cup C, & A \cap (B \cap C) &= (A \cap B) \cap C,\\
A \cup A &= A, & A \cap A &= A,\\
A \cup (A \cap B) &= A, & A \cap (A \cup B) &= A,\\
A \cup \emptyset &= A, & A \cap \emptyset &= \emptyset,\\
A \cup (B \cap C) &= (A \cup B) \cap (A \cup C), & &\\
A \cap (B \cup C) &= (A \cap B) \cup (A \cap C). & &
\end{aligned}
$$

Proof. For the proofs, we use the properties of the logical functors $\vee$ and $\wedge$. As an example, we prove the first equations:

$$A \cup B = \{x \mid x \in A \vee x \in B\} = \{x \mid x \in B \vee x \in A\} = B \cup A,$$

since the disjunction is commutative, while

$$A \cap B = \{x \mid x \in A \wedge x \in B\} = \{x \mid x \in B \wedge x \in A\} = B \cap A,$$

since the conjunction is commutative. □

These laws are called commutative laws for $\cup$ and $\cap$, associative laws, idempotent laws, distributive laws, and absorption laws. The empty set behaves with respect to $\cup$ like 0 for the addition of numbers and with respect to $\cap$ like 0 for multiplication. The following set operation has only one input.

Definition 1.3.5. Let E be a basic area and A an arbitrary subset of E, i. e., each element from A is also an element from E. Then the *complement* of A consists of exactly all elements of E which do not belong to A,

$$\neg A := \{x \mid x \in E \wedge x \notin A\}.$$

Instead of $\neg A$, one also uses A' or $\sim A$.

Using the complement, we define one more binary operation, namely the *difference of sets*.

Definition 1.3.6. Let E be a basic area for the variable x, and let A, B be subsets of E. Then the *difference set* of A and B is the set of all elements from E, belonging to A but not to B,

$$A \setminus B := \{x \mid x \in A \wedge \neg(x \in B)\}.$$

For the complement and difference, the following laws are satisfied:

Proposition 1.3.7. *Let A, B, C be arbitrary sets. Then:*

$$
\begin{aligned}
\neg(A \cup B) &= \neg A \cap \neg B, & \neg(A \cap B) &= \neg A \cup \neg B,\\
(A \cup B) \setminus C &= (A \setminus C) \cup (B \setminus C), & (A \cap B) \setminus C &= (A \setminus C) \cap (B \setminus C),\\
A \setminus (B \cup C) &= (A \setminus B) \cap (A \setminus C), & A \setminus (B \cap C) &= (A \setminus B) \cup (A \setminus C),\\
\neg(\neg A) &= A,\\
A \setminus B &= A \cap \neg B.
\end{aligned}
$$

Proof. We use the definitions and the corresponding logical equivalences, for example:

$$
\begin{aligned}
A \setminus (B \cap C) &= \{x \mid x \in A \wedge \neg(x \in (B \cap C))\}\\
&= \{x \mid x \in A \wedge (\neg(x \in B) \vee \neg(x \in C))\}\\
&= \{x \mid (x \in A \wedge (\neg(x \in B))) \vee (x \in A \wedge (\neg(x \in C)))\}\\
&= (A \setminus B) \cup (A \setminus C).
\end{aligned}
$$

The other proofs are similar. □

The first two rules are called *de Morgan rules.*

One more binary set operation is the Cartesian product of two sets.

Definition 1.3.8. Let E be a basic area for variables, and let A, B be subsets of E. Then the *Cartesian product* of A and B is the set of all ordered pairs (a, b), $a \in A$, $b \in B$,

$$A \times B := \{(a, b) \mid a \in A \wedge b \in B\}.$$

Ordered pairs are equal if the respective components are equal at the same time, i. e.,

$$(a, b) = (c, d) \Longleftrightarrow a = c \wedge b = d.$$

The definition of the Cartesian product can be generalized to more than two sets. This is also possible for the union and intersection. Because of the associative law for the union and intersection, each already obtained result can be combined step by step with one more set. For the Cartesian product, this is not possible since $A \times (B \times C)$ is different from $(A \times B) \times C$. The reason is that the elements $(a, (b, c))$ are different from the elements $((a, b), c)$. Therefore, the Cartesian product $A_1 \times A_2 \times \cdots \times A_n$ for $n \geq 2$ is defined as set of all n-tuples $(a_1, a_2, \ldots, a_n)$ with $a_i \in A_i$ for $i = 1, \ldots, n$. If all factors A_i are equal, then one has the nth Cartesian power A^n. The first Cartesian power of A is A. For $n = 0$, we define

$$A^0 := \{\emptyset\},\ A \neq \emptyset.$$

Therefore, $\emptyset^0$ is not defined.

If any element from set A is contained in set B, then A is said to be a *subset* of B, denoted $A \subseteq B$. Clearly,

$$A = B \Leftrightarrow A \subseteq B \wedge B \subseteq A.$$

We notice that for the definition of $A \subseteq B$ we use the implication.

Definition 1.3.9. $A \subseteq B :\Longleftrightarrow (x \in A \Rightarrow x \in B)$.

The relation denoted by $\subseteq$ is said to be the *inclusion*.

A nonempty subset of A, different from A, is said to be a *proper subset* of A, denoted $A \subset B$. Instead of $A \subseteq B$ and $A \subset B$, one often uses also $B \supseteq A$ and $B \supset A$, respectively, which are called the *superset* and *proper superset*, respectively.

These relations have the following properties:

Proposition 1.3.10.
(i) $A \subseteq A$, *reflexivity*,
(ii) $((A \subseteq B) \wedge (B \subseteq C)) \Rightarrow (A \subseteq C)$, *transitivity*,
(iii) $((A \subseteq B) \wedge (B \subseteq A)) \Rightarrow A = B$, *antisymmetry*,
(iv) $\emptyset \subseteq A$,
(v) $A \not\subset A$, *irreflexivity*,
(vi) $((A \subset B) \wedge (B \subset C)) \Rightarrow (A \subset C)$, *transitivity*,
(vii) $(A \subset B) \Rightarrow (B \not\subset A)$, *asymmetry*.

Proof. As an example, we prove (ii). Then we have:

$$\begin{aligned}((A \subseteq B) \wedge (B \subseteq C)) \Rightarrow A \subseteq C :\Leftrightarrow\ &((x \in A \Rightarrow x \in B) \wedge (x \in B \Rightarrow x \in C))\\ &\Rightarrow (x \in A \Rightarrow x \in C).\end{aligned}$$

The right-hand side follows from the properties of the implication. □

Notice that (i) and (v) are also true for $A = \emptyset$ because $x \in \emptyset \Rightarrow x \in \emptyset$ is, as an implication with false premise, a true statement.

An interesting set is the set of all subsets of a given set.

Definition 1.3.11. The *power set* of a set M is the set of all its subsets,

$$\mathcal{P}(M) := \{B \mid B \subseteq M\}.$$

For any set M, $\emptyset \in \mathcal{P}(M)$ and $M \in \mathcal{P}(M)$.

Example 1.3.12. The power set of $\{a,b,c,d\}$ consists of the empty set, $\{a,b,c,d\}$ itself, one-element sets $\{a\}, \{b\}, \{c\}, \{d\}$, two-element sets $\{a,b\}, \{a,c\}, \{a,d\}, \{b,c\}, \{b,d\}, \{c,d\}$, and three-element sets $\{a,b,c\}, \{a,b,d\}, \{a,c,d\}, \{b,c,d\}$. Altogether there are 16 sets.

As a generalization there holds

Proposition 1.3.13. *For any natural number n, the power set of an n-element set consists of* 2^n *elements.*

Proof. The proof is left to the reader to practice inductive proofs. Later on, we will come back to the proof and will see that this proposition is a special case of a more general theorem. In Chapter 2 we will come back to proofs by induction. □

1.4 Equivalence relations

Relations occur in various branches of economics. Comparing the results of production processes or investment projects and finding out which is "the best" belong to the most important activities of economists.

Definition 1.4.1. An n-ary relation R on a set M is a subset of the nth Cartesian power of M.

For $n = 2$, one has a binary relation, for $n = 3$, a ternary relation. Unary relations are subsets of M. The empty set is called an empty relation.

Example 1.4.2.
1. Let $M = \{0,1\}$ and $R_1 = \{(0,0),(0,1),(1,1)\}$ be a binary relation on M. Instead of $(0,0),(0,1),(1,1) \in R_1$, we can write $0 \le 0, 0 \le 1, 1 \le 1$.
2. Let

$$R_2 = \{(0,0,1,1),(0,1,0,1),(0,0,0,0),(1,1,1,1),(1,1,0,0),(1,0,1,0),\\ (1,0,0,1),(0,1,1,0)\}$$

be a quaternary relation on M. Obviously, R_2 can be written as

$$R_2 = \{(a,b,c,d) \mid a,b,c,d \in M \wedge a+b = c+d\},$$

where + is defined as addition modulo 2 by the table

+	0	1
0	0	1
1	1	0

Binary relations R can be given as *directed graphs* (see Chapter 5). The elements of M are the vertices of the graph. If two elements of M form a pair belonging to R then this pair is a directed edge of the graph, and conversely.

Example 1.4.3. Let $M = \{1,2,3,4,5\}$. The relation

$$R = \{(1,2),(1,4),(2,4),(2,3),(3,4),(3,1),(4,5)\}$$

corresponds to the directed graph of Figure 1.2.

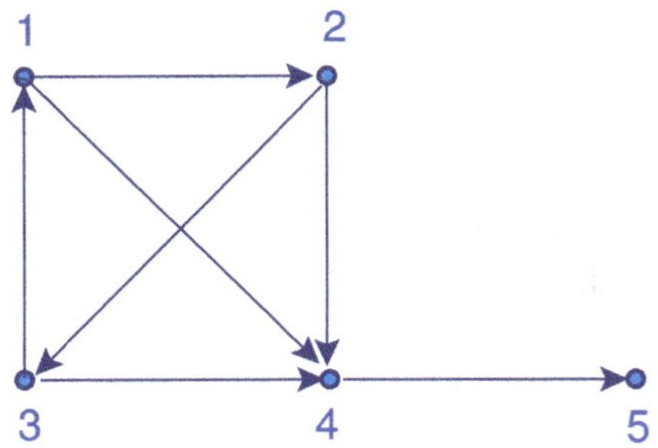

Figure 1.2: The directed graph of a binary relation.

We are especially interested in the properties of two important classes of binary relations, namely *equivalence* and *order relations.*

Definition 1.4.4. A binary relation R on a set M is said to be an equivalence relation on M if the following properties are satisfied:

(E1) R is reflexive, i. e., $\forall x \in M\ ((x,x) \in R)$,

(E2) R is symmetric, i. e., $\forall x,y \in M\ ((x,y) \in R \Rightarrow (y,x) \in R)$,

(E3) R is transitive, i. e., $\forall x,y,z \in M\ (((x,z) \in R \land (z,y) \in R) \Rightarrow (x,y) \in R)$.

Example 1.4.5.

1. Let

$$M = \{0,1,2\} \quad \text{and} \quad R = \{(0,0),(1,1),(2,2),(1,2),(2,1)\}.$$

Then R is reflexive, symmetric, and transitive.

2. Let m be a fixed positive integer and let x, y be any integers. We define

$$x \equiv y(m) :\Leftrightarrow \exists g \in \mathbb{Z}\ (x - y = gm).$$

Then $R_m := \{(x,y) \mid x,y \in \mathbb{Z} \land x \equiv y(m)\}$ is an equivalence relation on $\mathbb{Z}$, which is called *congruence modulo m.*

3. Equality of numbers is an equivalence relation.

Definition 1.4.6. Let R be an equivalence relation on M and $x \in M$. Then

$$[x]_R := \{y \mid y \in M \land (y,x) \in R\}$$

is said to be *the equivalence class generated by* x and

$$M/R := \{[x]_R \mid x \in M\}$$

is said to be the *factor* or *quotient set* of M by R.

Example 1.4.7.
1. The factor set for the first example is the set $M/R = \{[0]_R, [1]_R\}$ with $[0]_R = \{0\}$, $[1]_R = \{1, 2\}$.
2. In the second example, we have $\mathbb{Z}/m := \{[0]_m, \dots, [m-1]_m\}$ as the factor set.

The factor set partitions a given set M into a set of pairwise disjoint subsets.

Definition 1.4.8. Let $M \neq \emptyset$. Then a subset K of the power set $\mathcal{P}(M)$ is said to be a partition of M if $K \subseteq \mathcal{P}(M)$ and
1. $\forall X\ (X \in K \Rightarrow X \neq \emptyset)$ ($\emptyset$ does not belong to K);
2. $\forall m \in M\ \exists X \in K\ (m \in X)$ (each element of M occurs in a set of K);
3. $\forall X, Y \in K\ (X \cap Y \neq \emptyset \Rightarrow X = Y)$ (the sets of K are pairwise disjoint).

Equivalence relations and partitions are closely related to each other as the following "Main theorem for equivalence relations" states.

Theorem 1.4.9. *Let R be an equivalence relation on a set M. Then the factor set M/R forms a partition K_R of M. If, conversely, K is a partition of M, then there exists an equivalence relation R_K in M such that the equivalence classes with respect to this equivalence relation are exactly the sets of the partition. Moreover,*

$$K_{R_K} = K, \quad R_{K_R} = R,$$

i. e., the correspondence between equivalence relations and partitions of M is one-to-one.

Proof. We show that K_R is a partition of M if R is an equivalence relation in M. The equivalence classes of R are subsets of M. Because of reflexivity of R, i. e., $a \approx a$ for every $a \in M$, we get $a \in [a]_R$ and thus $\emptyset \notin K_R$ and $\forall a \in M (a \in [a]_R)$, i. e., every element belongs to one of the equivalence classes. We show that two different equivalence classes $[a]_R$, $[b]_R$ are disjoint. Suppose that $[a]_R \cap [b]_R \neq \emptyset$. Then there is an element $c \in M$ with $c \in [a]_R$, $c \in [b]_R$, i. e., with $(c, a) \in R$, $(c, b) \in R$. By the symmetry and transitivity of R, we have $(a, b) \in R$, i. e., $[a]_R = [b]_R$. Therefore, the equivalence classes with respect to R form a partition of M.

Conversely, let K be a partition of M. Then by

$$R_K := \{(a, b) \mid a \text{ and } b \text{ belong to the same set of the partition } K\}$$

we define an equivalence relation in M whose equivalence classes are the sets of the partition K.

Now we prove the both equations. If $X \in K$ and $x \in X$, then, by the definition of R_K, we have the equality $X = [x]_{R_K} \in K_{R_K}$, and conversely. This proves $K = K_{R_K}$. Moreover,

$$(x, y) \in R \Leftrightarrow (x, y) \in R_{K_R}$$

by the definition of R_{K_R}, i. e., $R = R_{K_R}$. □

The following example shows how the main theorem for equivalence relations can be applied to define concepts in any scientific area.

Example 1.4.10. Let E be an infinite set and let $M_1, M_2 \subset E$. We say that M_1 and M_2 are sets of equal quantity, written as $M_1 \sim M_2$, if there is a bijection from M_1 onto M_2. It is easy to see that $\sim$ is an equivalence relation on $\mathcal{P}(E)$, the power set of E. The cardinal number of a set $M \in \mathcal{P}(E)$ is defined as the equivalence class of all sets which have equal quantity as M:

$$\operatorname{card} M := [M] = \{M' \mid M' \in \mathcal{P}(E) \text{ and } M' \sim M\}.$$

Now, a natural number is defined as a cardinal number of a finite set, i. e., as $[M] = \{M' \mid M' \in \mathcal{P}(E) \text{ and } M' \sim M \text{ where } M \text{ is finite}\}$.

1.5 Partial order relations

One more important class of binary relations is the class of *order relations.*

Definition 1.5.1. A relation R on a set M (also written as $\leq$) is said to be *a partial order relation* on M if R is *reflexive, transitive,* and *antisymmetric*, i. e., if

$$\forall a, b, c \in M\ (a \leq a,\ (a \leq b \wedge b \leq c) \Rightarrow a \leq c,\ (a \leq b \wedge b \leq a) \Rightarrow a = b)$$

are satisfied. The pair $(M; \leq)$ is said to be *a partially ordered set*. Two elements a, b of a partially ordered set are called *comparable* if $a \leq b$ or $b \leq a$, otherwise *they are called incomparable*. The partially ordered set $(M; \leq)$ is said to be *linearly ordered* (or *totally ordered*) if any two elements of M are comparable.

Sometimes we will use $a \geq b :\Leftrightarrow b \leq a$ and say that "a is greater than or equal to b."

Example 1.5.2.
1. The nonnegative rational numbers $\mathbb{Q}^+$ form with respect to the usual partial order relation $\leq$ on $\mathbb{Q}^+$ a partially ordered set which is even linearly ordered.
2. The set $\mathbb{N}$ of all natural numbers forms together with the binary relation $|$ defined by

   $$a|b :\Leftrightarrow \exists c \in \mathbb{N}\ (b = ca)$$

   a partially ordered set $(\mathbb{N}; |)$. It is easy to check and left to the reader to show that

   $$\forall a, b, c \in \mathbb{N}\ \big(a|a, (a|b \wedge b|c) \Rightarrow a|c, (a|b \wedge b|a) \Rightarrow a = b\big)$$

   are satisfied.

The partially ordered set $(\mathbb{N}; |)$ is not linearly ordered since there are pairs (a, b) of natural numbers with $\neg(a \mid b)$.

We notice that $(\mathbb{Z}; |)$ is not a partially ordered set. For $a, b \in \mathbb{Z}$, $a|b \wedge b|a$ implies $a = b \vee a = -b$, i. e., $|$ is not antisymmetric.

If $\leq$ is a partial order relation, then

$$a < b :\Leftrightarrow a \leq b \wedge a \neq b$$

defines one more binary relation. Obviously, $<$ is transitive, but *irreflexive*, i. e., for all $a \in M$ we have $(a \nless a)$ (this means $\neg(a < a)$), and *asymmetric*, i. e.,

$$\forall a, b \in M\ (a < b \Rightarrow \neg(b < a)).$$

Definition 1.5.3. Let $(A; \leq)$ be a partially ordered set and $B \subseteq A$. An element $a \in A$ is said to be an *upper bound* of B if for all $b \in B$ the inequality $b \leq a$ is satisfied, and a *lower bound* of B if for all $b \in B$ the inequality $b \geq a$ is satisfied. An upper bound a is said to be *the supremum* of B if a is the least upper bound, i. e., if
1. $\forall b \in B\ (b \leq a)$ and
2. $\forall a' \in A\ \forall b \in B\ (b \leq a' \Rightarrow a \leq a')$

are satisfied.

A lower bound a is said to be the *infimum* of B if a is the greatest lower bound, i. e., if
1. $\forall b \in B\ (b \geq a)$ and
2. $\forall a' \in A\ \forall b \in B\ (b \geq a' \Rightarrow a \geq a')$

are satisfied. The supremum and infimum of a subset B of A are denoted by $a = \bigvee B$ and $a = \bigwedge B$, respectively.

Example 1.5.4. We determine the infimum and supremum of a two-element set $\{a, b\}$ of natural numbers with respect to the divisor relation $|$. The supremum and infimum c of a and b satisfy
1. $a|c, b|c$,
2. $\forall c' \in \mathbb{N}\ ((a|c' \wedge b|c') \Rightarrow c|c')$

for the supremum, and
1. $c|a, c|b$,
2. $\forall c' \in \mathbb{N}\ ((c'|a \wedge c'|b) \Rightarrow c'|c)$

for the infimum.

Then it becomes clear that the supremum is the least common multiple $\mathrm{lcm}(a, b)$ and the infimum is the greatest common divisor $\gcd(a, b)$, which can be determined by using the *Euclidean algorithm* (which will be explained in Chapter 3, Section 3.6).

Definition 1.5.5. Let $(A; \leq)$ be a partially ordered set, and let $B \subseteq A$ be a subset of A. Then $b \in B$ is said to be a *maximal element* of B if

$$\forall a \in A\ (b < a \Rightarrow a \notin B).$$

The element $b \in B$ is said to be a *minimal* element of B if

$$\forall a \in A\,(a < b \Rightarrow a \notin B).$$

The *greatest element* b of $B \subseteq A$ is defined by

$$\forall b' \in B\,(b' \leq b)$$

and the *least element* b of $B \subseteq A$ is defined by

$$\forall b' \in B\,(b \leq b').$$

A linearly ordered set $(A; \leq)$ is said to be *well ordered* if for every nonempty subset $B \subseteq A$ the least element exists. In this case, we say that the *well-ordering theorem* is satisfied. Moreover,

$$[a, b] := \{x \mid a \leq x \leq b\}$$

is said to be a *closed interval* and

$$(a, b) := \{x \mid a < x < b\}$$

is said to be an *open interval*. An element $b \in A$ is said to be *an upper neighbor* of $a \in A$, written as

$$a \prec b, \text{ if } [a, b] = \{a, b\}.$$

Finite partially ordered sets can be presented by Hasse diagrams, which are directed graphs (see Chapter 5). Figure 1.3 shows some examples of Hasse diagrams. The vertices are the elements of the ordered set. If $a \leq b$ then there is a directed edge from a to b. Instead of arrows, the direction is defined from the bottom to the top.

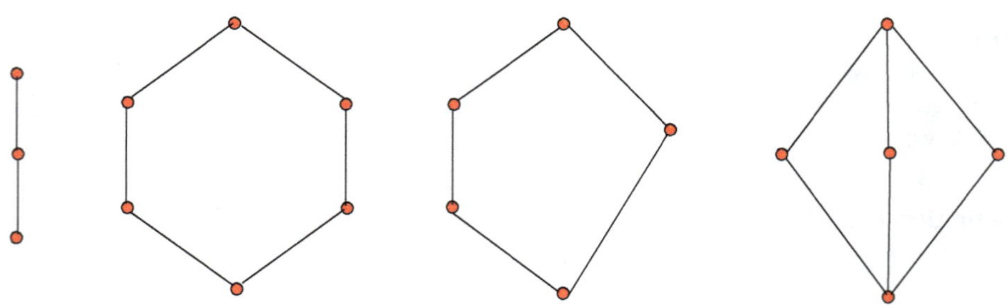

Figure 1.3: Hasse diagrams.

Paths of the form

$$a_1 \leq a_2 \leq \cdots \leq \cdots \leq a_n$$

in the Hasse diagram of a partially ordered set are also called *chains*. Infinite partially ordered sets can have infinite chains. *Zorn's lemma* gives a condition for the existence of a maximal (minimal) element of a partially ordered set:

Lemma 1.5.6 ([67]). *If each chain of a nonempty partially ordered set M has an upper (lower) bound, then M has a maximal (minimal) element.*

Zorn's lemma is equivalent to the axiom of choice and to the well-ordering theorem.

1.6 Partial order relations and efficiency of economic activities

The aim of an evaluation of production activities is to increase efficiency and effectiveness of economic performances. Effectiveness simply means doing the right things and efficiency is doing the right things in the right way. Efficiency can be conceived as the ratio of the output power of output products or services, related to the input power. Rational economic performances are characterized by an optimal relationship between input and output. The *maximum principle* is oriented towards the output and requires that by given amounts of inputs the amount of produced outputs is maximal. The *minimum principle* is input oriented and requires that for given amounts of produced outputs the amounts of inputs are minimal. To evaluate the efficiency, we classify the objects with respect to their desirability. The output objects can be desired, not desired, or neutral. The neutral objects have no further influence. The aim is to maximize the output of desired objects and to minimize the output of undesired things (like litter, used dirty water, etc.). On the input side, the aim consists in minimization of desired goods and maximization of undesired things.

Let

$$\begin{pmatrix} x_1 \\ \vdots \\ x_n \end{pmatrix}, \begin{pmatrix} y_1 \\ \vdots \\ y_n \end{pmatrix} \in (\mathbb{Q}^+)^n$$

be n-tuples. We think about amounts of input and output objects of a production system. Koopmans [40] introduced the *activity analysis* of production and defined an *activity* as the difference

$$\begin{pmatrix} z_1 \\ \vdots \\ z_n \end{pmatrix} = \begin{pmatrix} x_1 \\ \vdots \\ x_n \end{pmatrix} - \begin{pmatrix} y_1 \\ \vdots \\ y_n \end{pmatrix}.$$

We will come back to production systems in more details in Chapter 8. Here we are only interested in the evaluation of activities as an application of partial order relations. The following notation goes back to Koopmans.

Let $z_1, \dots, z_l$ be input objects and let $z_{l+1}, \dots, z_n$ be output objects of a production system. We set $z_1 = -x_1, \dots, z_l = -x_l, z_{l+1} = y_{l+1}, \dots, z_n = y_n$, $x_1, \dots, x_l, y_{l+1}, \dots, y_n \in \mathbb{Q}^+$. Then

$$\begin{pmatrix} -x_1 \\ \vdots \\ -x_l \\ y_{l+1} \\ \vdots \\ y_n \end{pmatrix} = \begin{pmatrix} 0 \\ \vdots \\ 0 \\ y_{l+1} \\ \vdots \\ y_n \end{pmatrix} - \begin{pmatrix} x_1 \\ \vdots \\ x_l \\ 0 \\ \vdots \\ 0 \end{pmatrix}$$

is said to be an *admissible activity* of the production system. Let T^{ad} be the set of all admissible activities of a given production system.

Let

$$z^1 = \begin{pmatrix} -x_{11} \\ \vdots \\ -x_{1l} \\ y_{1l+1} \\ \vdots \\ y_{1n} \end{pmatrix}, \quad z^2 = \begin{pmatrix} -x_{21} \\ \vdots \\ -x_{2l} \\ y_{2l+1} \\ \vdots \\ y_{2n} \end{pmatrix}$$

be two admissible activities. We want to compare z^1 and z^2 to find out which one is more efficient. We introduce a binary relation $\preceq$ on the set T^{ad}. For a desired kind k of output objects, $z^1 \preceq z^2$ should mean that $y_{1k} \leq y_{2k}$, and for a desired kind k of input objects, $z^1 \preceq z^2$ should follow $-x_{1k} \leq -x_{2k}$. For an undesired kind k of output objects, $z^1 \preceq z^2$ should mean that $y_{1k} \geq y_{2k}$, and for an undesired kind k of input objects, $z^1 \preceq z^2$ should have $-x_{1k} \geq -x_{2k}$.

Before giving the formal definition of the relation $\preceq$, we start with an example. Let

$$z^1 = \begin{pmatrix} -1 \\ -3 \\ 2 \end{pmatrix}, \quad z^2 = \begin{pmatrix} -3 \\ -5 \\ 2 \end{pmatrix}, \quad z^3 = \begin{pmatrix} -2 \\ -1 \\ 2 \end{pmatrix}$$

be three activities. By pairwise checking, we determine all comparable and all incomparable pairs of activities.

In lines 2, 3, 5, and 6, the reasoning for comparable pairs is given. The reader should check the details.

desired	undesired	comparable pairs	incomparable pairs
1, 2, 3	-	$z^2 \preceq z^1$, $z^2 \preceq z^3$	z^1, z^3
2, 3	1	$z^1 \preceq z^3$ $(-3 < -1, 2 \leq 2, -1 > -2)$	z^1, z^2, z^2, z^3
1, 3	2	$z^3 \preceq z^1$ $(-1 > -3, 2 \geq 2, -3 < -1)$	z^1, z^2, z^2, z^3
1, 2	3	$z^2 \preceq z^1$, $z^2 \preceq z^3$	z^1, z^3
3	1, 2	$z^1 \preceq z^2$, $z^3 \preceq z^2$	z^1, z^3,
2	1, 3	$z^1 \preceq z^3$ $(-3 < -1, -1 > -2, 2 \geq 2)$	z^1, z^2, z^2, z^3
1	2, 3	$z^3 \preceq z^1$ $(-1 > -2, -3 < -1, 2 \leq 2)$	z^1, z^2, z^2, z^3
-	1, 2, 3	$z^1 \preceq z^2$, $z^3 \preceq z^2$	z^3, z^1

Definition 1.6.1. Let

$$z^1 = \begin{pmatrix} -x_{11} \\ \vdots \\ -x_{1l} \\ y_{1l+1} \\ \vdots \\ y_{1n} \end{pmatrix}, \quad z^2 = \begin{pmatrix} -x_{21} \\ \vdots \\ -x_{2l} \\ y_{2l+1} \\ \vdots \\ y_{2n} \end{pmatrix}$$

be two admissible activities. Then

$$z^2 \preceq z^1 :\Leftrightarrow \begin{cases} -x_{2k} \leq -x_{1k} & \text{if } k \in \{1, \dots, l\},\ k \text{ is desired,} \\ y_{2k} \leq y_{1k} & \text{if } k \in \{l+1, \dots, n\},\ k \text{ is desired,} \\ -x_{2k} \geq -x_{1k} & \text{if } k \in \{1, \dots, l\},\ k \text{ is undesired,} \\ y_{2k} \geq y_{1k} & \text{if } k \in \{1+1, \dots, n\},\ k \text{ is undesired.} \end{cases}$$

Theorem 1.6.2. *The relation $\preceq$ is a partial order relation on the set T^{ad} of all admissible activities.*

Proof. We have to show that $\preceq$ is reflexive, antisymmetric, and transitive.

The relation $\preceq$ is reflexive since for all $z^1 \in T^{\text{ad}}, z^1 \preceq z^1$, i. e.,

$$\begin{pmatrix} -x_{11} \\ \vdots \\ -x_{1l} \\ y_{1l+1} \\ \vdots \\ y_{1n} \end{pmatrix} \preceq \begin{pmatrix} -x_{11} \\ \vdots \\ -x_{1l} \\ y_{1l+1} \\ \vdots \\ y_{1n} \end{pmatrix}$$

because of $-x_{1k} \leq -x_{1k}$ for $k \in \{1, \dots, l\}$ and $y_{1k} \leq y_{1k}$ for $k \in \{l+1, \dots, n\}$, if k is desired, and $-x_{1k} \geq -x_{1k}$ for $k \in \{1, \dots, l\}$ and $y_{1k} \geq y_{1k}$ for $k \in \{l+1, \dots, n\}$, if k is not desired.

The relation $\preceq$ is transitive, i. e.,

$$z^1 \preceq z^2, z^2 \preceq z^3 \Rightarrow z^1 \preceq z^3.$$

We use

$$y_{1k} \leq y_{2k} \leq y_{3k} \Rightarrow y_{1k} \leq y_{3k}$$

if $k \in \{l+1, \dots, y_n\}$ and k is desired, and

$$-x_{1k} \leq -x_{2k} \leq -x_{3k} \Rightarrow -x_{1k} \leq -x_{3k}$$

if $k \in \{1, \dots, l\}$ and k is desired. In the case that k is undesired, we conclude in the corresponding way.

Antisymmetry can be proved similarly. □

As in the general case, by

$$z^1 \prec z^2 :\Leftrightarrow z^1 \preceq z^2 \wedge z^1 \neq z^2$$

an irreflexive order relation can be defined. In this case, we say "activity z^2 dominates activity z^1."

Definition 1.6.3. An activity $z \in T^{\mathrm{ad}}$ is said to be *efficient* if z is a maximal element with respect to $\preceq$, i. e., if

$$\forall z' \in T^{\mathrm{ad}} \; (z \prec z' \Rightarrow z' \notin T^{\mathrm{ad}}).$$

The set of all maximal elements of T^{ad} is also called the *efficient edge* of T^{ad}.

1.7 Functions

The concept of a *function* is one of the most important mathematical concepts with many applications.

Definition 1.7.1. Let M_1 and M_2 be nonempty sets. A subset $f \subseteq M_1 \times M_2$ of the Cartesian product is said to be *function from M_1 into M_2* if the following conditions are satisfied:

(i) $\forall x \in M_1 \; \exists y \in M_2 \; ((x,y) \in f)$,

(ii) $\forall x \in M_1 \; \forall y_1, y_2 \in M_2 \; (((x,y_1) \in f \wedge (x,y_2) \in f) \Rightarrow y_1 = y_2)$, i. e., f is *unique*.

We will use the following notation:

$$\begin{array}{ll} f : M_1 \to M_2, y = f(x), & \text{if } (x,y) \in f, \; x \mapsto y, \\ \text{(image of } f) & \mathrm{Im}(f) := \{y \in M_2 \mid \exists x \in M_1 \; ((x,y) \in f)\}, \\ \text{(preimage of } f) & \mathrm{Preim}(f) := \{x \in M_1 \mid \exists y \in M_2 \; ((x,y) \in f)\}. \end{array}$$

Instead of preimage of f, one uses also *domain of f, denoted* $\mathrm{dom}(f)$. Function f is said to be a *partial function* from M_1 into M_2 if $\mathrm{Preim}(f) \subset M_1$. Assume that $\mathrm{Preim}(f) = M_1$. Then f is said to be

$$\begin{aligned} \textit{injective} &:\Leftrightarrow \forall x, x' \in M_1\ \forall y \in M_2\ (((x,y) \in f \wedge (x',y) \in f) \Rightarrow x = x') \\ &\quad (f \text{ is one-to-one}); \\ \textit{surjective} &:\Leftrightarrow \forall y \in M_2\ \exists x \in M_1\ ((x,y) \in f) \quad (f \text{ maps onto } M_2); \\ \textit{bijective} &:\Leftrightarrow f \text{ is injective and surjective.} \end{aligned}$$

Example 1.7.2.
1. The function $id_{M_1} := \{(x,x) \mid x \in M_1\}$, the identity function on M_1, is bijective.
2. The function $f = \{(n,2n) \mid n \in \mathbb{N}\}$ is bijective.

If f is a function $f: M_1 \to M_2$ and $A \subseteq M_1$, we define

$$f(A) := \{y \in M_2 \mid \exists x \in A\ ((x,y) \in f)\}.$$

The set $f(A)$ is said to be the *image* of A. If $B \subseteq M_2$, we define

$$f^{-1}(B) := \{x \in M_1 \mid \exists y \in B\ ((x,y) \in f)\}$$

and call $f^{-1}(B)$ the *preimage* or *original* of B.

By our definition, functions are sets of pairs, and as sets we can compare functions and define operations of functions:

(a) *Equality.* Let f and g be functions. Equality of functions is defined by

$$f = g :\Leftrightarrow \{(x, f(x)) \mid x \in \mathrm{dom}(f)\} = \{(y, f(y)) \mid y \in \mathrm{dom}(g)\},$$

i. e., f and g are equal if and only if $\mathrm{dom}(f) = \mathrm{dom}(g)$ and $f(x) = g(x)$ for all $x \in \mathrm{dom}(f)$.

(b) *Inclusion.* Let f and g be functions. Then

$$f \subseteq g :\Leftrightarrow \{(x, f(x)) \mid x \in \mathrm{dom}(f)\} \subseteq \{(y, f(y)) \mid y \in \mathrm{dom}(g)\},$$

i. e., $f \subseteq g$ if and only if $\mathrm{dom}(f) \subseteq \mathrm{dom}(g)$ and $f(x) = g(x)$ for all $x \in \mathrm{dom}(f)$.

(c) *Product.* Let $f: M \to N$ and $g: N \to P$ be functions from M into N and from N into P, respectively. Then $g \circ f: M \to P$ with $(g \circ f)(x) := g(f(x))$ is said to be the product of f and g.

Proposition 1.7.3. *The product $\circ$ of functions is associative, i. e.,*

$$(f \circ g) \circ h = f \circ (g \circ h),$$

assuming that both sides exist.

Proof. Let $x \in \operatorname{dom}(h)$, $h(x) \in \operatorname{dom}(g)$, and $g(h(x)) \in \operatorname{dom}(f)$. Then $h(x) \in \operatorname{dom}(f \circ g)$ and both sides exist. Moreover,

$$((f \circ g) \circ h)(x) = (f \circ g)(h(x)) = f(g(h(x))) = f((g \circ h)(x)) = (f \circ (g \circ h))(x)$$

shows the needed equality. □

The inverse of a function $f : M \to N$ is defined by

$$f^{-1} := \{(y, x) \mid (x, y) \in f\}.$$

Theorem 1.7.4. *The inverse of a function $f : M \to N$ from M into N is a function from N into M if and only if f is bijective. If f is bijective, then also f^{-1} is bijective. If both f and g are bijective, then also $f \circ g$ is bijective.*

The proof is left as an exercise to the reader.

Bijective mappings on a finite set M play an important role in several applications.

Definition 1.7.5. Let M be a finite set, different from the empty set. Without loss of generality, we may assume that $M = \{1, 2, \ldots, n\}$. A permutation of order n is a bijective function of M onto M. Let S_n be the set of all permutations of order n. The elements of S_n can be given in the form

$$s = \begin{pmatrix} 1 & 2 & \ldots & n \\ k_1 & k_2 & \ldots & k_n \end{pmatrix}, \quad k_i \in M, k_i \neq k_j \text{ for } i \neq j.$$

Another possibility is to write down permutations as products of disjoint cycles. Elements which are mapped to themselves do not occur in this presentation. We give some examples:

$$s = \begin{pmatrix} 1 & 2 & 3 \\ 1 & 3 & 2 \end{pmatrix} = (2 \quad 3), \quad \begin{pmatrix} 1 & 2 & 3 & 4 \\ 2 & 1 & 4 & 3 \end{pmatrix} = (1 \quad 2)(3 \quad 4),$$
$$\begin{pmatrix} 1 & 2 & 3 \\ 1 & 2 & 3 \end{pmatrix} = (1).$$

Here

$$(1 \quad 2)(3 \quad 4)$$

means that the permutation maps 1 to 2, 2 to 1, as well as 3 to 4 and 4 to 3. The identity permutation, which maps each element to itself, will be written as (1). Any two-element cycle is said to be a *transposition*.

We recall that $n!$ ("n-factorial") denotes the product $1 \cdots n$ if $n \geq 1$ and $0! := 1$.

Theorem 1.7.6. *There are exactly $n!$ permutations of order n.*

Proof. We give a proof by induction on n. Let $M = \{1\}$. Then S_1 consists of one element since $|S_1| = 1! = 1$. Now let $M = \{1, \dots, n+1\}$. By inductive hypothesis, there are $n!$ permutations for which 1 is mapped to itself, there are $n!$ permutations, for which 2 is fixed, etc., and $n!$ permutations, which fix $n+1$. Altogether these are $(n+1)n! = (n+1)!$ permutations. □

As bijective functions, permutations can be multiplied and the result is a uniquely determined permutation of the same order. Consider, for example,

$$s_1 = \begin{pmatrix} 1 & 2 & 3 \\ 2 & 1 & 3 \end{pmatrix}, \quad s_2 = \begin{pmatrix} 1 & 2 & 3 \\ 2 & 3 & 1 \end{pmatrix}.$$

We calculate $(s_1 \circ s_2)(x) := s_1(s_2(x))$ for all $x \in \{1, 2, 3\}$ and obtain

$$s_1 \circ s_2 = \begin{pmatrix} 1 & 2 & 3 \\ 1 & 3 & 2 \end{pmatrix}.$$

For the identical permutation, all elements can be written in such a way that elements in the upper and lower rows are in natural order. In each other permutation, there is at least one pair of columns such that the orders in the upper and lower rows are opposite:

$$s = \begin{pmatrix} \dots & i_\mu & \dots & i_\nu & \dots \\ \dots & j_\mu & \dots & j_\nu & \dots \end{pmatrix} \quad \text{with} \quad \begin{matrix} i_\mu < i_\nu, \\ j_\mu > j_\nu, \end{matrix}$$

or conversely.

A pair of columns $\binom{i_\mu}{j_\mu}, \binom{i_\nu}{j_\nu}$ with $\substack{i_\mu < i_\nu \\ j_\mu > j_\nu}$, or conversely, is called an *inversion*.

Definition 1.7.7. A permutation s is said to be even or odd depending on an even or odd number of inversions. Also

$$\operatorname{sgn}(s) := \begin{cases} 1 & \text{if } s \text{ even}, \\ -1 & \text{if } s \text{ odd}, \end{cases}$$

is said to be the *sign* of s.

Theorem 1.7.8. *The sign* $\operatorname{sgn} : S_n \to \{-1, +1\}$ *defined by* $s \mapsto \operatorname{sgn}(s)$ *for each permutation* $s \in S_n$ *is a function with*

$$\operatorname{sgn}(s_1 \circ s_2) = \operatorname{sgn}(s_1) \cdot \operatorname{sgn}(s_2).$$

Proof. For the proof, we consider the product of differences

$$\prod := \prod_{\mu<\nu} (x_\mu - x_\nu) = \begin{matrix} (x_1 - x_2) \cdots (x_1 - x_n) \\ \ddots \\ \cdot (x_{n-1} - x_n). \end{matrix}$$

For $s \in S_n$, let $s(\prod)$ be the product of differences arising from $\prod$ by permuting the indices using s:

$$s\Bigl(\prod\Bigr) := \prod_{\mu<\nu}(x_{s(\mu)} - x_{s(\nu)}).$$

For instance,

$$(12)\Bigl(\prod\Bigr) = \begin{array}{ccc} (x_2 - x_1)(x_1 - x_3)\cdots(x_1 - x_n) & & \\ & \ddots & \\ & & \cdot\,(x_{n-1} - x_n), \end{array}$$

and thus $(12)(\prod) = -\prod$.

Furthermore, one has

$$s\Bigl(\prod\Bigr) = \prod \Leftrightarrow s \text{ even}, \quad s\Bigl(\prod\Bigr) = -\prod \Leftrightarrow s \text{ odd}.$$

Then:

- for $\operatorname{sgn}(s_1) = \operatorname{sgn}(s_2) = +1$, one has $(s_1 \circ s_2)(\prod) = s_1(s_2(\prod)) = \prod$ and $\operatorname{sgn}(s_1 \circ s_2) = +1 = \operatorname{sgn}(s_1) \cdot \operatorname{sgn}(s_2)$,
- for $\operatorname{sgn}(s_1) = \operatorname{sgn}(s_2) = -1$, one has $(s_1 \circ s_2)(\prod) = s_1(s_2(\prod)) = s_1(-\prod) = -(-\prod) = \prod$ and $\operatorname{sgn}(s_1 \circ s_2) = +1 = (-1) \cdot (-1) = \operatorname{sgn}(s_1) \cdot \operatorname{sgn}(s_2)$,
- for $\operatorname{sgn}(s_1) = +1$ and $\operatorname{sgn}(s_2) = -1$, one has $(s_1 \circ s_2)(\prod) = s_1(s_2(\prod)) = s_1(-\prod) = -\prod$ and $\operatorname{sgn}(s_1 \circ s_2) = -1 = (+1) \cdot (-1) = \operatorname{sgn}(s_1) \cdot \operatorname{sgn}(s_2)$,
- for $\operatorname{sgn}(s_1) = -1$ and $\operatorname{sgn}(s_2) = +1$, one has $(s_1 \circ s_2)(\prod) = s_1(s_2(\prod)) = s_1(\prod) = -\prod$ and $\operatorname{sgn}(s_1 \circ s_2) = -1 = (-1) \cdot (+1) = \operatorname{sgn}(s_1) \cdot \operatorname{sgn}(s_2)$.

This finishes the proof. □

The following observation is easy to check: If the permutation s, as a product of disjoint cycles, consists of r elements, then $\operatorname{sgn}(s) = (-1)^{r-l}$.

Example 1.7.9. Consider $s = (12)(345)$. Then $r = 5$, $l = 2$, $\operatorname{sgn}(s) = (-1)^{5-2} = -1$.

Moreover, we have

Corollary 1.7.10. *The set S_n contains as many even as odd permutations.*

Proof. Without loss of generality, we assume that there are more even than odd permutations. Let A_n be the set of all even permutations of S_n. Let s_u be an arbitrary odd permutation from S_n. We form the set of all products $\{s_g \circ s_u \mid s_g \in A_n\}$. From the properties of S_n, there follows that these products are pairwise different. Thus, there are as many products as we have elements in A_n. Each of these products is odd. This means that there are more odd than even permutations, a contradiction, and so the numbers of even and odd permutations are equal. □

The set S_n of all permutations of order n is closed under multiplication, i. e., the product of two permutations of order n is a uniquely determined permutation of order n. The inverse of a permutation of order n is a permutation of order n and the identity permutation (1) of order n satisfies $(1)\circ s = s\circ(1) = s$ for all $s \in S_n$. Sets with a binary operation having these properties are called *groups*. The set A_n is also closed with respect to multiplication. The identity permutation is even. It is not difficult to see that the inverse of an even permutation is even. This shows that $(A_n;\circ)$ is also a group. We speak of the *alternating* subgroup of the group S_n. Furthermore, A_n contains $\frac{n!}{2}$ elements since there are $n!$ permutations of order n. In Chapters 3 and 7 we will come back to the important structural concept of a group.

The following concept connects functions with equivalence relations.

Definition 1.7.11. Let $f : M \to N$ be a function. The kernel of f is defined by

$$\ker(f) := \{(x,y) \in M \times M \mid f(x) = f(y)\}.$$

Proposition 1.7.12. *The kernel* $\ker(f)$ *is an equivalence relation on the set* M.

Proof. By definition, $\ker(f)$ is a binary relation on M. We have to prove that $\ker(f)$ is reflexive, symmetric, and transitive. For each $x \in M$, the pair (x,x) belongs to M since $f(x) = f(x)$. Let $(x,y),(y,z) \in \ker(f)$. Then $f(x) = f(y) = f(z)$ and therefore $(x,z) \in \ker(f)$. This shows that $\ker(f)$ is transitive. Symmetry can be similarly proved. □

We apply Definition 1.4.6 and form the factor set $M/\ker(f)$. By Theorem 1.4.9, $M/\ker(f)$ defines a partition of set M. The sets of this partition are the equivalence classes $[x]_{\ker(f)}$. By the definition of the kernel, the equivalence classes $[x]_{\ker(f)}$ consist of all elements which are mapped to the same element of N. Since each element of M belongs to precisely one equivalence class with respect to $\ker(f)$, each $x \in M$ can be uniquely mapped to one equivalence class $[x]_{\ker(f)}$. This gives a function

$$\mathrm{nat}(\ker(f)) : M \to M/\ker(f), \quad x \mapsto [x]_{\ker(f)}.$$

Then we have the following result which will be used in Chapters 2 and 6.

Theorem 1.7.13. *Let* $f : M \to N$ *be a mapping from set* M *onto set* N, *i. e.,* f *is surjective. Then there exists a uniquely determined bijective mapping* $f' : M/\ker(f) \to N$ *with* $f = f' \circ \mathrm{nat}(\ker(f))$. *(Here* $f = f' \circ \mathrm{nat}(\ker(f))$ *means the "commutativity" (=) of the diagram.)*

Proof. We define $f' := f \circ (\mathrm{nat}(\ker(f)))^{-1}$ and show that f' has the required properties:
1. $f' : M/\ker(f) \to N$ is a function, i. e., is well defined,
2. f' is the unique function with the property $f = f' \circ \mathrm{nat}(\ker(f))$,

3. f is injective,
4. f is surjective.

1. Obviously, f' is a function from $M/\ker(f)$ into N satisfying

$$f' \circ \mathrm{nat}(\ker(f)) = f,$$

since $\mathrm{dom}(f') = M/\ker(f)$ and because of the surjectivity of $\mathrm{nat}(\ker(f))$.

Because of

$$\begin{aligned}&\forall c_1, c_2 \in M/\ker(f)\ \exists a_1, a_2 \in M\\ &(\mathrm{nat}(\ker(f))(a_1) = c_1, \quad \mathrm{nat}(\ker(f))(a_2) = c_2),\end{aligned}$$

and thus

$$\begin{aligned}&c_1 = c_2 \Leftrightarrow \mathrm{nat}(\ker(f))(a_1) = \mathrm{nat}(\ker(f))(a_2) \Leftrightarrow\\ &\quad (a_1, a_2) \in \ker(\mathrm{nat}(\ker(f))) \Rightarrow (a_1, a_2) \in \ker f\\ &\Rightarrow f(a_1) = f(a_2) \Rightarrow f((\mathrm{nat}(\ker(f)))^{-1}(c_1)) = f((\mathrm{nat}(\ker(f)))^{-1}(c_2))\end{aligned}$$

and

$$(f \circ \mathrm{nat}(\ker(f))^{-1})(c_1) = f'(c_1) = f'(c_2) = (f \circ \mathrm{nat}(\ker(f))^{-1})(c_2),$$

and so f' is well defined.

2. Each mapping $f'' : M/\ker(f) \to N$ with $f'' = f \circ \mathrm{nat}(\ker(f))^{-1}$ is equal to $f' = f \circ \mathrm{nat}(\ker(f))^{-1}$, since from $f' \circ \mathrm{nat}(\ker(f)) = f$ and $f'' \circ \mathrm{nat}(\ker(f)) = f$, i. e., $f' \circ \mathrm{nat}(\ker(f)) = f'' \circ \mathrm{nat}(\ker(f))$, there follows for all $a \in M$ that

$$\begin{aligned}&(f' \circ \mathrm{nat}(\ker(f)))(a) = (f'' \circ \mathrm{nat}(\ker(f)))(a), \text{ i. e.,}\\ &f'(\mathrm{nat}(\ker(f))(a)) = f''(\mathrm{nat}(\ker(f))(a)).\end{aligned}$$

The surjectivity of $\mathrm{nat}(\ker(f))$ implies $f(c) = f'(c)$ for all $c \in M/\ker(f)$ and thus $f = f'$.

3. Because of the surjectivity of $\mathrm{nat}(\ker(f))$, each pair of elements c_1, c_2 can be written as $c_1 = \mathrm{nat}(\ker(f))(a_1)$, $c_2 = \mathrm{nat}(\ker(f))(a_2)$. Then from $f(c_1) = f(c_2)$, there follows $c_1 = c_2$ and thus f is injective.

4. Since f is surjective, $f' \circ \mathrm{nat}(\ker(f)) = f$ is surjective and for every $b \in N$ there exists an element $a \in M$ with $(f' \circ \mathrm{nat}(\ker(f)))(a) = b$, and then for all $b \in N$ there exists an element $c \in M/\ker(f)$, namely $c = \mathrm{nat}(\ker(f))(a)$, with $f'(c) = b$. The functions f, f' and $nat(ker(f))$ can be seen in Figure 1.4. □

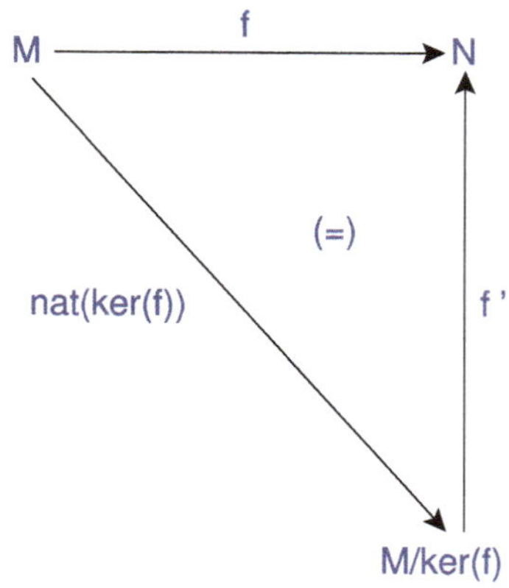

Figure 1.4: Diagram lemma.

1.8 Historical remarks

The traditional formal logic dates back to *Aristotle* (384–322 BC). As a young man, he went to Athens and studied at Plato's academy. Later, around 343 BC, he became the teacher of Alexander the Great. Returning back to Athens in 335 BC, he founded a philosophical school in the Lyceum. He was a researcher and teacher in many subjects. His interests covered a broad range of areas, from philosophy, logic, linguistic and natural sciences, economics, politics, psychology to arts. Aristotle developed the first theory of logic, known as the organon. He insisted that thinking has to be based on several principles. The first principle is *noncontradiction*. In his book *Metaphysics*, he wrote "it is impossible for the same thing to be both affirmed and denied of the same thing, at the same time, and in the same way" (*Metaphysics, IV, 3, 1005b 19–20)*. This principle cannot be proved, we would say, it is an *axiom*. A demonstration (*proof*) is a deduction that shows why something is necessarily so. Aristotle formulated principles for valid inferences and also explored inductive reasoning, beginning from observations. His aim was to understand the reality by correct observations and thinking.

Mathematical logic is based on formal logic, but differs from the traditional formal logic in methods and surpasses it in power. Its beginnings are placed with *George Boole* (1815–1864). He introduced the Boolean functions which allow, using the principle of extensionality, to calculate whether a logical sentence is true or false if the truth values of the inputs are known. The work of *Leibniz, Peirce, Schröder, Frege, Peano, Whitehead, Russel,* and their successors made mathematical logic reach the state of a reputable branch of knowledge and an important theoretical foundation of computer science.

Set theory is closely connected with mathematical logic by the axiom of comprehension, saying that for any statement form $H(x)$ there exists exactly one set $\{x \mid H(x)\}$ consisting of all elements x for which $H(x)$ is true. Then the empty set $\emptyset$ and the set operations can be defined by logical sentences as we did in Section 1.3.

According to *Georg Cantor* (1845–1918) and *Felix Hausdorff* (1868–1942), sets are collections of different objects which are the elements of the set. This “definition” is insufficient and led to contradictions, the so-called *antinomies* of the set theory. One of these antinomies is Russel’s paradox connected with “the set of all sets,” which is not a set, otherwise it would be an element of itself, which is a self-contradiction. At the beginning of the twentieth century, these antinomies resulted in the foundational crisis of mathematics. The crisis was resolved by the axiomatic foundations of set theory, for instance, by the *Zermelo–Frankel* system of axioms.

2 Commodities and money – a relational model

A commodity, also called a ware or a good, is an object which satisfies with its properties human needs and desires. But its use value alone does not show the real character of a commodity. It is the result of human labor and is not produced to satisfy the producer's own needs. At the historical beginning of the capitalist mode of production, commodities were exchanged on the ware market for other commodities, and later for money. Since the emergence of capitalism is based on commodities and since the wealth of our capitalist societies presents itself as an enormous accumulation of commodities, it is important to understand this concept and its historical development. Karl Marx analyzes this economical basic concept in the first part of his work *"Capital" ("Das Kapital")* [51]. He finds that his analytic way "makes the reader of the first chapters rather arduous" (see the preface of the French edition). In our opinion, the reason is that the author consequently follows the epistomological way to define concepts, using equivalence relations, equivalence classes, and abstraction. Therefore, it is not difficult to create a model based on the mathematical concepts of an equivalence relation on a set, equivalence classes, the factor set, and the kernel of a function [50, 22, 20]. This approach can help understand the formal aspects behind Marx' long text easier, at least for those who understood the mathematical ideas of Chapter 1 of this book. On the other hand, we present an application of some mathematical concepts and ideas from Chapter 1. Clearly, our short exposition cannot and will not replace Part I of "Capital."

https://doi.org/10.1515/9783112234433-002

2.1 Use and exchange values of a commodity

From the very origin of the capitalist mode of production, the exchange of commodities prevails more and more. The creation of commodities is based on and connected with the social division of labor and on the private ownership of the means of production. In merchandise management, the use value is the carrier of the exchange value (in short, we will call it "value") of the goods. The value appears as quantitative proportion under which the use value of a commodity is exchanged for the use value of another commodity. Certain amounts of one good will be equalized with amounts of another good:

$$x \text{ units of commodity } A \sim y \text{ units of commodity } B.$$

This "equality" only makes sense if both sides have a common basis, and what they have in common is not usefulness, since use values of different goods are not comparable. What both sides have in common is that they are the results of human labor.

The reason for the twofold character of a commodity to have a use value and an exchange value is the twofold character of the labor embodied. The concrete labor of a taylor, carpenter, or baker pursues a certain purpose and creates the use value. All the different use values reflect the social division of labor, but the division of labor is also a necessary condition to produce commodities. The converse is not always true, i. e., the production of commodities is not necessary for the division of labor. Marx mentions primitive Indian communities with a social division of labor which did not produce commodities. What do different concrete labor activities have in common? The expenditure of human labor. Forgetting about the concrete kind of labor, by abstraction one obtains abstract human labor. The value of a commodity represents the abstract human labor embodied in it. In this way, the socially necessary labor arises as an abstraction of concrete labor and can be measured by a mean value, the socially necessary labor time. Human labor becomes also a commodity that is traded on the labor market. Its price, expressed in money, measures the value of commodities traded on the ware market. The price of a commodity is per se not fixed, but becomes nearly fixed if supply and demand are in equilibrium. We assume for our considerations that equilibrium is reached. For more information on value and capital see e. g. [35].

2.2 Elementary and expanded form of the value of a commodity

The "equality"

$$x \text{ units of commodity } A \sim y \text{ units of commodity } B$$

means that on the ware market x units of ware A are exchanged for y units of ware B. The value of commodity A is expressed by the value of commodity B, so that x units of commodity A have the same value as y units of commodity B. This means that the value of A is expressed relative to the value of B or is in relative form, and the value of B is in equivalent form. But the value of commodity B can also be expressed by the value of A:

$$y \text{ units of commodity } B \sim x \text{ units of commodity } A.$$

Now B is in a relative form and A is in an equivalent form. This means that the binary relation expressed by pairs

$$(x \text{ units of commodity } A, y \text{ units of commodity } B),$$
$$(y \text{ units of commodity } B, x \text{ units of commodity } A)$$

is symmetric.

The *expanded* form of value is characterized by the fact that x units of commodity A can be expressed by the value of many other commodities:

$$x \text{ units of commodity } A \sim \begin{cases} & y_1 \text{ units of commodity } B_1, \\ \text{or} & y_2 \text{ units of commodity } B_2, \\ \vdots & \\ \text{or} & y_n \text{ units of commodity } B_n. \end{cases}$$

Thus x units of commodity A are then the expanded form of value since it can be expressed relative to y_1 units of commodity B_1, or relative to y_2 units of commodity B_2, …, or relative to y_n units of commodity B_n. But if x units of commodity A have the value of y_1 units of commodity B_1 and also the value of y_2 units of commodity B_2, then, by symmetry, y_1 units of commodity B_1 have the value of x units of commodity A and then also the value of y_2 units of commodity B_2. This means the relation is transitive. During the historical development of commodity exchange, certain amounts of gold became a general equivalent which is much more comfortable, but gold is also a commodity. Later on, money became a general equivalent which was at first not a commodity. We will see why this also works. On the ware market, a commodity will not be exchanged for exactly the same commodity because x units of A cannot be exchanged for $y \neq x$ units of A since the values are different – to exchange x units of A with x units of A makes no sense. As a consequence, the binary relation above, is not reflexive.

2.3 The main theorem for the exchange of commodities

In this section we will set up the mathematical model of the "exchange of commodities," by defining the concept of the *value of a commodity*, and will prove the main result. At first we fix the notation. We require that the commodity market (ware market) is considered in equilibrium between supply and demand, i. e., the market is assumed to be balanced, meaning that "supply = demand."

Let $C := \{W_1, W_2, \ldots, W_n\}$ be a finite set (commodities) and let $\underline{a} := (a_1, \ldots, a_n) \in (\mathbb{Q}^+)^n$ be a finite sequence of positive rational numbers. Then $a_i W_i$ means "a_i units of commodity W_i." Let $\underline{a}C := \{a_1 W_1, a_2 W_2, \ldots, a_n W_n\}$ and

$$\text{lab} : \underline{a}C \to \mathbb{Q}^+$$

be a function which assigns to each element of $\underline{a}C$ a positive rational number q_i, which we denote by $a_i W_i \mapsto q_i \in \mathbb{Q}^+ \setminus \{0\}$. (Here we think of the value of a_i units of commodity W_i expressed by the amount of socially necessary labor, embodied in a_i units of commodity W_i.)

Definition 2.3.1. We say that $C = \{W_1, \ldots, W_n\}$ is a set of commodities if the following conditions are satisfied:

(i) There is a function

$$\text{lab} : \underline{a}C \to \mathbb{Q}^+.$$

(ii) For each i, $1 \leq i \leq n$, there exists an index j, $1 \leq j \leq n$ with $i \neq j$, such that

$$\text{lab}(a_i W_i) = \text{lab}(a_j W_j).$$

Condition (i) means that in each unit of commodity W_i some labor is embodied, which makes W_i a useful object.

Condition (ii) means that there is a market, where a_i units of W_i can be exchanged for a_j units of W_j (generally, the equivalent W_j means the general equivalent money).

The equation

$$\text{lab}(a_i W_i) = \text{lab}(a_j W_j)$$

means that the socially necessary labor which is embodied in a_i units of W_i and in a_j units of W_j is equal. Therefore, the values of the commodities should be equal.

Definition 2.3.2. Let $\sim_{\text{SIR}}$ be the binary relation defined by

$$a_i W_i \sim_{\text{SIR}} a_j W_j :\Leftrightarrow \text{lab}(a_i W_i) = \text{lab}(a_j W_j) \text{ and } i \neq j.$$

Interpretation: $a_i W_i \sim_{\text{SIR}} a_j W_j$ means that a_i units of W_i can be exchanged for a_j units of W_j since the socially necessary human labor embodied in $a_i W_i$ and in $a_j W_j$ is equal. Because $i \neq j$, the commodities W_i and W_j are different, but a_i can be equal a_j since $\underline{a} = (a_1, \ldots, a_n)$ is a sequence.

Lemma 2.3.3. *The relation $\sim_{\text{SIR}}$ is symmetric and irreflexive.*

Proof. Let $a_i W_i \sim_{\text{SIR}} a_j W_j$, i. e., $\text{lab}(a_i W_i) = \text{lab}(a_j W_j)$ and $1 \leq i \neq j \leq n$. Since the equality is symmetric, we also have $a_j W_j \sim_{\text{SIR}} a_i W_i$, since $j \neq i$. Thus relation $\sim_{\text{SIR}}$ is symmetric. Since, by the definition of $\sim_{\text{SIR}}$, the indices i and j are different, the relation $\sim_{\text{SIR}}$ is irreflexive. □

Remark 2.3.4.
1. Relation $\sim_{\text{SIR}}$ is not transitive. Indeed, if $a_iW_i \sim_{\text{SIR}} a_jW_j$, $a_jW_j \sim_{\text{SIR}} a_kW_k$, then $\text{lab}(a_iW_i) = \text{lab}(a_jW_j)$ and $\text{lab}(a_jW_j) = \text{lab}(a_kW_k)$, $i \neq j$ and $j \neq k$. Then $\text{lab}(a_iW_i) = \text{lab}(a_kW_k)$, but i and k can be equal and thus $a_iW_i \nsim_{\text{SIR}} a_kW_k$.
2. On the ware market, a_i units of commodity W_i will not be exchanged for a_i units of W_i. This corresponds to irreflexivity of $\sim_{\text{SIR}}$ in the model. If $a_i \neq a_j$ then on the ware market a_i units of W_i will not be exchanged for a_j units of W_i. By the model, this is also impossible.
3. The model does not consider the use value of a commodity since the use values are not measurable.

Two different commodities are exchangeable if and only if the socially necessary labor is equal. On the other hand, the market shows that they are exchangeable if and only if they have the same value. But what is the exchange value of a commodity? Up to now, this concept is not defined. What we know is that we could measure its value by the labor embodied. In the genesis of these concepts, at first a general equivalent for commodities with the same value appeared, certain amounts of gold or silver. These general equivalents are also commodities. Then money took over the role of gold and silver. But money is not a commodity. To use money, our model needs a set different from $\underline{a}C$.

For a better understanding of our next step, we offer the following example. If small children learn the numbers, for instance, the concept of the natural number denoted by 3, they consider many different sets consisting of three elements. By a process of abstraction, they forget the concrete kind of the elements of these sets, three balls or three apples, or three pieces of chocolate, etc. There is a one-to-one mapping between two three-element sets. On the collection of all finite sets, these one-to-one mappings define an equivalence relation. Indeed, each three-element set is related to itself since the identity mapping is one-to-one and maps a three-element set onto itself. This gives reflexivity. If the one-to-one mapping f maps the three-element set A to the three-element set B, then its inverse f^{-1}, which is also one-to-one, maps B to A, and this leads to the symmetry of the relation. If f_1 maps A to B and f_2 maps B to C, then the composition $f_1 \circ f_2$ is a one-to-one mapping which maps A to C, hence the transitivity. Using the main theorem of equivalence relations, our equivalence relation produces a partition of the collection of all finite sets and the equivalence classes are precisely the sets of the partition. Now one defines:

The cardinal number three (denoted by 3) is the equivalence class of all three-element sets. In Chapter 3 we will come back to these ideas when we define natural numbers in general. Now we follow this epistemological way to use common properties of objects (equivalence relation) to define concepts.

Definition 2.3.5. The subsets

$$[a_iW_i]_{\text{SIR}} := \{a_jW_j \mid a_jW_j \sim_{\text{SIR}} a_iW_i\}$$

of $\underline{a}C$ are said to be SIR-classes.

Let $\Delta(\underline{a}C) := \{(a_iW_i, a_iW_i) \mid i \in \{1,\dots,n\}\}$ and let ker(lab) be the kernel of the function lab.

Proposition 2.3.6. *The* SIR-*classes have the following properties:*
(i) $a_iW_i \notin [a_iW_i]_{\text{SIR}}$ *for each* i, $1 \le i \le n$.
(ii) $[a_iW_i]_{\text{SIR}} \ne \emptyset$ *for each* i, $1 \le i \le n$.
(iii) *If* $\text{lab}^{-1}(q)$ *contains at least two elements for every* $q \in \text{Im(lab)}$, *then* $\sim_{\text{SIR}} \cup \Delta(\underline{a}C) = \text{ker(lab)}$.

Proof. (i) Since $a_iW_i \nsim_{\text{SIR}} a_iW_i$ for each $i \in \{1,\dots,n\}$, we have $a_iW_i \notin [a_iW_i]_{\text{SIR}}$.

(ii) By Definitions 2.3.1(ii) and 2.3.2, for each $i \in \{1,\dots,n\}$ there exists a $j \ne i \in \{1,\dots,n\}$ with $a_iW_i \sim_{\text{SIR}} a_jW_j$.

(iii) If $(a_iW_i, a_jW_j) \in \sim_{\text{SIR}}$, then $\text{lab}(a_iW_i) = \text{lab}(a_jW_j)$ and thus $(a_iW_i, a_jW_j) \in \text{ker(lab)}$. Moreover, $\Delta(\underline{a}C) \subseteq \text{ker(lab)}$. Therefore, $\sim_{\text{SIR}} \cup \Delta(\underline{a}C) \subseteq \text{ker(lab)}$.

Since $\text{lab}^{-1}(q)$ contains at least two elements for every $q \in \text{Im(lab)}$, there exist $i \ne j$ with $(a_iW_i, a_jW_j) \in \text{ker(lab)}$, i. e., $\text{lab}(a_iW_i) = \text{lab}(a_jW_j)$ and $(a_iW_i, a_jW_j,) \in \sim_{\text{SIR}}$. If $i = j$, then $(a_iW_i, a_jW_j,) \in \Delta(\underline{a}C)$. Therefore, $(a_iW_i, a_jW_j) \in \sim_{\text{SIR}} \cup \Delta(\underline{a}C)$ for all $i,j \in \{1,\dots,n\}$. Therefore, $\text{ker(lab)} \subseteq \sim_{\text{SIR}} \cup \Delta(\underline{a}C)$ and altogether we have equality. □

To define the value of a commodity, one needs an equivalence relation. The relation $\sim_{\text{SIR}}$ cannot be used since it is irreflexive. But, as we have seen in Chapter 1, Section 1.7, the kernel of a function is an equivalence relation. Applying the definition (Chapter 1, Section 1.4) of the kernel of a function to the function $\text{lab} : \underline{a}C \to \mathbb{Q}^+$, we have

$$\text{ker(lab)} = \{(a_iW_i, a_jW_j) \mid \text{lab}(a_iW_i) = \text{lab}(a_jW_j)\}.$$

This ker(lab) is an equivalence relation on $\underline{a}C$ and the factor set $\underline{a}C/\text{ker(lab)}$ is the corresponding partition of the set $\underline{a}C$ (see Chapter 1, Theorem 1.7.13). Moreover, by the same theorem, there is a one-to-one mapping

$$f' : \underline{a}C \to \text{Im(lab)} \subseteq \mathbb{Q}^+.$$

Now we formulate

Definition 2.3.7. The value of a_iW_i (a_i units of commodity W_i), denoted by $\text{val}(a_iW_i)$, is the equivalence class $[a_iW_i]_{\text{ker(lab)}}$ of a_iW_i with respect to ker(lab) and is given by the positive rational number $f'([a_iW_i]_{\text{ker(lab)}})$, where f' is the bijection

$$f' : \underline{a}C/\text{ker(lab}^*) \to \text{Im(lab)} \subseteq \mathbb{Q}^+.$$

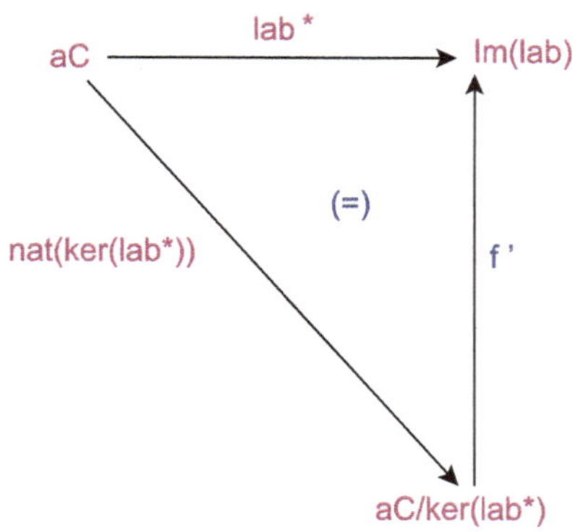

Figure 2.1: Value of a commodity.

Since $\text{lab}^* : \underline{a}C \to \text{Im}(\text{lab}) \subset \mathbb{Q}^+$ is surjective, there exists a uniquely determined bijective function $f' : \underline{a}C/\ker(\text{lab}) \to \text{Im}(\text{lab})$ with

$$f' \circ \text{nat}(\ker(\text{lab})) = \text{lab}^* \quad \text{(Figure 2.1)}.$$

Here f' maps $\text{val}(a_i W_i) = [a_i W_i]_{\ker(\text{lab})}$ to a nonnegative rational number (which is not yet the price of the commodity). In this way, the value of a commodity is uniquely determined (at a certain fixed time which depends on the equilibrium of supply and demand). The number $f'(\text{val}(a_i W_i))$ can be used to calculate the price of the human labor embodied in a_i units W_i, or the price of commodity W_i.

The relationship between the irreflexive and symmetric relation $\sim_{\text{SIR}}$ and the equivalence relation ker(lab) allows using the main theorem for equivalence relations.

Altogether, we have the *main theorem for the exchange of commodities*:

Theorem 2.3.8. *Let $C = \{W_1, \ldots, W_n\}$ be a set of commodities and $\underline{a} = (a_1, \ldots, a_n) \in (\mathbb{Q}^+)^n$. Let $\underline{a}C = \{a_1 W_1, \ldots, a_n W_n\}$ (where $a_i W_i$ means a_i units of commodity W_i). Let $\Delta(\underline{a}C)$ be the identity relation (diagonal) of $\underline{a}C$ and let*

$$\text{lab} : \underline{a}C \to \mathbb{Q}^+$$

be a function. Then

$$\sim_{\text{SIR}} = \ker(\text{lab}) \setminus \Delta(\underline{a}C)$$

if and only if $\text{lab}^{-1}(q)$ contains at least two elements for every $q \in \text{Im}(\text{lab})$.

Proof. Clearly,

$$\sim_{\text{SIR}} = \ker(\text{lab}) \setminus \Delta(\underline{a}C) \Longleftrightarrow \sim_{\text{SIR}} \cup \Delta(\underline{a}C) = \ker(\text{lab}).$$

By Proposition 2.3.6(iii), if $\text{lab}^{-1}(q)$ contains at least two elements for every $q \in \text{Im}(\text{lab})$, then $\sim_{\text{SIR}} \cup \Delta(\underline{a}C) = \ker(\text{lab})$.

If, conversely, $\sim_{\text{SIR}} = \ker(\text{lab}) \setminus \Delta(\underline{a}C)$, then $\sim_{\text{SIR}} \subseteq \ker(\text{lab})$ and thus $[a_iW_i]_{\text{SIR}} \subseteq [a_iW_i]_{\ker(\text{lab})}$. By Proposition 2.3.6, $[a_iW_i]_{\text{SIR}} \neq \emptyset$, i. e., $[a_iW_i]_{\text{SIR}}$ contains at least one element which is different from a_iW_i since $\sim_{\text{SIR}}$ is irreflexive. Then $[a_iW_i]_{\ker(\text{lab})}$ contains at least one element which is different from a_iW_i and, moreover, $a_iW_i \in [a_iW_i]_{\ker(\text{lab})}$ since ker(lab) is reflexive. Then $[a_iW_i]_{\ker(\text{lab})}$ contains for every $i \in \{1,\dots,n\}$ at least two elements. By definition of ker(lab), the function lab maps all elements of $[a_iW_i]_{\ker(\text{lab})}$ to the same element $q \in \mathbb{Q}^+$. Therefore, $\text{lab}^{-1}(q)$ contains at least two elements for every $q \in \text{Im}(\text{lab})$. □

Remark 2.3.9. The main theorem for equivalence relations says that there is a one-to-one correspondence between equivalence relations on a set and partitions of this set. The sets of each partition are precisely the equivalence classes of the corresponding equivalence relation. Epistemologically, equivalence classes are used to define concepts as we did in Definition 2.3.7 with the concept *value of a commodity*. The essential binary relation for commodities is the exchange relation $\sim_{\text{SIR}}$ which is related to the equivalence relation ker(lab) as we formulated in Theorem 2.3.8. This is the reason to call Theorem 2.3.8 the "main theorem for the exchange of commodities."

2.4 Historical remarks

Karl Marx (1818–1883) was born in 1818 in Trier, studied in Berlin and Bonn, and received his doctorate in philosophy from the University of Jena. His philosophical work was influenced by Immanuel Kant, Georg Friedrich Wilhelm Hegel, and Ludwig Feuerbach. Together with his friend Friedrich Engels, he was the founder of the dialectical and historical materialism, a great achievement in scientific thinking up to our days. The economics part of his work is based on the economic analysis of the bourgeois society by Adam Smith (1723–1790) and David Ricardo (1772–1823). In his main work "Das Kapital" Karl Marx analyzed the capitalist mode of production. The three volumes appeared between 1867 and 1894. His deep analysis convinced him that capitalism produces internal tensions which would lead to its self-destruction and replacement by a completely new and better system. In 1872 Marx wrote a preface to the French edition answering a letter written by the French citizen Maurice Lachâtre, who suggested to publish "Das Kapital" as a serial, to become more accessible to working class people. Marx answered that he is very much interested in this idea since he wanted as many people as possible to understand his analysis of the capitalist mode of production. He also mentioned the difficulties. His scientific method, first applied in such a form in the economics literature, makes reading the first chapters rather difficult for many people. On the other hand, he wrote

> There is no royal road to science, and only those who do not dread the fatiguing climb of its steep paths have a chance of gaining its luminous summits.

3 Numbers

The concept of a number is one of the basic concepts of human culture. Numbers allow us to express quantities, to count, and to measure. Numerals, i. e., symbols for numbers, are among the earliest witnesses of writing. Numbers have become an indispensable part of practical human activity and form the foundation of any science, in particular economics. Conversely, the origin and development of numbers have influenced the growth of all branches of mathematics. For references, see [25, 26, 16, 42, 46–48, 19].

https://doi.org/10.1515/9783112234433-003

3.1 Natural numbers

There are several ways towards a mathematical foundation of natural numbers. One is based on the theory of finite sets and another is a recursive definition, let 0 be a natural number and for each natural number n, let its successor n' be a natural number. We mentioned already in Chapter 2 how little children learn to understand the concepts of small natural numbers considering different sets consisting of the same number of elements. Actually, children compare two sets by assigning each element of the first to an element of the second, and conversely. This simple idea leads to the following definition.

Definition 3.1.1. Set M_1 is *equinumerous* to set M_2, written as $M_1 \sim M_2$, if there is a bijective function $f : M_1 \to M_2$.

Equinumerosity is a reflexive, transitive, and symmetric binary relation on a system of sets. Indeed, for any sets M_1, M_2, M_3 of a system $\mathcal{M}$ of sets, we have $M_1 \sim M_1$ (using id_{M_1} as bijection),

$$M_1 \sim M_2,\ M_2 \sim M_3 \implies M_1 \sim M_3$$

(using the bijective composition $f_1 \circ f_2$ of two bijections), and

$$M_1 \sim M_2 \implies M_2 \sim M_1$$

(using the bijective inverse f^{-1} of a bijection $f : M_1 \to M_2$). The equivalence relation $\sim$ defines a partition into classes of equinumerous sets of a system of sets. All sets consisting of the same number of elements belong to the same class.

Definition 3.1.2. Let $\mathcal{M}$ be a system of sets. The cardinality (cardinal number) of $M \in \mathcal{M}$ is the class of all sets which are equinumerous to M, written as $[M]$ or $\operatorname{card} M$,

$$\operatorname{card} M := \{M' \mid M' \in \mathcal{M} \wedge M' \sim M\}.$$

Now natural numbers shall be defined as cardinalities of sets with an appropriate property. The needed property is *finiteness*. What is a finite set? A definition cannot use natural numbers which we only want to define. We use *R. Dedekind's* (1831–1916) definition.

Definition 3.1.3. A set M is said to be finite if there is no proper subset $N \subset M$ which is equinumerous to M, i. e.,

$$M \text{ finite } :\Leftrightarrow \neg\exists N \subset M \wedge N \sim M.$$

If a set M is not finite it is said to be *infinite*. The existence of an infinite set is fixed by the corresponding axiom (see Chapter 1, Section 1.3).

Another idea to define finite sets goes back to *B. Russel* (1872–1970). The mathematician *G. Cantor* (1845–1918) from Halle (Germany) contributed also a lot to the theory of natural numbers.

Any finite set can be obtained starting with the empty set and then successively forming the union with pairwise different one-element sets. Using Russel's axiomatic system of set theory, one can prove:

Proposition 3.1.4. *Let A, B be finite sets. Then the union $A \cup B$, the intersection $A \cap B$, and the Cartesian product $A \times B$ of A and B are finite sets.*

Definition 3.1.5. Let X be an infinite set and let $M \subset X$ be finite. Then $\operatorname{card} M$ is said to be a natural number (finite cardinality).

We mention that this definition is independent on the choice of the infinite set X.

Remark 3.1.6. The empty set $\emptyset$ is defined to be finite and $\operatorname{card}\emptyset := 0$. If $\{\alpha\}$ is a one-element set, then, by Proposition 3.1.4, $\emptyset \cup \{\alpha\}$ is finite and $\operatorname{card}(\emptyset \cup \{\alpha\}) := 1$. If $\operatorname{card} M = n$ and $\alpha \notin M$, then $\operatorname{card}(M \cup \{\alpha\}) = n'$ is the *successor* of n.

The other approach by "continued counting" can be formulated as a system of axioms, called *Peano axioms* (*G. Peano* (1858–1932)) from which all propositions of this theory can be derived. In our approach, Peano's axioms are theorems which can and have to be proved.

Theorem 3.1.7 (Peano theorems). *The system $\mathbb{N}$ is a Peano structure, i. e., a triple* $(\mathbb{N}, 0, \sigma)$, *with the following properties:*

(PI) *There exists an element $0 \in \mathbb{N}$.*

(PII) *There exists a function $\sigma : \mathbb{N} \to \mathbb{N}$ defined by $n \mapsto n'$ for all $n \in \mathbb{N}$; n' is said to be the successor of n and σ is said to be the successor function.*

(PIII) *σ is injective, i. e., $\forall m, n\ (m' = n' \implies m = n)$.*

(PIV) *0 is not a successor.*

(PIV) *Each subset M of $\mathbb{N}$, which contains 0 together with n and n', is equal to $\mathbb{N}$.*

Proof. (PI) Since $\emptyset$ is finite and we defined $\operatorname{card}\emptyset = 0$, we have $0 \in \mathbb{N}$.

(PII) To prove that σ is a function, we have to show that from

$$n = [M] = [M'], \quad \alpha \notin M,\ \alpha' \notin M' \text{ there follows } [M \cup \{\alpha\}] = [M' \cup \{\alpha'\}].$$

By assumption, there exists a bijective function $f : M \to M'$. This f can be extended by $\alpha \to \alpha'$ to a bijective function $g : M \cup \{\alpha\} \to M' \cup \{\alpha'\}$.

(PIII) Let $n' = [M \cup \{\alpha\}]$ and $m' = [N \cup \{\beta\}]$ with $\alpha \notin M, \beta \notin N$, and $n' = m'$. Then there exists a bijection $f : M \cup \{\alpha\} \to N \cup \{\beta\}$. We consider the following cases: (1) $f(\alpha) = \beta$ and (2) $f(\alpha) \neq \beta$. Then $f^{-1}(\beta)$ will be mapped by f to β. By exchanging $f(\alpha)$ with β, the second case can be changed to the first. Then there exists a bijection $g : A \to B$ and $m = n$.

(PIV) Because $a \in M \cup \{a\}$, there is $m' = [M \cup \{a\}] \neq [\emptyset]$.

(PV) Let $M \subseteq \mathbb{N}$ be an arbitrary subset of $\mathbb{N}$ satisfying both assumptions of (PV). We consider $\mathcal{M}^* = \{N \mid N \in \mathcal{E} \wedge [N] \in M\}$ (where $\mathcal{E}$ is the system of all finite sets of X and X is infinite).

This $\mathcal{M}^*$ is a so-called inductive system of sets (for the definition of inductive systems of sets, see [19]):

(i) $\emptyset \in \mathcal{M}^*$ because $0 = [\emptyset] \in M$ and
(ii) from $A \in \mathcal{M}^*$ and $a \in X$ there follows $n = [A] \in M$, i. e., $n' = [A \cup \{a\}] \in M$, and so $A \cup \{a\} \in \mathcal{M}^*$.

From $n = [A] \in \mathbb{N}$, i. e., $A \in \mathcal{E}$, there follows that A is contained in each inductive set system over X and then also in $\mathcal{M}$, i. e., $n = [A] \in M$. This means that $\mathbb{N}$ is a subset of $M : \mathbb{N} \subseteq M$. Since $M \subseteq \mathbb{N}$ is also satisfied, we have the equality $M = \mathbb{N}$. □

Corollary 3.1.8.

(i) *The image of σ is the set of all natural numbers different from* 0.
(ii) *For all $n \in \mathbb{N}$, there holds $n \neq n'$.*
(iii) *Any two Peano structures, $(N_1, 0_1, \sigma_1)$ and $(N_2, 0_2, \sigma_2)$, are in a one-to-one correspondence using a bijective function $\varphi : N_1 \to N_2$, with $\varphi(0_1) = 0_2$ and $\varphi(\sigma_1(n)) = \sigma_2(\varphi(n))$ for all $n \in N_1$.*

For the proof, see [19].

Theorem (PV) can be used to prove statements of the form

$$\forall n \in \mathbb{N}\,(H(n)),$$

where $H(n)$ is a predicate form which contains a variable n for any natural number.

Let $M \subseteq \mathbb{N}$ be the set of all natural numbers for which $H(n)$ is true. To prove that $M = \mathbb{N}$, by (PV), one has to show that $0 \in M$, i. e., that $H(0)$ is true and if $n \in M$, i. e., if $H(n)$ is true, then $n' \in M$, i. e., that $H(n')$ is true. Then by (PV), there follows that $M = \mathbb{N}$. This means that $\forall n \in \mathbb{N}\,(H(n))$ is true.

We will give a simple example of a *proof by induction.*

Example 3.1.9. We want to prove that for each natural number the sum of the first $n+1$ natural numbers, i. e., $0 + 1 + \cdots + n$ can be calculated by

$$\sum_{i=1}^{n} i = \frac{n(n+1)}{2}.$$

Indeed, for $n = 0$ the left-hand side is 0 and the right-hand side is $\frac{0 \cdot 1}{2} = 0$, i. e., $H(0)$ is true. Suppose that $H(n)$ is satisfied, i. e., suppose that $\sum_{i=1}^{n} i = \frac{n(n+1)}{2}$. If we add $n + 1$ on both sides, we obtain

$$0 + 1 + \cdots + n + (n+1) = \frac{n(n+1)}{2} + (n+1) = \frac{n(n+1)}{2} + \frac{2(n+1)}{2}$$

$$= \frac{(n+1)(n+2)}{2}.$$

Therefore, $H(n')$ is true.

In Chapter 1, Section 1.1, we mentioned already that on the set $\mathbb{N}$ of natural numbers addition and multiplication are always defined:

$$+, \cdot : \mathbb{N} \times \mathbb{N} \longrightarrow \mathbb{N}$$

by

$$\forall a, b \in \mathbb{N}\ (a, b) \mapsto a + b \quad \text{and} \quad (a, b) \mapsto a \cdot b, \text{ respectively.}$$

Let A be an arbitrary nonempty set. A function $A \times A \longrightarrow A$ is said to be a *binary operation* on A. We mentioned also that the binary operations $+, \cdot$ satisfy the associative and commutative laws, and that $+$ and $\cdot$ satisfy the distributive law. We know this from our experiences with calculations in $\mathbb{N}$. We know also that 0 is a *zero element* and 1 is a *unit element (identity element)* of $\mathbb{N}$. This means that these elements satisfy

$$\forall a \in \mathbb{N}\ (a + 0 = 0 + a = 0 \text{ and } a \cdot 1 = 1 \cdot a = a).$$

After having defined addition and multiplication of natural numbers, we are able to prove

Theorem 3.1.10. *If on* $\mathbb{N}$ *sum* $+$ *and product* $\cdot$ *are defined by*

$$\begin{aligned} [M] + [N] &:= [M \cup N], \quad \textit{if}\, M \cap N = \emptyset, \\ [M] \cdot [N] &:= [M \times N], \end{aligned}$$

then $+$ *and* $\cdot$ *are commutative and associative, and the distributive laws*

$$\forall a, b, c \in \mathbb{N}\ \big(a \cdot (c + d) = a \cdot c + a \cdot d, (a + b) \cdot c = a \cdot c + b \cdot c\big)$$

are satisfied. Further, 0 *satisfies* $0 + a = a + 0 = a$ *and* 1 *satisfies* $1 \cdot a = a \cdot 1 = a$ *for all* $a \in \mathbb{N}$ *(*0 *and* 1 *are called the neutral or identity elements with respect to addition and multiplication, respectively).*

Proof. We prove at first that $+, \cdot$ are binary operations on $\mathbb{N}$. Since natural numbers are defined as equivalence classes of finite sets, we have to prove the independence of the choice of the representative which we use from each equivalence class.

Let $[M] = [M']$ and $[N] = [N']$ with $M \cap N = \emptyset$ and $M' \cap N' = \emptyset$. Then there exist bijections $f : M \to M'$ and $g : N \to N'$. This gives a bijection $h : M \cup N \to M' \cup N'$ if we set

$$h(x) = \begin{cases} f(x) & \text{if } x \in M, \\ g(x) & \text{if } x \in N. \end{cases}$$

Therefore $[M \cup N] = [M' \cup N']$. By Proposition 3.1.4, $M \cup N, M' \cup N'$ are finite, and $[M \cup N]$ and $[M' \cup N']$ are natural numbers. Since $M \cap N = \emptyset$ and $(M \cup N) \cap P = \emptyset$ are satisfied if and only if $N \cap P = \emptyset$ and $M \cap (N \cup P) = \emptyset$, the sum $([M] + [N]) + [P]$ is defined if and only if the sum $[M] + ([N] + [P])$ is defined. Both sums are equal because

$$[M \cup N] + [P] = [(M \cup N) \cup P] = [M \cup (N \cup P)] = [M] + [N \cup P].$$

The addition is commutative because of

$$[M] + [N] = [M \cup N] = [N \cup M] = [N] + [M],$$

and $M \cap N = \emptyset$ if and only if $N \cap M = \emptyset$. Also

$$[M] + [\emptyset] = [M \cup \emptyset] = [M]$$

shows that $0 = [\emptyset]$ is a zero element.

Now we show that the multiplication is a binary operation on $\mathbb{N}$. From $[M] = [M']$ and $[N] = [N']$, there follows that there are bijections $f : M \to M'$ and $g : N \to N'$. From these bijections, we obtain a bijection $h : M \times N \to M' \times N'$ by setting

$$h((\alpha, \beta)) = (f(\alpha), g(\beta)) =: (f \times g)((\alpha, \beta)).$$

Therefore, $[M \times N] = [M' \times N']$. Since M and N are finite by Proposition 3.1.4, $M \times N$ is finite. There follows

$$([M] \cdot [N]) \cdot [P] = [M] \cdot ([N] \cdot [P]),$$

since the mapping $(M \times N) \times P \to M \times (N \times P)$ with $((a, b), c) \mapsto (a, (b, c))$ is bijective. Thus, the multiplication is associative.

Because $M \times N \to N \times M$ with $(a, b) \mapsto (b, a)$, the multiplication is also commutative, i. e., $[M] \cdot [N] = [N] \cdot [M]$.

Also $[M] \cdot [\{a\}] = [M]$ is satisfied since $M \times \{a\} \to M$ with $(a, a) \mapsto a$ is bijective. This means that $[\{a\}]$ is an identity element.

Because $N \cap P = \emptyset$ if and only if $(M \times N) \cap (M \times P) = \emptyset$, $[M]([N] + [P])$ is defined if and only if $[M] \cdot [N] + [M] \cdot [P]$ is defined. Because $M \times (N \cup P) = (M \times N) \cup (M \times P)$, both terms are equal. □

The following propositions can be proved in a similar way (see [19])

Corollary 3.1.11.

1. $m + n = 0$ *if and only if* $m = 0$ *and* $n = 0$.

2. $m + n = m$ *if and only if* $n = 0$.
3. $m \cdot n = 0$ *if and only if* $m = 0$ *or* $n = 0$.
4. $m \cdot n = 1$ *if and only if* $m = 1$ *and* $n = 1$.
5. $m + n = m^* + n \Longrightarrow m = m^*$ *and* $m \cdot n = m^* \cdot n \Longrightarrow m = m^*$ *(cancellation property for addition and multiplication).*

If we want to express that we are not only interested in the natural numbers as elements of $\mathbb{N}$, but also in the binary operations on $\mathbb{N}$, we write $(\mathbb{N}; +, \cdot)$. If we also want to say that the binary operations satisfy the associative, commutative, and distributive laws we use the concept of a *semiring*.

Definition 3.1.12. We say that $\mathcal{A} = (A; +, \cdot)$ is a semiring if the operations $+$ and $\cdot$ are associative, the addition is commutative, the distributive laws

$$\forall a, b, c \in \mathbb{N}\ (a \cdot (c + d) = a \cdot c + a \cdot d, (a + b) \cdot c = a \cdot c + b \cdot c)$$

are satisfied, and if there is an element $0 \in A$ satisfying

$$\forall a \in A\ (a + 0 = a, a \cdot 0 = 0 \cdot a = 0).$$

A semiring is said to be commutative if the multiplication $\cdot$ is commutative. The semiring $\mathcal{A}$ can have an identity element, i. e., a neutral element with respect to multiplication satisfying $1 \cdot a = a \cdot 1 = a$.

Then we have

Theorem 3.1.13. *The system $(\mathbb{N}; +, \cdot)$ is a commutative semiring (with the zero element 0) with the identity element 1, which has the cancellation properties with respect to $+$ and $\cdot$:*

$$\forall a, x_1, x_2\ (a + x_1 = a + x_2 \Rightarrow x_1 = x_2 \wedge a \cdot x_1 = a \cdot x_2 \Rightarrow x_1 = x_2).$$

Proof. All properties follow from Theorem 3.1.10 and Corollary 3.1.11. □

In Chapter 1, Section 1.1 (also in Section 1.5), we mentioned that on $\mathbb{N}$ the usual order relation $<$ is an irreflexive linear order which is compatible with addition and multiplication. Now we define the relation $<$ on $\mathbb{N}$ and prove its properties.

Definition 3.1.14. We set $n < m :\Leftrightarrow \exists x \in \mathbb{N} \setminus \{0\}\ (n + x = m)$ ("n is less than m").

Theorem 3.1.15. *Relation $<$ is an irreflexive linear order relation on $\mathbb{N}$ which is compatible with respect to addition and multiplication.*

Proof. For all $m \in \mathbb{N}$, the natural number m is not smaller than m, since otherwise there would be a natural number $x \neq 0$ with $m + x = m$, which contradicts Corollary 3.1.11.

Relation $<$ is asymmetric, i. e., if m is less than n, then n is not less than m. Indeed, from $m + x = n$ and $n + y = m$ with $x, y \neq 0$ one has $n + x + y = n$ with $x + y \neq 0$, which contradicts Corollary 3.1.11.

Relation $<$ is transitive, since from $m < n$ and $n < p$, i. e., $m + x = n$ and $n + y = p$ with $x, y \neq 0$, there follows $m + (x + y) = n + y$ and then $m + (x + y) = p$, and so $m < p$ because of $x + y \neq 0$.

Relation $<$ is linear, i. e., for all $m, n \in \mathbb{N}$ there is $m < n$ or $n < m$, or $m = n$. We consider $H(n) : \forall m\, (m < n \vee n < m \vee m = n)$ and give a proof by induction on n.

First, $H(0)$ is true, since either $m = 0$ or $m \neq 0$. From the latter follows $0 < m$ because of $0+m = m$. Now we show that $\forall n\, (H(n) \Rightarrow H(n+1))$, where $H(n+1) : \forall m\, (m < n+1 \vee m = n + 1 \vee n + 1 < m)$. By the inductive hypothesis, we have three cases:

(i) $m = n$. Then $m + 1 = n + 1$ and with $1 = 0' \neq 0$ we get $m < n + 1$.

(ii) $m < n$. Then $m + x = n$ and with $x \neq 0$ we get $m + 1 + x = n + 1$ and from $1 + x \neq 0$ there follows $m < n + 1$.

(iii) $n < m$. Then $n + x = m$ with $x \neq 0$. Therefore, there is an element $y \in \mathbb{N}$ with $y' = x$.

Then from $n + y + 1 = m$ there follows $n + 1 = m$ (resp. $n + 1 < m$) if $y = 0$ (resp. $y \neq 0$).

Now $n < m$ implies $n + l < m + l$ for any $l \in \mathbb{N}$, i. e., $<$ is compatible with respect to addition. Indeed, from $n + x = m$ with $x \neq 0$, we obtain $n + l + x = m + l$, i. e., $n + l < m + l$.

Relation $<$ is also compatible with respect to multiplication since from $n < m$ and $l \neq 0$ we get $n + x = m$ with $x \neq 0$, and thus $(n + x) \cdot l = ml$ with $xl \neq 0$ because $x \neq 0$ and $l \neq 0$ (Corollary 3.1.11). Therefore $nl < ml$. □

We notice that $<$ defined in Definition 3.1.14 is said to be the *natural order* and that from $<$ the reflexive order relation $\leq$ can be defined by

$$n \leq m :\Leftrightarrow n < m \vee n = m.$$

If $n \leq m$ and $m \leq n$ then, using the asymmetry of $<$, one has $m = n$.

The structural properties of the natural numbers can be summarized by the concept of an ordered semiring as a quadruple $(A; <, +, \cdot)$, where $(A; +, \cdot)$ is a semiring and where the irreflexive order relation $<$ is compatible with respect to addition and multiplication.

Corollary 3.1.16. *The system* $\underline{\mathbb{N}} := (\mathbb{N}; <, +, \cdot)$ *is a naturally ordered semiring.*

Definition 3.1.17. An ordered algebraic structure with binary operations consists of a nonempty set C, an irreflexive order relation $<_C$ on C, and binary operations $\circ_C$ on C such that $<_C$ is compatible with respect to the binary operations $\circ_C$. Let $\mathcal{A}$ and $\mathcal{B}$ be two ordered algebraic structures with two binary operations. A function $\varphi : A \to B$ from A to B is said to be an *o*-homomorphism from $\mathcal{A}$ to $\mathcal{B}$, if the following conditions are satisfied:

(i) $\varphi(a \circ_A b) = \varphi(a) \circ_B \varphi(b)$ for all $a, b \in A$ and all binary operations $\circ_A$ of A.

(ii) $a <_A b \Leftrightarrow \varphi(a) <_B \varphi(b)$ for all $a, b \in A$.

Then $\varphi(\mathcal{A}) \subseteq B$ is said to be an o-homomorphic image of $\mathcal{A}$. If φ is bijective, φ is said to be an o-isomorphism onto $\mathcal{B}$ and $\mathcal{A}$ and $\mathcal{B}$ are said to be o-isomorphic (order-isomorphic).

The following result shows the importance of the natural numbers as an ordered semiring among all other similar ordered algebraic structures. For the proof, see [19].

Theorem 3.1.18. *Let $\mathcal{A} = (A; <, +, \cdot)$ be an algebraic structure with a binary relation and two binary operations. Then the following statements are equivalent:*
(i) *$\mathcal{A}$ is o-isomorphic to $\underline{\mathbb{N}} := (\mathbb{N}; <, +, \cdot)$.*
(ii) *$\mathcal{A}$ is a minimal naturally ordered semiring with zero and identity elements.*

(Here $\mathcal{A}$ being minimal means that $\mathcal{A}$ has no proper substructure with these properties.)

3.2 Representation of natural numbers

For calculations, it is important that any natural number can be represented in a positional numeral system. The most important positional numeral systems are the decimal system and the dyadic system.

The following theorem allows the representation in any positional system.

Theorem 3.2.1 (Remainder theorem). *Let n and m be any natural numbers, and let $m \neq 0$. Then there exist unique natural numbers q and r (remainder) such that $n = qm + r$ $(0 \leq r < m)$.*

Proof. We define $M := \{x \mid x \in \mathbb{N} \text{ and } xm \leq n\}$. Because $0m = 0 \leq n$ for all n, we have $0 \in M$, and thus $M \neq \emptyset$. From $1 \leq m$, there follows $x = x1 \leq xm \leq n$. Therefore, set M has an upper bound. We use the well-known fact that every nonempty set of natural numbers which has an upper bound has a maximum $q := \max M$. This maximum satisfies $qm \leq n$, but $(q+1)m > n$. This shows the existence of a natural number r with $qm + r = n$ and $n \geq 0$. Because $qm + r < (q+1)m = qm + m$, there follows $r < m$, since from $r \geq m$ we would obtain $qm + r \geq qm + m$. This shows the existence of the representation.

Suppose there are two representations $n = q_1m + r_1 = q_2m + r_2$ with $0 \leq r_1, r_2 < m$. Without loss of generality, we assume that $q_2 \leq q_1$. Then $0 \leq (q_1 - q_2)m = r_2 - r_1 < m$ and also $0 = q_1 - q_2$, therefore $0 = r_2 - r_1$, i. e., $q_1 = q_2$ and $r_1 = r_2$. □

Theorem 3.2.2 (Representation theorem). *Any natural number $n \geq 1$ has precisely one representation $n = a_k10^k + a_{k-1}10^{k-1} + \cdots + a_110 + a_0$ in the decimal system, where $a_k, \ldots, a_0$ are natural numbers with $0 \leq a_i \leq 9$, $i = 0, 1, 2, \ldots, k$ and $a_k \neq 0$. We write $n = a_ka_{k-1}\ldots a_0$.*

Proof. Dividing n by 10 and with the remainder, one has $n = q_010 + a_0$, $0 \leq a_0 \leq 9$. Dividing q_0 by 10 with the remainder gives $q_0 = q_110 + a_1$, $0 \leq a_1 \leq 9$. After finitely many steps, the procedure stops with $q_k = 0$, since $n > q_1 > q_2 > \cdots > q_k = 0$. Substituting the term obtained for q_0 in the first equation, the term for q_1 in the second equation, etc., one obtains

$$n = a_k 10^k + a_{k-1} 10^{k-1} + \cdots + a_1 10 + a_0.$$

The uniqueness of this representation can be proved going from $n = q_0 10 + a_0 = q_0' 10 + a_0'$ to $a_0 = a_0'$, $q_0 = q_0'$, then from $q_0 = q_0'$ to $a_1 = a_1'$, $q_1 = q_1'$, etc., and finally, to $a_k = a_k'$. □

This procedure can be generalized for each natural number $m > 1$ as basis (m-adic positional system).

Theorem 3.2.3 (Representation in arbitrary positional system). *Each natural number $n \geq 1$ has precisely one representation $n = a_k m^k + a_{k-1} m^{k-1} + \cdots + a_1 m + a_0$ in the m-adic positional system, where $a_k, \ldots, a_0$ are natural numbers with $0 \leq a_i \leq m-1$, $i = 0, 1, 2, \ldots, k$ and $a_k \neq 0$ (m-adic digits). This representation will be written as $n = (a_k a_{k-1} \ldots a_0)_m$.*

The representation of natural numbers in the dyadic system, i. e., $m = 2$, is especially important. There are only two digits $0, 1$ which correspond to the two states of an electrical system.

3.3 Nonnegative rational numbers

In $\mathbb{N}$, addition and multiplication are always defined, but not division. There are natural numbers a, b such that $a : b$ is not a natural number. Our aim is to extend $\mathbb{N}$ to a number system in which division, except for division by 0, is always possible. Moreover, we want to extend $(\mathbb{N}; <, +, \cdot)$ in such a way that no property of $(\mathbb{N}; <, +, \cdot)$ is lost. This concerns the relation $<$, operations $+, \cdot$, but also the laws which are satisfied in $(\mathbb{N}; <, +, \cdot)$. Algebraically, this means that we want to construct an algebraic structure which contains $(\mathbb{N}; <, +, \cdot)$ as a substructure and in which division, except for division by 0, is always possible. It makes sense to look for a minimal upper structure with these properties. We will solve this problem for any commutative semiring $(A; +, \cdot)$ having the cancellation property

$$\forall a, x, y \in A\ (a + x = a + y \Rightarrow x = y).$$

We also need the zero element of the semiring

$$\forall a \in A\ (a \cdot 0 = 0)$$

since the semiring of natural numbers has both properties. We will use the following method. On the set $A \times A \setminus \{0\}$, we will define an equivalence relation $\sim$, form the factor set $(A \times A \setminus \{0\})/\sim$, and define addition and multiplication on the factor set in such a way that our new algebraic structure fulfils all the requirements. To learn how we have to define ~, +, and ·, we assume at first that the upper structure which we want to have exists, and then we prove the existence and uniqueness of the new structure.

Definition 3.3.1. An algebraic structure $\mathcal{A} = (A; +, \cdot)$ is said to be a semifield if $\mathcal{A} = (A; +, \cdot)$ is a semiring and if the multiplication is invertible (except at 0), i. e.,

$$\forall a, b \in A\ \exists x, y \in A\ (ax = b \wedge ya = b),$$

where $a \neq 0$.

Theorem 3.3.2 (Structure theorem). *Let $\mathcal{H}$ be a minimal upper semifield of a (nontrivial) commutative semiring $\mathcal{A} = (A, +, \cdot)$ satisfying the cancellation rule. Then $\mathcal{H}$ is also commutative and each element $x \in \mathcal{H}$ can be written as a quotient $x = ab^{-1} = \frac{a}{b}$, $a, b \in A$, $b \neq 0$. Calculations in $\mathcal{H}$ reduce to calculations in $\mathcal{A}$ using the following rules:*

$$\begin{array}{ll} \text{I.} & \dfrac{a}{b} = \dfrac{a'}{b'} \quad \textit{if and only if} \quad ab' = a'b, \\ \text{II.} & \dfrac{a}{b} + \dfrac{c}{d} = \dfrac{ad + bc}{bd}, \\ \text{III.} & \dfrac{a}{b} \cdot \dfrac{c}{d} = \dfrac{ac}{bd}. \end{array}$$

Proof. In $\mathcal{H}$, each equation of the form $xb = a$, $a, b \in A$, $b \neq 0$ is uniquely solvable and, because of the commutativity of $\mathcal{A}$, the uniquely determined solution can be written as $x = ab^{-1} = \frac{a}{b}$. (From $ab = ba$, by multiplying with $b^{-1} \in H$ in $\mathcal{H}$ there follows $a = bab^{-1}$ and then $b^{-1}a = ab^{-1}$.) Let

$$B := \left\{ \frac{a}{b} \,\middle|\, a, b \in A \wedge b \neq 0 \right\}$$

be the set of these quotients.

Now we show that $\mathcal{B} \subseteq \mathcal{H}$ is a semifield which contains $\mathcal{A}$ and is equal to $\mathcal{H}$ because of the minimality of $\mathcal{H}$.

At first, we prove that rules I, II, and III are satisfied for all elements of B:

I.
$$ab^{-1} = a'b'^{-1} \implies ab^{-1}bb' = a'b'^{-1}b'b \implies ab' = a'b \quad \text{and}$$
$$ab' = a'b \implies ab'b'^{-1}b^{-1} = a'bb^{-1}b'^{-1} \implies ab^{-1} = a'b'^{-1}.$$

We remark that all calculations are in $\mathcal{H}$ and therefore correctly defined.

II. In $\mathcal{H}$, there holds $ab^{-1} + cd^{-1} = add^{-1}b^{-1} + cbb^{-1}d^{-1} = (ad + bc)(bd)^{-1}$.

III. In $\mathcal{H}$, one has $b^{-1}c = cb^{-1}$, since

$$ab^{-1} \cdot cd^{-1} = acb^{-1}d^{-1} = (ac)(bd)^{-1}.$$

Rules II and III show that the addition and multiplication give results in B, thus $\mathcal{B}$ is a subsemiring of $\mathcal{H}$. The commutativity also follows from rules II and III.

Furthermore, $e_\mathcal{B} := \frac{d}{d}$ $(d \in A,\ d \neq 0)$ is an identity element of $\mathcal{B}$, since by rules II and I, for all $\frac{a}{b} \in B$, $\frac{a}{b} \cdot \frac{d}{d} = \frac{ad}{bd} = \frac{a}{b}$. Here we can choose any $d \in A$, $d \neq 0$, since by rule I, $e_B = \frac{d}{d} = \frac{d'}{d'}$ for all $d, d' \in A$ with $d, d' \neq 0$.

In $\mathcal{B}$, there exists an inverse for each $\frac{a}{b}$ with $a \neq 0$, $b \neq 0$, namely $\frac{b}{a}$, since from rule III there follows $\frac{a}{b} \cdot \frac{b}{a} = \frac{ab}{ba} = \frac{ab}{ab} = e_B$. This means that all elements of $\mathcal{B}$ which are different from the zero element have inverses. Indeed, $\frac{0}{b}$ is for any $b \neq 0$ the zero element of $\mathcal{B}$ because

$$\frac{a}{c} + \frac{0}{b} = \frac{ab + 0c}{cb} = \frac{ab}{cb} = \frac{a}{c}.$$

Moreover, we have $\frac{0}{b} = \frac{0}{b'}$, due to rule I.

There holds $\mathcal{A} \subseteq \mathcal{B}$ because

$$a = a(bb^{-1}) = (ab)b^{-1} = \frac{ab}{b}, \quad \text{for each } a \in A \text{ and } b \neq 0.$$

Consequently, $\mathcal{B}$ is a subsemifield of $\mathcal{H}$, which contains $\mathcal{A}$ and hence $\mathcal{B} = \mathcal{H}$. □

Definition 3.3.3. A commutative upper semifield of a semiring $\mathcal{A}$, whose elements can be written as quotients $ab^{-1} = \frac{a}{b}$ of elements a and $b \neq 0$ from A, is said to be a quotient semifield of $\mathcal{A}$.

In a quotient semifield $\mathcal{H}$ of $\mathcal{A}$, each equation $xb = a$ with $a \in A$ and $b \in A$, $b \neq 0$, has exactly one solution from $\mathcal{H}$, and $\mathcal{H}$ consists of all these solutions. Therefore, no proper subset of H containing A can be a semifield.

Theorem 3.3.4. *Each minimal upper semifield of a (nontrivial) commutative semiring $\mathcal{A}$ which satisfies the cancellation rule is a quotient semifield of $\mathcal{A}$, and conversely.*

Now we got some ideas how a minimal upper semifield of a commutative semiring with a zero element and cancellation rule looks like and can show its existence.

Theorem 3.3.5 (Existence theorem). *For each (nontrivial) commutative semiring $\mathcal{A}$ in which the cancellation rule is satisfied, there exists a quotient semifield $\mathcal{H}$.*

Proof. Let $A^* = A \setminus \{0\}$ where A is a nonempty set with at least two elements. Our construction set is the set $A \times A^*$ of all ordered pairs (a, b), $a, b \in A$, $b \neq 0$. On this set we define a binary relation $\sim$ by

$$(a, b) \sim (a', b') \quad \text{if and only if} \quad ab' = a'b$$

(looking at the equality of quotients).

This relation is:

- reflexive because of $ab = ab$;
- symmetric, since from $ab' = a'b$ there follows $a'b = ab'$; and

- transitive, since from $ab' = a'b$ and $a'b'' = a''b'$ there follows $ab''b' = a'b''b = a''b'b$ and then, since $b \neq 0$ and due to the cancellation rule in $\mathcal{A}$, we get $ab'' = a''b$.

Let $A \times A^*/\sim$ be the set of all equivalence classes with respect to the equivalence relation $\sim$, i. e., the factor set. Let $[a, b]$ be the class of the pair (a, b). We recall that the equality of equivalence classes corresponds to the equivalence of the pairs. We define addition and multiplication on $A \times A^*/\sim$ by looking at the operations for quotients:

I. $[a, b] = [a', b']$ if and only if $(a, b) \sim (a', b')$, if and only if $ab' = a'b$.
II. $[a, b] + [c, d] := [ad + bc, bd]$;
III. $[a, b] \cdot [c, d] := [ac, bd]$.

To show that these definitions really give operations on $A \times A^*/\sim$, we have to prove that the sum and product of the classes are independent of the choice of the representatives (a, b) and (c, d). Indeed, from $[a, b] = [a', b']$ and $[c, d] = [c', d']$, i. e., from $ab' = a'b$ and $cd' = c'd$, there follows

$$\begin{aligned} (ad + bc)b'd' &= (a'd' + b'c')bd, \quad \text{i. e.,} \\ [ad + bc, bd] &= [a'd' + b'c', b'd'], \end{aligned}$$

for the sum, and similarly one concludes for the product. We notice that the addition can be equivalently defined by

$$[a, b] + [c, b] = [a + c, b].$$

With these definitions, $A \times A^*/\sim$ becomes a commutative semiring with a zero element, in which the cancellation rule is satisfied since

$$\begin{aligned} ([a, b] + [c, d]) + [e, f] &= [ad + bc, bd] + [e, f] \\ &= [(ad + bc)f + bde, bdf] \\ &= [adf + b(cf + de), bdf] \\ &= [a, b] + ([c, d] + [e, f]); \\ ([a, b] \cdot [c, d]) \cdot [e, f] &= [ac, bd] \cdot [e, f] = [ace, bdf] \\ &= [a, b] \cdot [ce, df] \\ &= [a, b]([c, d] \cdot [e, f]); \\ [a, b] \cdot [c, d] &= [ac, bd] = [ca, db] = [c, d] \cdot [a, b]; \\ [a, b] \cdot ([c, d] + [e, f]) &= [a, b] \cdot [cf + de, df] = [a(cf + de), bdf] \\ &= [acbf + bdae, bdbf] = [a, b] \cdot [c, d] + [a, b] \cdot [e, f]; \end{aligned}$$

$$\begin{aligned} [a, b] \cdot [x, y] = [a, b] \cdot [u, v] &\Longrightarrow [ax, by] = [au, bv] \Longrightarrow \\ axbv = byau &\Longrightarrow abxv = abyu \Longrightarrow xv = yu \Longrightarrow [x, y] = [u, v], \end{aligned}$$

for all $[a, b] \neq [0, b]$, and here $[0, b]$ is the zero element because

$$[c,d] + [0,b] = [cb + d0, db] = [cb, db] = [c,d].$$

For each $d \in A$ with $d \neq 0$, the class $[d,d]$ is, due to

$$[a,b] \cdot [d,d] = [ad, bd] = [a,b],$$

the one element, and $[b,a]$ for $b \neq 0$, i. e., $[a,b]$ different from the zero element, is, due to

$$[a,b] \cdot [b,a] = [ab, ba] = [ab, ab],$$

the inverse of $[a,b]$. Thus, $A \times A^*/\sim$ becomes a semifield. Unfortunately, $\underline{A \times A^*}/\sim$ is not an upper semifield of $\mathcal{A}$. (Recall that $A \times A^*/\sim$ consists of equivalence classes of ordered pairs of elements from A.) Fortunately, $A \times A^*/\sim$ contains a semiring $\mathcal{A}'$ which is isomorphic to $\mathcal{A}$. This is the set

$$A' := \{[ab, b] \mid a, b \in A \wedge b \neq 0\}.$$

Indeed, $a \mapsto \bar{a} = [ab, b]$ defines an isomorphism of $\mathcal{A}$ onto $\mathcal{A}'$ as was shown in [19]. Also $\mathcal{A}'$ is a commutative subsemiring of $\underline{A \times A^*}/\sim$ with the cancellation rule, which is isomorphic to $\mathcal{A}$. Furthermore, A' can be removed from $A \times A^*/\sim$ and substituted by A. This was proved in [19].

Since

$$[a,b] = [ab, b] \cdot [b, bb] = [ab, b] \cdot [bb, b]^{-1},$$

each element of $A \times A^*$ can be written as a quotient of elements from A'. By this construction, $\underline{A \times A'}/\sim$ is a quotient semifield of the commutative semiring $\mathcal{A}'$ with the cancellation rule, and the replacement and substitution procedure finishes the proof. □

It is left to prove that our construction produces a uniquely determined minimal upper semifield of a commutative semiring with the zero element and cancellation rule.

Theorem 3.3.6. *Each semifield $\mathcal{H}^*$ which contains the (nontrivial) commutative semiring $\mathcal{A}$ with the cancellation property contains exactly one minimal upper semifield $\mathcal{H}$ of $\mathcal{A}$. Two minimal upper semifields $\mathcal{H}_1$, $\mathcal{H}_2$ of $\mathcal{A}$ are isomorphic as quotient semifields of $\mathcal{A}$.*

Proof. Let $\mathcal{H}$ be the intersection of all upper semifields of $\mathcal{A}$ contained in $\mathcal{H}^*$. One can show that $\mathcal{H}$ itself as an intersection is a uniquely determined upper semifield of $\mathcal{A}$ [19]. If $\mathcal{H}_1^*$ and $\mathcal{H}_2^*$ are two different upper semifields of $\mathcal{A}$, then the minimal upper semifields $\mathcal{H}_1$ and $\mathcal{H}_2$, respectively, contained in $\mathcal{A}$, can be different. By the structure theorem, H_1 consists of all quotients $x_1 = ab^{-1} = \frac{a}{b} \in H_1$, $b \neq 0$, of elements from A constructed in $\underline{H_1^*}$ and H_2 consists of all quotients $x_2 = ab^{-1} = \frac{a}{b} \in H_2$, $b \neq 0$, of elements from A constructed in $\mathcal{H}_2^*$. Calculations in $\mathcal{H}_1$ and $\mathcal{H}_2$ follow the rules I, II, III, and are determined by calculations in $\mathcal{A}$. By $x_1 \mapsto x_2$ an isomorphism of $\mathcal{H}_1$ onto $\mathcal{H}_2$ can be defined. □

Now we apply our results to the semiring of natural numbers and obtain

Corollary 3.3.7. *To the semiring* $\underline{\mathbb{N}} = (\mathbb{N};\, +,\, \cdot)$ *of natural numbers there exists up to isomorphism exactly one minimal upper semifield, the quotient semifield* $\underline{Q(\mathbb{N})}$*, which consists of all quotients*

$$ab^{-1} = \frac{a}{b}, \quad a, b \in \mathbb{N} \text{ and } b \neq 0.$$

Calculations with quotients are uniquely determined by rules I, II, *and* III.

The structural properties of the set $\mathbb{Q}^+$ of nonnegative rational numbers can be summarized as follows:

Definition 3.3.8. The quotient semifield $Q(\underline{\mathbb{N}})$ of the commutative semiring $\underline{\mathbb{N}} = (\mathbb{N};\, +,\, \cdot)$ with the cancellation rule is said to be the semifield of nonnegative rational numbers $\underline{\mathbb{Q}^+} = (\mathbb{Q}^+; +, \cdot)$.

By this definition, $\underline{\mathbb{Q}^+}$ consists of all quotients $\frac{m}{n}$, $m, n \in \mathbb{N}$, $n \neq 0$. The calculation in $\underline{\mathbb{Q}^+}$ follows the rules

$$\text{I.} \quad \frac{m}{n} = \frac{m'}{n'} \quad \text{if and only if } mn' = m'n,$$

$$\text{II.} \quad \frac{m}{n} + \frac{p}{q} := \frac{mq + np}{nq},$$

$$\text{III.} \quad \frac{m}{n} \cdot \frac{p}{q} := \frac{mp}{nq}.$$

By our construction, $\underline{\mathbb{Q}^+}$ is a minimal upper semifield of the semiring $\underline{\mathbb{N}}$. This means that for each upper semifield $\underline{H}^*$ of $\underline{\mathbb{N}}$ with $\underline{\mathbb{N}} \subset \underline{H}^* \subseteq \underline{\mathbb{Q}^+}$, one has $\underline{H}^* = \underline{\mathbb{Q}^+}$. Conversely, each minimal upper semifield of $\underline{\mathbb{N}}$ is a quotient semifield of $\underline{\mathbb{N}}$ and therefore isomorphic to $\underline{\mathbb{Q}^+}$.

As for natural numbers, also on the set of nonnegative rational numbers we want to define an irreflexive order relation.

Definition 3.3.9. Let $\frac{m}{n}, \frac{m'}{n'} \in \mathbb{Q}^+$. Then we define a binary relation $<$ on $\mathbb{Q}^+$ by

$$\frac{m}{n} < \frac{m'}{n'} \quad \text{if and only if } mn' < m'n.$$

Theorem 3.3.10. *The relation* $<$ *on* $\mathbb{Q}^+$ *is an irreflexive order relation in* $\mathbb{Q}^+$*, which is monotone with respect to addition and multiplication in* $\mathbb{Q}^+$*, which means that the addition and multiplication are compatible with* $<$*. Altogether,* $\underline{\mathbb{Q}^+} = (\mathbb{Q}^+; <, +, \cdot)$ *is a naturally ordered semifield.*

Proof. At first, we show that the definition of $<$ is independent of the representatives.

From

$$\frac{m}{n} = \frac{m'}{n'}, \quad \frac{p}{q} = \frac{p'}{q'}, \quad \text{and} \quad \frac{m}{n} < \frac{p}{q},$$

there follows, using rule I and the definition of <,

$$mn' = m'n, \quad pq' = p'q, \quad \text{and} \quad mq < pn,$$
$$mn'qq' = m'nqq', \quad mqn'q' < pnn'q' \quad (q \neq 0,\ q' \neq 0) \ \text{(by monotonicity)}$$

and

$$pq'nn' = p'qnn', \quad \text{thus } m'nqq' < p'qnn', \text{ i. e., } m'q'nq < p'n'nq.$$

Now, because $n \neq 0$, $q \neq 0$, we get $m'q' < p'n'$, and then by definition,

$$\frac{m'}{n'} < \frac{p'}{q'}.$$

Since $mn \not< mn$ in $\mathbb{N}$, we obtain $\frac{m}{n} \not< \frac{m}{n}$ for all $\frac{m}{n} \in \mathbb{Q}^+$, and so < is irreflexive.

Moreover, < is asymmetric, since from $\frac{m}{n} < \frac{p}{q}$ and $\frac{p}{q} < \frac{m}{n}$, we have $mq < pn$ and $pn < mq$, which contradicts the asymmetry of the order in $\mathbb{N}$.

Relation < is transitive, since $\frac{m}{n} < \frac{m'}{n'}$, i. e., $mn' < m'n$, and $\frac{m'}{n'} < \frac{p}{q}$, i. e., $m'q < n'p$, imply $mn'q < m'nq$ and $m'nq < n'pn$, and then, by the transitivity of the order in $\mathbb{N}$, $mqn' < pnn'$ and, because of $n' \neq 0$, $mq < pn$, i. e., $\frac{m}{n} < \frac{p}{q}$.

Relation < is a linear order, i. e., for all $\frac{m}{n}, \frac{p}{q} \in \mathbb{Q}^+$ we have $\frac{m}{n} < \frac{p}{q}$, $\frac{p}{q} < \frac{m}{n}$, or $\frac{m}{n} = \frac{p}{q}$. This follows easily from the linearity of the order in $\mathbb{N}$. From $\frac{m}{n} < \frac{p}{q}$ there follows

$$\frac{m}{n} + \frac{m'}{n'} < \frac{p}{q} + \frac{m'}{n'}$$

for arbitrary $\frac{m'}{n'} \in \mathbb{Q}^+$, since from $mq < pn$ there follows

$$mn'qn' < pn'nn' \quad \text{and} \quad mn'qn' + m'nqn' < pnn'n' + m'qnn'$$

by the monotonicity of the addition in $\underline{\mathbb{N}}$, thus

$$(mn' + m'n)qn' < (pn' + m'q)nn',$$

and then

$$\frac{mn' + m'n}{n'n} < \frac{pn' + m'q}{qn'}, \quad \text{i. e.,} \quad \frac{m}{n} + \frac{m'}{n'} < \frac{p}{q} + \frac{m'}{n'}.$$

Let $\frac{m}{n} < \frac{p}{q}$ and $\frac{m'}{n'}$ be different from the zero element in $\mathbb{Q}^+$, i. e., $m' \neq 0$. Then $mm'qn' < pm'nn'$ ($m' \neq 0$, $n' \neq 0$) from $mq < pn$ and the compatibility of $<$ with respect to the multiplication in $\underline{\mathbb{N}}$. This means

$$\frac{mm'}{nn'} < \frac{pm'}{qn'} \quad \text{and} \quad \frac{m}{n} \cdot \frac{m'}{n'} < \frac{p}{q} \cdot \frac{m'}{n'},$$

and we obtain the compatibility of $<$ with respect to the multiplication. For the proof that $<$ is a natural order, see [19]. For the proof of the transitivity, we used that $m < p$ in $\mathbb{N}$ follows from $mn' < pn'$ ($n' \neq 0$). Otherwise, if $m \geq p$, then, by the definition of the order in $\mathbb{N}$, there would be an element $x \in \mathbb{N}$ with $m = p + x$, and then $mn' = pn' + xn'$, and therefore $mn' \geq pn'$. □

It is important to notice that the order relation defined in $\mathbb{Q}^+$ extends the order in $\mathbb{N}$.

Proposition 3.3.11. *The order defined in $\mathbb{Q}^+$ extends the order in $\mathbb{N}$ and is the only possible natural order relation on $\mathbb{Q}^+$ which extends the order in $\mathbb{N}$.*

Proof. The natural numbers m, p can be written in $\mathbb{Q}^+$ as $\frac{mn}{n}$ and $\frac{pn}{n}$ with $n \neq 0$. Then $\frac{mn}{n} < \frac{pn}{n}$ in $\mathbb{Q}^+$ if and only if $mnn < pnn$, which means $m < p$ in $\mathbb{N}$.

Let $<\cdot$ be another natural order relation in $\mathbb{Q}^+$, which extends the order in $\mathbb{N}$. Then

$$\frac{m}{n} <\cdot \frac{p}{q} \Longrightarrow \frac{m}{n} \cdot \frac{nq}{1} <\cdot \frac{p}{q} \cdot \frac{nq}{1},$$

since $n \neq 0$ and $q \neq 0$ and, moreover, $mq <\cdot pn$, i. e., $mq < pn$ in $\mathbb{N}$, and thus, by our definition, $\frac{m}{n} < \frac{p}{q}$ in $\mathbb{Q}^+$, and conversely. □

Finally, we can prove the following algebraic characterization of the semifield of all nonnegative rational numbers.

Theorem 3.3.12. *The semifield $\underline{\mathbb{Q}^+} = (\mathbb{Q}^+; <, +, \cdot)$ is a minimal, naturally ordered semifield, and every such semifield $\underline{H}$ is o-isomorphic (Definition 3.1.17) to $\mathbb{Q}^+$.*

Proof. Each ordered semifield $\underline{U}$ of $\underline{\mathbb{Q}^+}$ is equal to $\mathbb{Q}^+$, since from $1 \in \underline{U}$ there follows $\underline{\mathbb{N}} \subseteq \underline{U}$ and, by the structure theorem, we get $\underline{\mathbb{Q}^+} \subseteq \underline{U}$, thus $\underline{\mathbb{Q}^+} = \underline{U}$.

If, conversely, $\underline{H}$ is a minimal naturally ordered semifield, then $\underline{H}$ contains an identity element e and therefore a minimal ordered subsemiring $\underline{N'}$ with e as the identity element, namely the intersection of all subsemirings of $\underline{H}$ which contain e. Then $\underline{H}$ is, as a minimal naturally ordered semifield, also a minimal ordered upper semifield of $\underline{N'}$ ($\underline{H} = Q(\underline{N'})$). Since $\underline{N'}$ is o-isomorphic to $\underline{\mathbb{N}}$, the proof is finished. □

Measuring in economy concerns the national income, domestic investment, net export, gross domestic product (GDP), meaning national economy's total output, monetary measure, etc. There is always a huge amount of data, which are mostly given by positive rational numbers with a unit, sometimes by negative rational numbers, but never by irrational numbers, i. e., by infinite nonperiodic decimal fractions. In the following section, we will come to the representation of nonnegative rational numbers.

3.4 Representation of nonnegative rational numbers

Definition 3.4.1. An (infinite) nonnegative decimal fraction

$$a = a_0.a_1a_2\dots$$

is a sequence of natural numbers $a_0, a_1, a_2, \dots$ with $0 \le a_i \le 9$ for $i \ge 1$ and $a_0 \in \mathbb{N}$. The numbers $a_1, a_2, \dots$ are said to be *the decimal digits* of a. Also

$$s_n = a_0 + \frac{a_1}{10} + \dots + \frac{a_n}{10^n}$$

and $s'_n = s_n + \frac{1}{10^n}$, $n = 0, 1, 2, \dots$, are said to be the *lower and upper approximation fractions* of a, respectively. The number a is said to be *periodic* if a has the form

$$a = a_0.a_1\dots a_l a_{l+1}\dots a_{l+k}a_{l+k+1}\dots a_{l+2k}\dots,$$

in short

$$a = a_0.a_1\dots a_l\overline{a_{l+1}\dots a_{l+k}},$$

where

$$\begin{aligned} a_{l+1} &= a_{l+k+1} = a_{l+2k+1}, \dots \\ a_{l+2} &= a_{l+k+2} = a_{l+2k+2}, \dots, \\ &\vdots \end{aligned}$$

i. e., if from index l onwards, k digits in the same order are repeated. Also $a_1 \dots a_l$ is said to be the *preperiod of length* l and $a_{l+1}\dots a_{l+k}$ *the period of length* k. Moreover, we introduce the following notions:
- if $k = 1$, $a_{l+1} = 0$, then a is said to be *a finite decimal fraction,*
- if $l = 0$, then a is said to be *a purely periodic decimal fraction, and*
- if $k = 1$, $a_{l+1} = 9$, then a is said to be *a decimal fraction with* 9 *as its period.*

We mention that $\mathbb{Q}^+$ is Archimedean ordered [19], i. e.,

$$\forall a, b \in \mathbb{Q}^+, b' \in \mathbb{Q}^+ \setminus \{0\}\ \exists n \in \mathbb{N}\ (nb > a).$$

Corollary 3.4.2. *The approximation fractions of a decimal fraction have the following properties:*

1. (s_n) *is a monotonically increasing sequence.*
2. (s'_n) *is a monotonically decreasing sequence.*
3. *For all* $n, m \in \mathbb{N}$ *there holds* $s_n < s'_m$.
4. *For each* $t \in \mathbb{Q}^+ \setminus \{0\}$, *there exists an* $n \in \mathbb{N}$, *such that* $s'_n - s_n < t$.

Proof. 1. One has $s_{n+1} - s_n = \frac{a_{n+1}}{10^{n+1}} \geq 0$ for $n = 0, 1, 2, \ldots$

2. Also

$$s'_n - s'_{n+1} = s_n + \frac{1}{10^n} - s_{n+1} - \frac{1}{10^{n+1}} = \frac{10-1}{10^{n+1}} - \frac{a_{n+1}}{10^{n+1}} = \frac{9 - a_{n+1}}{10^{n+1}} \geq 0$$

for $n = 0, 1, 2, \ldots$

3. If $n \leq m$, then $s_n \leq s_m < s'_m$ because of claim 1, and if $n \geq m$, then $s_n < s_n \leq s_m$ because of claim 2.

4. Since $\mathbb{Q}^+$ is Archimedean ordered, there exists an n with $1 < nt$. Together with $n < 10^n$, then $1 < 10^n t$ and thus

$$s'_n - s_n = \frac{1}{10^n} < t. \qquad \square$$

Everybody knows how to get the decimal fraction from the fraction $\frac{n}{m}$, $m \neq 0$, simply by division. Having a closer look at the division algorithm helps in obtaining a deeper insight and in understanding the mathematical background.

Theorem 3.4.3. *Applying the division algorithm DA, one obtains from* $s = \frac{p}{q}$, $p \in \mathbb{N}$, $q \in \mathbb{N} \setminus \{0\}$ *a uniquely determined decimal fraction* $DA(s) = a_0.a_1a_2\ldots$, *which is independent of the representative of* $\frac{p}{q}$.

DA consists in the following repeated division with remainder:

$$\begin{aligned} p &= a_0 q + r_1, \quad 0 \leq r_1 < q, \\ 10r_1 &= a_1 q + r_2, \quad 0 \leq r_2 < q, \\ 10r_2 &= a_2 q + r_3, \quad 0 \leq r_3 < q, \\ &\vdots \\ 10r_n &= a_n q + r_{n+1}, \quad 0 \leq r_{n+1} < q, \end{aligned}$$

DA gives the following uniquely determined representation of s:

$$s = a_0 + \frac{a_1}{10} + \cdots + \frac{a_n}{10^n} + \sigma_n, \quad a_i \in \mathbb{N},\ 0 \leq a_i \leq 9 \textit{ for } i > 0,\ 0 \leq \sigma < \frac{1}{10^n}.$$

Proof. From the given equations, we obtain:

$$\frac{p}{q} = a_0 + \frac{r_1}{q} = a_0 + \frac{a_1}{10} + \frac{1}{10}\frac{r_2}{q}$$

$$= a_0 + \frac{a_1}{10} + \frac{a_2}{10^2} + \frac{1}{10^2}\frac{r_3}{q}$$

$$\vdots$$

$$= a_0 + \frac{a_1}{10} + \cdots + \frac{a_n}{10^n} + \frac{1}{10^n}\frac{r_{n+1}}{q}.$$

Here we have $0 \le a_i \le 9$ for $i > 0$,

$$a_i q = 10r_i - r_{i+1} \le 10r_i < 10q,$$

and, with $\sigma_n := \frac{1}{10^n}\frac{r_{n+1}}{q}$, we get $0 \le \sigma_n < \frac{1}{10^n}$ because of $0 \le \frac{r_{n+1}}{q} < 1$.

Assume that

$$t = b_0 + \frac{b_1}{10} + \cdots + \frac{b_n}{10^n} + \tau_n, \quad b_i \in \mathbb{N},\ 0 \le b_i \le 9 \text{ for } i > 0,\ 0 \le \tau_n < \frac{1}{10^n}.$$

We show that $s = t$ if and only if for all n,

$$a_0 = b_0, \ldots, a_n = b_n, \quad \text{and} \quad \sigma_n = \tau_n.$$

Let $n = 0$, then $s = t$ means, that $a_0 + \sigma_0 = b_0 + \tau_0$. If $a_0 > b_0$, we have $0 < a_0 - b_0 = \tau_0 - \sigma_0 < 1$, a contradiction, and thus $a_0 = b_0$ and $\sigma_0 = \tau_0$. (Similarly, $a_0 < b_0$ leads to a contradiction.)

Inductively assume that the statement is true for n, which means

$$s_n + \frac{a_{n+1}}{10^{n+1}} + \sigma_{n+1} = t_n + \frac{b_{n+1}}{10^{n+1}} + \tau_{n+1}.$$

Without loss of generality, assume that $a_{n+1} > b_{n+1}$. Then from the inductive hypothesis, we obtain

$$a_{n+1} + 10^{n+1}\sigma_{n+1} = b_{n+1} + 10^{n+1}\tau_{n+1} \quad \text{and then} \quad 0 < a_{n+1} - b_{n+1} < 1.$$

This is a contradiction and therefore $a_{n+1} = b_{n+1}$ and $\sigma_{n+1} = \tau_{n+1}$. This shows also the independence of the decimal fraction $DA(s)$ from the representation of $\frac{p}{q}$ of s. □

We mention that for the approximation fractions of $DA(s)$ there holds

$$\forall n\,(s_n \le s \le s_n').$$

Note that s is the only nonnegative number with this property.

The division algorithm DA maps each fraction of the form $\frac{p}{q}$ to a uniquely determined decimal fraction $DA(\frac{p}{q})$.

Let $\mathbb{Q}^+$ be the set of all decimal fractions and let Dec $\mathbb{Q}^{+(p)}$ be the set of all periodic decimal fractions without 9 as their period.

Theorem 3.4.4. *The division algorithm defines a bijective function from* $\mathbb{Q}^+$ *onto* $\mathrm{Dec}\,\mathbb{Q}^{+(p)}$, $DA\colon \mathbb{Q}^+ \to \mathrm{Dec}\,\mathbb{Q}^{+(p)}$ *by* $s = \frac{p}{q} \mapsto DA(s)$. *Therefore, we can identify* $s \in \mathbb{Q}^+$ *and* $DA(s) \in \mathrm{Dec}\,\mathbb{Q}^{+(p)}$ *such that the following conditions are satisfied:*
(i) $\frac{A}{10^l} \mapsto a_0.a_1 \dots a_l$,
(ii) $\frac{B}{10^k-1} \mapsto 0.\overline{b_1 \dots b_k}$,
(iii) $\frac{A}{10^l} + \frac{B}{10^l(10^k-1)} \mapsto a_0.a_1 \dots a_l\overline{b_1 \dots b_k}$,

where A is the greatest natural number which is smaller than $s \cdot 10^l$ *and B is the greatest natural number which is smaller than* $s \cdot 10^{l+k} - A \cdot 10^k$.

Proof. We show that $DA(s)$ is a periodic decimal fraction without 9 as its period. If $a_n = 9$, then, by Theorem 3.4.3,

$$10r_n = 9q + r_{n+1} \quad \text{with } 0 \le r_{n+1} < q.$$

If $r_{n+1} \ge r_n$, we have

$$9q + r_{n+1} \ge 9q + r_n > 9r_n + r_n = 10r_n,$$

i. e., $r_n > r_{n+1}$. Suppose that from a certain index l onward we have $a_n = 9$ for all $n > l$, then the following inequalities are true: $r_{l+1} > r_{l+2} > r_{l+3} > \cdots$. Since there are only finitely many natural numbers less than r_{l+1}, $DA(s)$ cannot be a periodic decimal fraction with 9 as its period. But $DA(s)$ is a periodic decimal fraction. Because $0 \le r_n < q$ for r_n, there are only finitely many possibilities, i. e., there exist indices l and k with $r_{l+1} = r_{l+k+1}$.

Then in the next steps of the algorithm DA, the remainders are also equal, i. e.,

$$r_{l+2} = r_{l+k+2}, \dots$$

From

$$10r_{l+1} = a_{l+1}q + r_{l+2} \quad \text{and} \quad 10r_{l+k+1} = a_{l+k+1}q + r_{l+k+2},$$

there follows

$$a_{l+1} = a_{l+k+1}, \dots, a_{l+k} = a_{l+k+k},$$

and thus $DA(s)$ is periodic, i. e.,

$$DA(s) = a_0.a_1 \dots a_l\overline{a_{l+1} \dots a_{l+k}}.$$

The map DA is injective, since from $DA(s) = DA(t)$ there follows that s and t have equal approximation fractions. From our remark before the theorem, we obtain $s = t$. Also the map DA is surjective, i. e., it maps $\mathbb{Q}^+$ onto $\mathrm{Dec}\,\mathbb{Q}^{+(p)}$. Obviously, for decimal fractions from $\mathrm{Dec}\,\mathbb{Q}^{+(p)}$, there are precisely the possibilities given by (i)–(iii):

(i) Suppose a is finite, $a = a_0.a_1a_2 \dots a_l$. Then by the theorem before,

$$s = a_0 + \frac{a_1}{10} + \dots + \frac{a_l}{10^l} = \frac{a_0 10^l + a_1 10^{l-1} + \dots + a_l}{10^l} = \frac{A}{10^l}.$$

(ii) If a is purely periodic, $a = 0.\overline{b_1 \dots b_k}$. Then $s = \frac{B}{10^k-1}$. Because

$$B = b_1 10^{k-1} + \dots + b_k < 9(10^k - 1 + \dots + 1) = 9 \cdot \frac{10^{k-1}}{10-1} = 10^k - 1,$$

an application of the division algorithm DA gives

$$B = b_1 b_2 \dots b_k = 0 \cdot (10^k - 1) + B.$$

From

$$b_1 b_2 \dots b_k \cdot 0 = b_1 \cdot (10^k - 1) + b_2 \dots b_k \cdot 0 + b_1,$$

there follows

$$b_1 b_2 \dots b_k \cdot 0 = b_1(10^k - 1) + b_2 \dots b_k b_1 \quad \text{with } b_2 \dots b_k b_1 < 10^k - 1.$$

In this way, one obtains:

$$b_2 \dots b_k b_1 = b_2 \cdot (10^k - 1) + b_3 \dots b_k b_1 b_2,$$

$$\vdots$$

$$b_k b_1 \dots b_{k-1} 0 = b_k \cdot (10^k - 1) + b_1 b_2 \dots b_{k-1},$$

where the remainder each time is less than $10^k - 1$. There holds $DA(s) = a$.

(iii) If a is mixed periodic, then $a = a_0.a_1 \dots a_l \overline{b_1 \dots b_k}$ and so

$$s = \frac{A}{10^l} + \frac{B}{10^l(10^k - 1)}.$$

Now, on the one hand, we have

$$s = a_0 + \frac{a_1}{10} + \dots + \frac{a_l}{10^l} + \sigma_l \quad \text{with } \sigma_l = \frac{B}{10^l(10^k - 1)} < \frac{l}{10^l},$$

and, on the other hand, the division algorithm gives

$$s = a_0' + \frac{a_1'}{10} + \dots + \frac{a_l'}{10^l} + \sigma_l' \quad \text{with } \sigma_l' = \frac{l}{10^l}\frac{r_{l+1}'}{q} < \frac{l}{10^l}.$$

Because of the uniqueness of the division algorithm,

$$a_0 = a_0', \ldots, a_l = a_l', \quad \text{and} \quad \frac{B}{10^k - 1} = \frac{r_{l+1}'}{q} =: s'.$$

Applying the division algorithm to s', because of the independence of s' from the representation as fraction, by (ii) we obtain

$$10r_{l+1}' = b_1 q + r_{l+2}$$

and, continuing the application to s,

$$10r_{l+1}' = a_{l+1}' q + r_{l+2}'.$$

From the division with remainder in $\mathbb{N}$, we get

$$\begin{aligned} b_1 &= a_{l+1}' \quad \text{and} \quad r_{l+2} = r_{l+2}' \\ &\vdots \\ b_k &= a_{l+k}' \quad \text{and} \quad r_{l+k+1} = r_{l+k+1}' = r_{l+1}', \text{ etc.} \end{aligned}$$

It is left to prove the statements on A and B.

Since (i) and (ii) are special cases, we consider

$$s = \frac{A}{10^l} + \frac{B}{10^l(10^k - 1)}.$$

From $s \cdot 10^l = A + \frac{B}{10^k-1}$ and $\frac{B}{10^k-1} < 1$ follows that A is the greatest natural number which is smaller than $s \cdot 10^l$, and from

$$\begin{aligned} s \cdot 10^l(10^k - 1) &= A \cdot (10^k - 1) + B, \\ s \cdot 10^{l+k} - A \cdot 10^k &= (s \cdot 10^l - A) + B, \\ \text{and} \quad s \cdot 10^l - A &< 1, \end{aligned}$$

there follows that B is the greatest natural number which is smaller than $s \cdot 10^{l+k} - A \cdot 10^k$. □

We notice that all our observations can be repeated in an arbitrary positional system, i. e., with any basic digit $d > 1$.

3.5 Integers

In the semiring $\underline{\mathbb{N}}$, subtraction is not always possible. Indeed, not for any natural numbers a, b there exists a natural number x with $a + x = b$. Therefore, we ask for an extension of $\underline{\mathbb{N}}$ such that this equation is solvable.

For integers, i. e., in $\mathbb{Z}$, addition is invertible, but not multiplication, as it is in $\underline{\mathbb{Q}}^+$. Addition and multiplication of integers satisfy the following laws:
1. $+$ and $\cdot$ are associative.
2. $+$ is commutative.
3. $\forall a, b \in \mathbb{Z}\, \exists x \in \mathbb{Z}\, (a + x = b)$, i. e., $+$ is invertible.
4. The distributive laws

$$\forall a, b, c \in \mathbb{Z}\, (a \cdot (b + c) = a \cdot b + a \cdot c, (a + b) \cdot c = a \cdot c + b \cdot c)$$

   are satisfied.

More generally, we define an algebraic structure $\mathcal{A} = (A; +, \cdot)$ to be a *ring* if the following conditions are satisfied:

Definition 3.5.1. Let $\mathcal{A} = (A;\ +,\ \cdot)$ be an algebraic structure with two binary operations

$$+, \cdot : A \times A \to A.$$

Then $\mathcal{A}$ is said to be a ring if $+, \cdot$ are associative, $+$ is commutative, the distributive laws

$$\forall a, b, c \in A\, (a \cdot (b + c) = a \cdot b + a \cdot c, (a + b) \cdot c = a \cdot c + b \cdot c)$$

are satisfied, and $+$ is invertible, i. e., if

$$\forall a, b \in A\, \exists x \in A\, (a + x = b)$$

is satisfied.

Clearly, $\mathbb{Z}$ is a ring (with some additional properties). In every ring, we have the following

Corollary 3.5.2. *Let $\mathcal{A} = (A;\ +,\ \cdot)$ be a ring. Then*
(i) *In $\mathcal{A}$, there exists a zero element 0, i. e., 0 satisfies $a + 0 = 0 + a = 0$ and, for any $a \in A$, there exists an element $-a \in A$ such that $a + (-a) = (-a) + a = 0$.*
(ii) $\forall a \in A\, (a \cdot 0 = 0 \cdot a = 0)$.
(iii) $\forall a, b \in A\, (a(-b) = (-a)b = -ab;\ (-a)(-b) = ab)$.

Proof. (i) For all $c \in A$, there exists an element $y \in A$ such that $y + c = c$. Moreover, for all $a \in A$, there exists an $x \in A$ such that $c + x = a$. Then there follows $y + a = y + (c + x) = (y + c) + x = c + x = a$. Thus there is an element $y := 0_l$ with $y + a = 0_l + a = a$ for all $a \in A$. In a similar way, one shows that there is an element 0_r with $a + 0_r = a$ for all $a \in A$. Then $0_l = 0_l + 0_r = 0_r$ is a uniquely determined element $0 := 0_l = 0_r$ with $a + 0 = 0 + a = a$ for all $a \in A$. The existence of a uniquely determined element $-a \in A$ for any $a \in A$ can be proved similarly; this is left as an exercise to the reader.

(ii) From

$$a \cdot 0 + 0 = a \cdot 0 = a \cdot (0 + 0) = a \cdot 0 + a \cdot 0$$

follows by subtraction of $a \cdot 0$ from the left that $a \cdot 0 = 0$. Analogously, one proves $0 \cdot a = 0$. (We can also say that we applied the cancellation rule for + which is satisfied).

(iii) Also $a(-b) = -ab$ holds because of

$$ab + a(-b) = a\big(b + (-b)\big) = a \cdot 0 = 0$$

and $ab + (-ab) = 0$ by the cancellation rule. Furthermore, $(-a)b = -ab$ follows from

$$ab + (-a)b = \big(a + (-a)\big)b = 0 \cdot b$$

and $ab + (-ab) = 0$ also by cancellation rule. Likewise, $(-a)(-b) = ab$ follows from

$$a(-b) + (-a)(-b) = \big(a + (-a)\big)(-b) = 0 \cdot (-b) = 0$$

and $a(-b) + ab = 0$ because of $a(-b) = -ab$. □

The ring $\underline{\mathbb{Z}} = (\mathbb{Z}; +, \cdot)$ has some more properties which are not satisfied in arbitrary rings:

1. $\underline{\mathbb{Z}}$ is commutative (i. e., the multiplication is commutative).
2. $\underline{\mathbb{Z}}$ has an identity element e (satisfying $\forall a \in \mathbb{Z}\ (a \cdot e = e \cdot e = a)$), this is $e = 1$.
3. $\underline{\mathbb{Z}}$ satisfies the cancellation rule $\forall a, x_1, x_2 \in \mathbb{Z}\ (a \neq 0 \wedge a \cdot x_1 = a \cdot x_2 \Rightarrow x_1 = x_2)$.

The cancellation rule with respect to multiplication is equivalent to another possible property of a ring which is connected with the question of whether a product can be 0 if none of the factors is equal to 0.

Definition 3.5.3. Let $\mathcal{A}$ be a nontrivial ring (i. e., $\mathcal{A}$ is different from the ring consisting of the zero element only). An element $a \in A$ is said to be a left zero-divisor, if $a \neq 0$ and if an element $b \in A, b \neq 0$, exists with $ab = 0$. In this case, b is said to be a right zero-divisor. If a is a left and a right zero-divisor, then a is called a zero-divisor (for instance, if $\mathcal{A}$ is a commutative ring). A ring $\mathcal{A}$ is said to be zero-divisor-free, if $\mathcal{A}$ has neither a right nor a left zero-divisor. Any commutative, zero-divisor-free ring is said to be an integral domain.

Theorem 3.5.4. *A ring $\mathcal{A} = (A; +, \cdot)$ is zero-divisor-free if and only if $\mathcal{A}$ satisfies the cancellation rule:*

$$\begin{aligned} \underline{A} \text{ zero-divisor-free} \quad \Leftrightarrow \quad & \forall a, x_1, x_2 \in A\big((a \neq 0 \wedge ax_1 = ax_2) \Rightarrow x_1 = x_2\big) \wedge \\ & \forall a, y_1, y_2 \in A\big((a \neq 0 \wedge y_1 a = y_2 a) \Rightarrow y_1 = y_2\big). \end{aligned}$$

Proof. Let $\mathcal{A}$ be zero-divisor-free and $a \in A$ be an arbitrary element from A, different from 0. Assume that $a \cdot x_1 = a \cdot x_2$. Then $a \cdot x_1 - a \cdot x_2 = a \cdot (x_1 - x_2) = 0$ and, from $a \neq 0$ and since $\mathcal{A}$ is zero-divisor-free, we obtain $x_1 = x_2$. Similarly, one shows that from $a \neq 0$ and $y_1 a = y_2 a$ there follows $y_1 = y_2$.

Conversely, we assume that the cancellation rule is satisfied and that $ab = 0$ for $a \neq 0$. Then from $a \cdot 0 = 0$ (Corollary 3.5.2(ii)) there follows $ab = a \cdot 0$, and thus $b = 0$, i. e., $\mathcal{A}$ is zero-divisor-free. □

Corollary 3.5.5. *The ring* $\underline{\mathbb{Z}} = (\mathbb{Z}; +, \cdot)$ *is an integral domain.*

In a similar way as we constructed $\underline{\mathbb{Q}^+}$ from the semiring $\underline{\mathbb{N}}$, we can construct the integral domain $\underline{\mathbb{Z}}$. Pairs written as $a - b, a, b, \in \mathbb{N}$, which are called *differences*, take over the role of quotients. As in the case of nonnegative rational numbers, a structure theorem helps to recognize the way of our construction.

Theorem 3.5.6 (Structure theorem). *Let $\mathcal{R}$ be a minimal upper ring of a semiring $\mathcal{A}$. Then each element $x \in R$ can be written as a difference*

$$x = a + (-b) = a - b$$

of elements $a, b \in A$. Calculation in $\mathcal{R}$ follows the rules

$$\begin{aligned} &\text{I.} && a - b = a' - b' \quad \textit{if and only if} \quad a + b' = a' + b, \\ &\text{II.} && (a - b) + (c - d) = (a + c) - (b + d), \\ &\text{III.} && (a - b) \cdot (c - d) = (ac + bd) - (ad + bc). \end{aligned}$$

If $\mathcal{A}$ is commutative, also $\mathcal{R}$ is such.

Proof. Indeed, since $\mathcal{R}$ is a ring, the differences $a - b$ are elements of R and the rules I, II, and III can be proved from the axioms of a ring. Therefore, the set of all such differences forms a subring $\mathcal{D}$ of $\mathcal{R}$. Now D contains A, since each element a of A can be written as $a = a + (b - b) = a - (b - b)$, i. e., as a difference of elements from A. Since $\mathcal{R}$ is a minimal upper ring of $\mathcal{A}$, we have $\mathcal{D} = \mathcal{R}$. The commutativity of $\mathcal{R}$ follows from the commutativity of $\mathcal{A}$ since

$$(a - b) \cdot (c - d) = (ac + bd) - (ad + bc) = (ca + db) - (cb + da) = (c - d) \cdot (a - b). \quad \square$$

Definition 3.5.7. An upper ring of a semiring $\mathcal{A}$, whose elements can be written as differences

$$a + (-b) = a - b, \quad a, b \in A,$$

is said to be *the difference ring* of $A : D(A)$.

Theorem 3.5.8. *Each minimal upper ring of a semiring $\mathcal{A}$ is a difference ring, and conversely.*

Proof. In a difference ring $\mathcal{R}$ of $\mathcal{A}$, every equation $b + x = a$ with $a, b \in A$ has exactly one solution $x \in R$, and R consists precisely of all these solutions. Therefore, no proper subset of R which contains A can be a ring. □

Theorem 3.5.9 (Existence theorem). *For every semiring $\mathcal{A}$, there exists a difference ring $\mathcal{R}$.*

The proof consists of the following steps:

(i) Definition of an equivalence relation $\sim$ on $A \times A$ by

$$(a, b) \sim (a', b') \Leftrightarrow a + b' = a' + b$$

and formation of the factor set $A \times A/\sim$.

(ii) Definition of addition and multiplication on this factor set, corresponding to rules II and III of the structure theorem, especially, proof of the independence of the representatives.

(iii) Proof that $A \times A/\sim$ is a ring.

(iv) Showing that $A \times A/\sim$ contains a subsemiring which is in one-to-one correspondence with $\mathcal{A}$ (isomorphic to $\mathcal{A}$).

(v) Using replacement and substitution to show that $\mathcal{R} := \underline{A \times A/\sim}$ is the difference ring of $\mathcal{A}$.

The details of the proof are left as an exercise to the reader. This construction gives a uniquely determined result.

Theorem 3.5.10. *Every ring $\mathcal{R}^*$ which contains a semiring $\mathcal{A}$ contains exactly one minimal upper ring $\mathcal{R}$ of $\mathcal{A}$. Two minimal upper rings of $\mathcal{A}$ are isomorphic as difference rings of $\mathcal{A}$.*

The proof is similar to the uniqueness proof of the quotient semifield of a commutative semiring with the cancellation rule and so is left as an exercise to the reader.

For the construction of the difference ring the cancellation rule was not needed. But if the cancellation rule is satisfied in a semiring, it can be satisfied in its difference ring. For the proof of the following theorem, see [19].

Theorem 3.5.11. *If in a semiring $\mathcal{A}$ the cancellation rule is satisfied and if for any two different elements a and b of A at least one of the equations $b + x = a$ or $a + y = b$ has a solution in $\mathcal{A}$, then the cancellation rule is satisfied in the difference ring $D(\mathcal{A})$.*

From our results, we obtain

Theorem 3.5.12. *For the semiring $\underline{\mathbb{N}}$ of natural numbers there exists a minimal upper ring which is uniquely determined up to isomorphism, namely the difference ring $\underline{\mathbb{Z}} := D(\underline{\mathbb{N}})$. This $D(\underline{\mathbb{N}})$ consists of all differences*

$$a + (-b) = a - b; \quad a, b \in \mathbb{N}.$$

Calculations in $\underline{\mathbb{Z}}$ *are uniquely determined by the rules* I, II, *and* III.

Definition 3.5.13. The difference ring $\underline{\mathbb{Z}} = \underline{D(\mathbb{N})}$ is said to be the ring of integers.

Now we define the order of the elements in $\mathbb{Z}$.

Definition 3.5.14. Let $m-n$, $m'-n' \in \mathbb{Z}$. The binary relation $<$ on $\mathbb{Z}$ is defined as follows: $m - n < m' - n'$ if and only if $m + n' < n + m'$. Here $m + n' < n + m'$ is the irreflexive order relation on $\mathbb{N}$.

We used the same symbol for the order on $\mathbb{N}$ as for the order on $\mathbb{Z}$. Indeed, the binary relation $<$ is the extension of the order relation on $\mathbb{Z}$.

Theorem 3.5.15. *The binary relation defined above is an irreflexive order relation on* $\mathbb{Z}$, *which is monotone with respect to addition and multiplication in* $\mathbb{Z}$ *(compatible with respect to addition and multiplication in* $\underline{\mathbb{Z}}$*) and which extends the order in* $\mathbb{N}$.

For the proof, see [19].

Altogether we obtain the following algebraic characterization of the integers.

Theorem 3.5.16. *The ring* $\underline{\mathbb{Z}}$ *of integers is a naturally ordered ring with identity element and each minimal ordered ring with an identity element is o-isomorphic to* $\underline{\mathbb{Z}}$.

Proof. It is easy to check that $a < b$ if and only if there exists an element $x \neq 0$ in $\mathbb{Z}$ such that $a + x = b$, i. e., $<$ is a natural order relation.

Let $\mathcal{R} \subseteq \mathbb{Z}$ be a minimal ordered subring with an identity element e, $e \neq 0$. From $a = a \cdot e = a \cdot 1$, there follows for all $a \neq 0$ from $\mathbb{Z}$, using the cancellation rule of $\mathbb{Z}$, that $e = 1$. Then also all multiples of 1 belong to R, i. e., $\mathcal{R} = \mathbb{Z}$, and thus $\mathbb{Z}$ is a minimal ordered ring with an identity element.

If, conversely, $\mathcal{R}$ is a nontrivial minimal ordered ring with an identity element e, then $\mathcal{R}$ contains a minimal ordered subsemiring $\mathcal{U}$ with e as an identity element, namely the intersection of all subsemirings of $\mathcal{R}$ which contain e.

Then $\mathcal{R}$ is, as a minimal ordered ring, a minimal ordered upper ring of this semiring $\mathcal{U}$. Since $\mathcal{U}$ is o-isomorphic to $\underline{\mathbb{N}}$, the proof is finished. □

3.6 Elementary number theory

In this section we will present the reader some basic concepts of number theory. We start with *divisibility* as an example of an order relation on $\mathbb{Z}$ which is different from $<$. Then we come to prime numbers and the fundamental theorem of elementary number theory, and mention some old solved and unsolved problems of number theory. The equivalence modulo m is another binary relation on $\mathbb{Z}$ which is compatible with addition and multiplication. This leads to linear congruences, diophantine equations, and their solutions which will be applied to consumer theory in Chapter 4.

In a naturally ordered semiring (also in a naturally ordered ring), the multiplication is not always invertible. Therefore, it is a special situation when m is divisible by n, and we define:

Definition 3.6.1. Let m and n be two integers. Then m is said to be a divisor of n, written as $m|n$, if there is an integer q with $n = mq$:

$$m \text{ divisor of } n :\Leftrightarrow \exists q \in \mathbb{Z}\,(n = mq).$$

If m is not a divisor of n, we write $m \nmid n$. If $n = mq$, then n is called a multiple of m and of q. Clearly, q is uniquely determined (this can be proved using the cancellation rule).

Theorem 3.6.2. *The following properties of the divisibility are easy consequences of the definition and of properties of the semiring* $\underline{\mathbb{N}}$:
1) $1|n$; $n|0$; $0|n \Rightarrow n = 0$,
2) *let* $m, n \in \mathbb{N}$, *then* $m|n \wedge n \neq 0 \Rightarrow m \leq n$,
3) $n|n$, *i. e.,* $|$ *as a binary relation is reflexive,*
4) $m|n$ *and* $n|s \Rightarrow m|s$, *i. e.,* $|$ *is transitive,*
5) $m|n$ *and* $n|m \Rightarrow m = n$, *i. e.,* $|$ *is antisymmetric,*
6) $m|n \Rightarrow am|an$,
7) $am|an \wedge a \neq 0 \Rightarrow m|n$,
8) $s|n \wedge s|m \Rightarrow s|(an + bm)$, *where* $m, n, q, s, a, b, \in \mathbb{Z}$.

Each reflexive, transitive, and antisymmetric relation on a set M defines a partial order relation on M.

Corollary 3.6.3. *The semiring* $\underline{\mathbb{N}}$ *is a partially ordered set with respect to* $|$.

The semiring $\underline{\mathbb{N}}$ is not an ordered semiring with respect to $|$ since from $m \mid n$ does not follow $m + s|n + s$ (e. g., $2|8$, but $2 + 3 = 5 \nmid 11 = 8 + 3$).

In $\mathbb{Z}$, the equality $m = n$ does not necessarily follow from $m \mid n$ and $n \mid m$ since $m = -n$ is also possible. This leads, for an arbitrary ring $\mathcal{R}$ with identity element e, to the following definition: the *units* of $\mathcal{R}$ are the divisors of the identity element e. Note that 1 and -1 are both units of $\underline{\mathbb{Z}}$.

The units of a ring are precisely those ring elements which have an inverse. The units are divisors of each element of the ring.

Definition 3.6.4. Let $a, a' \in \mathbb{Z}$. Then a is said to be an associate to a', denoted $a \cong a'$, if $a' = a$ or $a' = -a$ (generally, if there is a unit ε with $a\varepsilon = a'$).

Corollary 3.6.5. *Let* a *and* a' *be integers. Then*
1. *a is an associate to a' if a is divisor of a' and a' is divisor of a.*
2. *$\cong$ is an equivalence relation on $\mathbb{Z}$.*
3. $a \cong a' \wedge b \cong b' \wedge b \mid a \Rightarrow b' \mid a'$.

Proof. 1. If $a' = a$ or $a' = -a$, then $a|a'$ with $q = 1$, respectively with $q = -1$. If, conversely, $a|a'$ and $a' \mid a$, i. e., if there is an element $q \in \mathbb{Z}$ with $a' = a \cdot q$ and an element $q' \in \mathbb{Z}$ with $a = a' \cdot q'$, then $a' = a'(q'q)$ follows and, because of $a' = a' \cdot 1$ and from the cancellation rule in $\mathbb{Z}$, one has $q'q = 1$ if $a' \neq 0$. Now $q'q = 1$ is only possible if $q = q' = -1$ or if $q = q' = 1$. Thus $a \cong a'$. From $a' = 0$, we obtain $a = 0$. In this case, we also have $a \cong a'$.

2. The reflexivity of $\cong$ is clear. If $a' = a$ or $a' = -a$, then $a = a'$ or $a = -a'$, i. e., $\cong$ is symmetric. If $a' = a$ or $a' = -a$ and $a'' = a'$ or $a'' = -a'$, we get $a'' = a$ or $a'' = -a$, and $a'' \cong a$, and so $\cong$ is transitive.

3. If $a' = a$ or $a' = -a$ and $b' = b$ or $b' = -b$ and if b divides a, i. e., there is an element $q \in \mathbb{Z}$ with $bq = a$, then $b'q' = a'$ with $q' = q$ or $q' = -q$, i. e., b' is a divisor of a'. □

The equivalence relation $\cong$ defines a partition of $\mathbb{Z}$, $\mathbb{Z} = \cup\{a, -a\}$. The nonnegative representative of the class $\{a, -a\}$ is said to be the *absolute amount* or *norm* of a:

$$|a| = \begin{cases} a & \text{if } a > 0, \\ -a & \text{if } a < 0, \\ 0 & \text{if } a = 0. \end{cases}$$

Each class, except for the class of 0, consists of two elements. Property 3 shows that all statements related to divisibility are valid also for the associate elements. Therefore all propositions concerning divisibility are already satisfied in the semiring $\underline{\mathbb{N}}$. The relation $\cong$ allows classifying the set of possible divisors of an integer.

Definition 3.6.6. A number $b \in \mathbb{Z}$ is said to be a *trivial* divisor of $a \in \mathbb{Z}$ if $b \in \{1, -1\}$ or if b is an associate to a. Also b is said to be a *proper* divisor of a if b is a divisor of a, but b is not an associate to a, written as $b \| a$.

For instance, $3, -3, 2, -2$ are all nontrivial divisors of 6, the proper divisors of 6 are $3, -3, 2, -2, 1, -1$. Numbers which have only trivial divisors, i. e., the *prime numbers*, play an important role. Each integer can be written as a product of prime numbers. Prime numbers can be written as a product only in a trivial way.

Definition 3.6.7. A natural number $p > 1$ is said to be *a prime number* if it is divisible only by 1 and itself. An integer $n > 1$, which is not a prime number, i. e., which has no (positive) divisors except for 1 and itself, is said to be a composite number.

Corollary 3.6.8. *If $n \in \mathbb{N}$ is not a prime number and $n \notin \{0, 1\}$, then n can be written as a product of two natural numbers different from* 1.

Proof. By definition, n has a nontrivial divisor m, $m \neq 1$. Then the complementary divisor q must be different from 1, $n = mq \wedge m \neq 1 \wedge q \neq 1$. □

The sequence of prime numbers begins with $2, 3, 5, 7, 11, \ldots$ The prime number 2 is the only even prime number.

All prime numbers less than or equal to a given prime number n can be written down using a procedure which is called the *sieve of Eratosthenes (Eratosthenes of Cyrene)* (275–194 BC). The following small propositions help understand this procedure.

Lemma 3.6.9. *The smallest divisor of an integer $n > 1$, which is different from 1, is a prime number.*

Proof. Let m be the smallest divisor of $n > 1$ which is different from 1. Suppose m is a composite number. Then m has a divisor m_1 with $m_1 \leq m$. Now m_1 is also a divisor of n. This contradicts our choice of the number m. □

Lemma 3.6.10. *The smallest divisor of a composite integer n which is different from 1 is not greater than $\sqrt{n}$.*

Proof. Let m be a divisor of n, which satisfies the conditions of the lemma. Then by the previous lemma, m is a prime number. If $n = mq$, then $q \geq m$ and so $n = mq \geq m^2$, i. e., $m \leq \sqrt{n}$. □

Sieve of Eratosthenes. Let n be a natural number. If all prime numbers $\leq \sqrt{n}$ are known, then we can find all prime numbers $\leq n$ in the following way: We write down the integers

$$1, 2, \dots, n.$$

In this sequence, 2 is the first integer which is greater than 1 and is a prime number. Now we delete all multiples of 2, except for 2. Now 3 is the first integer following 2 which was not deleted. Thus 3 is only divisible by 1 and 3 and, by definition, it is a prime. Now we delete all multiples of 3, except for 3. Once all the multiples of each discovered and canceled prime have been deleted as composites, the remaining not deleted integers are primes. This procedure has to be continued up to $\sqrt{n}$. Let p be the greatest prime $\leq \sqrt{n}$. If we delete in the sequence all multiples of p, except for p, then all not deleted integers are primes.

The following easy to understand but, nevertheless, up to now unsolved problems are connected with the infinite sequence of all primes:

Goldbach hypotheses (Ch. Goldbach (1690–1764)):

1. Each even natural number $n > 5$ can be written as a sum of two odd primes.
2. Each odd natural number $n > 8$ can be written as a sum of three odd primes.

If the first statement is true, then also the second. Unfortunately, the converse is not satisfied. *I. M. Winogradow* (1891–1983) has proved that the second statement, the so-called weak Goldbach conjecture, is true for all integers which are large enough (greater than 10^{1350}). Up to now, the first statement was neither proved nor disproved, but verified for numbers less than $4 \cdot 10^{18}$.

In the sequence of prime numbers, there are so-called prime twins. These are pairs $(p, p+2)$ of primes with the difference 2. The first twins less than 100 are

$$(3,5),\ (5,7),\ (11,13),\ (17,19),\ (29,31),\ (41,43),\ (59,61),\ (71,73).$$

If $p > 3$, then the number in-between must be divisible by 3 since from $p, p+1, p+2$ at least one is divisible by 3. It is easy to prove, and we will do it, that there are infinitely many primes. But the following problem is open:

Problem. Is the set of all twins infinite?

The *twin conjecture* says that there are infinitely many twins.

Theorem 3.6.11 (Euclid). *The set of all prime numbers is infinite.*

Proof. Let $p_1 = 2, p_2 = 3, p_3 = 5, \ldots$ be the sequence of all prime numbers and suppose that the set $P = \{p_1, \ldots, p_k\}$ with $p_1 < p_2 < \cdots < p_k$ of all primes is finite. Let $n = p_1 p_2 \cdots p_k + 1$ be the product of all primes plus 1. Obviously, $n > 1$ and $p_i \nmid n$ for each of the numbers $p_i \in P$. But there exists a prime p, which is a divisor of n. Otherwise n would be a prime greater than p_k. Then $p_i < p$, $i = 1, \ldots, k$, a contradiction. Therefore the set P cannot be finite. □

In the theory of divisibility, the theorem of the unique representation of an arbitrary integer different from $-1, 0, 1$ as a product of prime factors plays an important role.

Theorem 3.6.12 (Existence and uniqueness of prime factorization). *Each integer different from $0, \pm 1$ has a prime factorization which is uniquely determined up to units and the order of the factors.*

Proof. Suppose there is an integer which has no factorization and let n be the least one having this property. Since n is not a prime (otherwise n would be a factorization), it must be a composite number. Then n has two nontrivial divisors $s, t \in \mathbb{N}$ with $1 < s$, $t < n$, and $n = st$. Since the numbers s, t are positive and less than n, they have prime factorizations. But then also n has a prime factorization. This contradicts our assumption that there is no factorization of n.

To prove the uniqueness, suppose there were two factorizations of $a \in \mathbb{Z}$, $a = p_1 p_2 \cdots p_n = q_1 q_2 \cdots q_m$, where p_i, q_j are primes or their opposites. Now one shows (a useful exercise for the reader) by induction that $n = m$ and $p_i \cong q_j$ for some order of the primes. □

If in the prime factorization of an integer the same prime occurs several times, it can be written as a prime power. If $p_1, p_2, \ldots, p_n$ is the ordered finite part of the infinite sequence of all pairwise different prime numbers and if the natural numbers $\alpha_1, \alpha_2, \ldots, \alpha_n$ show how often the corresponding primes from $p_1, p_2, \ldots, p_n$ occur in the prime factorization of the integer n, then we obtain

$$n = (-1)^{\varepsilon(n)} p_1^{a_1} p_2^{a_2} \cdots p_k^{a_k},$$

where

$$\epsilon(n) = \begin{cases} 0 & \text{if } n > 0, \\ 1 & \text{if } n < 0. \end{cases}$$

We can also use

$$n = (-1)^{\epsilon(n)} \prod_p p^{\alpha(p)}, \quad \text{with } \alpha(p) = \begin{cases} a_i & \text{if } p = p_i, \\ 0 & \text{if } p \nmid n, \end{cases} \tag{*}$$

where the product is over the set of all prime numbers and $\alpha(p)$ shows how often the prime p occurs in the product. This $\alpha(p)$ is also called the *order of p in n*, written as $\operatorname{ord}_p n$. Then $\operatorname{ord}_p n = 0$ if and only if $p \nmid n$. Moreover, we define $\operatorname{ord}_p 0 := \infty$. Obviously, $\alpha(n) \geq 0$, and $\alpha(n) > 0$ only for finitely many primes.

For instance, if $n = -360$, then

$$-360 = (-1)2^3 3^2 5, \quad \text{i. e., } \epsilon(-360) = 1,\ \alpha(2) = 3,\ \alpha(3) = 2,\ \alpha(5) = 1,$$

and $\alpha(p) = 0$ for all $p \notin \{2, 3, 5\}$.

For 0 and 1, we set

$$1 = \prod_p p^0 \quad \text{and} \quad 0 = \prod_p p^{\infty}.$$

This writing has the following advantages:

1. The multiplication of elements from $\mathbb{Z}$ corresponds to the componentwise addition of its exponent systems:

$$a \cdot b = \left((-1)^{\varepsilon(a)} \prod_p p^{\alpha(p)}\right)\left((-1)^{\varepsilon(b)} \prod_p p^{\beta(p)}\right) = (-1)^{\varepsilon(ab)} \prod_p p^{\alpha(p)+\beta(p)}.$$

2. For

$$a = (-1)^{\varepsilon(a)} \prod_p p^{\alpha(p)} \quad \text{and} \quad b = (-1)^{\varepsilon(b)} \prod_p p^{\beta(p)},$$

we have $b|a$ if and only if $\beta(p) \leq \alpha(p)$ for all p.

We will use the prime power factorization to calculate the greatest common divisor and the least common multiple.

Definition 3.6.13. Let $a_1, a_2, \ldots, a_k$ be integers different from 0. The integer d is said to be *greatest common divisor (gcd)* of $a_1, a_2, \ldots, a_k$,

$$d := (a_1, a_2, \ldots, a_k) \quad \text{or} \quad d := \gcd(a_1, a_2, \ldots, a_k),$$

if the following conditions are satisfied:
(g1) $d|a_1, d|a_2, \ldots, d|a_k$ and
(g2) each divisor t of $a_1, a_2, \ldots, a_k$ is also a divisor of d, i. e.,

$$\forall t \in \mathbb{Z}\,((t|a_1, t|a_2, \ldots, t|a_k) \Rightarrow t|d).$$

The integer f is said to be *least common multiple (lcm)* of $a_1, a_2, \ldots, a_k$,

$$f := [a_1, a_2, \ldots, a_k] \quad \text{or} \quad f := \operatorname{lcm}(a_1, a_2, \ldots, a_k),$$

if the following conditions are satisfied:
(v1) $a_1|f, a_2|f, \ldots, a_k|f$ and
(v2) if $a_1|w, a_2|w, \ldots, a_k|w$, then $f|w$, i. e.,

$$\forall w \in \mathbb{Z}\,((a_1|w, a_2|w, \ldots, a_k|w) \Rightarrow f|w).$$

We remark that the greatest common divisor and the least common multiple of

$$a = (-1)^{\epsilon(a)} \prod_p p^{\alpha(p)} \quad \text{and} \quad b = (-1)^{\epsilon(b)} \prod_p p^{\beta(p)}$$

from $\mathbb{Z}$ are given by

$$d = \gcd(a, b) \cong \prod_p p^{\delta(p)}, \quad \text{with } \delta(p) = \min(\alpha(p), \beta(p)),$$

and

$$f \cong \prod_p p^{\eta(p)}, \quad \text{with } \eta(p) = \max(\alpha(p), \beta(p)).$$

Definition 3.6.14. The integers $a_1, \ldots, a_k$ are said to be *relatively prime* if their greatest common divisor is an associate to 1, i. e., if $\gcd(a_1, \ldots, a_k) \cong 1$.

Lemma 3.6.15. *If $a, b, c \in \mathbb{Z}$ with $a|bc$ and $\gcd(a, b) \cong 1$, then $a|c$.*

Proof. Because of $\gcd(a, b) \cong 1$, one can show that the equation $ua + vb = 1$ for some integers u, v is satisfied. This gives $uac + vbc = c$. Now $a|bc$ implies that a divides the left-hand side of this equation. Thus $a|c$. ☐

The lemma is not true, if $\gcd(a, b) \not\cong 1$, as the following example shows: For $a = 4$, $b = 6$, and $c = 2$, one has $\gcd(4, 6) = 2$, $4|12$, but $4 \nmid 2$ and $4 \nmid 3$.

Corollary 3.6.16. *Let p be a prime and $b, c \in \mathbb{Z}$. If $p|bc$, then $p|b$ or $p|c$.*

Proof. By definition of a prime, we have $\gcd(p, b) \cong 1$ or $p \mid b$. For $p|b$, the proof is finished. Otherwise, $\gcd(p, b) \cong 1$ and, consequently, $p|c$ because of $p|bc$ and Lemma 3.6.15. ☐

If d_1 and d_2 are two greatest common divisors of $\gcd(a_1, a_2, \ldots, a_k)$, then $d_1|d_2$ and $d_2|d_1$. Therefore, $d_2 = \pm d_1$ and the gcd is uniquely determined up to units. A corresponding statement holds for the lcm.

The prime factorization is applicable to the representation of elements from $\mathbb{Q}^+$. Each element a from $\mathbb{Q}^+$ is uniquely representable as a shortened fraction $a = \frac{m}{n}$ with $m, n \in \mathbb{N}$, $n \neq 0$, $\gcd(m, n) = 1$, since from $a = \frac{r}{s}$, $s \neq 0$, $r, s \in \mathbb{N}$ follows, dividing by $\gcd(r, s)$, a "reduced" representation as a fraction, and this is uniquely determined up to units. Indeed, from $a = \frac{m}{n} = \frac{m'}{n'}$ with $\gcd(m, n) = \gcd(m', n') = 1$ follows $mn' = nm'$ since $n'|nm'$, and then $n'|n$ and $n|mn'$, and from this $n|n'$. Therefore, $n = n'$ and so $m = m'$.

For the following three decimal fractions, one has: If $a = \frac{m}{n}$ is the shortened fraction and $0 \leq a < 1$, i. e., $0 \leq m < n$, then

1. $a = 0.a_1 \ldots a_l \Leftrightarrow n = 2^{\alpha_1} 5^{\alpha_2}$, $\alpha_1, \alpha_2 \in \mathbb{N}$,
2. $a = 0.\overline{b_1 \ldots b_k} \Leftrightarrow \gcd(n, 10) = 1, \wedge n|10^k - 1$,
3. $a = 0.a_1 \ldots a_l \overline{b_1 \ldots b_k} \Leftrightarrow n = 2^{\alpha_1} 5^{\alpha_2} m$ with $1 < m < n$, $\gcd(m, 10) = 1$, $\alpha_1, \alpha_2 \in \mathbb{N}$ and $m|10^k - 1$.

For the proof, see [48].

Using the prime power factorization of an integer, one can calculate the greatest common divisor and the least common multiple of two or more integers. Another method to calculate the gcd is the Euclidean algorithm

Theorem 3.6.17 (Euclidean algorithm). *For any two integers a, b, there exists the greatest common divisor $d \cong \gcd(a, b)$. It can be calculated by the following algorithm and written as a linear combination of a and b.*

Proof. If $a = b = 0$, then $d = \gcd(a, b) = 0$, $0|0$ and $t|0$ for all $t \in \mathbb{Z}$. Therefore we may assume that one of elements is different from 0. Let $b \neq 0$. We will apply the division algorithm from Section 3.4 of this chapter and notice that in the case of $r_j < 0$ we can use the absolute amounts. By continued division, we obtain:

$$\begin{aligned}
a &= bq_0 + r_1 && \text{with } 0 < |r_1| < |b|, \\
b &= r_1 q_1 + r_2 && \text{with } 0 < |r_2| < |r_1|, \\
r_1 &= r_2 q_2 + r_3 && \text{with } 0 < |r_3| < |r_2|, \\
&\ldots \\
r_{n-2} &= r_{n-1} q_{n-1} + r_n && \text{with } 0 < |r_n| < |r_{n-1}|.
\end{aligned}$$

The algorithm stops, since the r_j form a strictly decreasing sequence of positive integers. Finally, we obtain

$$r_{n-1} = r_n q_n.$$

The last remainder which is different from 0, i. e., r_n, is the gcd of a and b. Indeed, from the last equation, there follows $r_n | r_{n-1}$ and, because of $r_n | r_n$ and $r_n | r_{n-1} q_{n-1}$, we obtain $r_n | r_{n-2}$. Continuing in this way, we have $r_n | a$ and $r_n | b$.

If t is a common divisor of a and b, then by the first equation t divides r_1, by the second one, t divides r_2, etc. Finally, one has $t | r_n$.

By the first equation, r_1 is linear combination of a and b. Substituting this linear combination of r_1 into the second equation, r_2 can be represented as a linear combination of a and b. Continuing in this way, finally one obtains r_n as linear combination of a and b. □

If we calculate the greatest common divisor for any two integers by the Euclidean algorithm, we get a result which is up to units uniquely determined.

Example 3.6.18. Let $a = 36$ and $b = 21$. Then the Euclidean algorithm gives:

$$\begin{aligned} 36 &= 21 \cdot 1 + 15, \\ 21 &= 15 \cdot 1 + 6, \\ 15 &= 6 \cdot 2 + 3, \\ 6 &= 3 \cdot 2. \end{aligned}$$

Therefore, $d = \gcd(36, 21) \cong 3$ and

$$\begin{aligned} 15 &= 36 \cdot 1 + 21 \cdot (-1), \\ 6 &= 21 \cdot 1 + 15 \cdot (-1) = 21 \cdot 1 + \big(36 \cdot 1 + 26(-1)\big)(-1) = 36 \cdot (-1) + 21 \cdot 2, \\ 3 &= 15 \cdot 1 + 6 \cdot (-2). \end{aligned}$$

Thus,

$$3 = 36 \cdot 1 + 21 \cdot (-1) + \big(36 \cdot (-1) + 21 \cdot 2\big) \cdot (-2) = 36 \cdot 3 + 21 \cdot (-5) = 3 \cdot 36 + (-5) \cdot 21.$$

For more than two integers, by associativity of the gcd, one has

$$d = \gcd(a_1, \ldots, a_n) = \gcd(\gcd(\ldots \gcd(\gcd(a_1, a_2), a_3) \ldots), a_{n-1}), a_n).$$

If the integers a and b lead to the same remainder when we divide both by the natural number m, then each divisor of a and m is also a divisor of b and m, and conversely. This observation leads to the following definition:

Definition 3.6.19. Let a and b be integers and let m be a natural number with $m > 1$. Then a is congruent to b modulo m if there is an integer g such that $a - b$ is g times m:

$$a \equiv b(m) :\Leftrightarrow \exists g \in \mathbb{Z}\ (a - b = gm) \quad (\Leftrightarrow m | (a - b)).$$

Clearly, $a \equiv b(m)$ if and only if a and b have the same remainder when we divide both by m. Indeed, assume that a and b have the same remainder r when we divide both by m, then

$$a = km + r, \quad b = lm + r, \quad 0 < |r| < m$$

and $a - b = (k - l)m$, thus $a \equiv b(m)$.

Now a and b can be written as

$$a = km + r_1, \quad b = lm + r_2, \quad 0 < |r_1| < m,\ 0 < |r_2| < m.$$

If $a \equiv b(m)$, i. e., $a - b = gm$ for some $g \in \mathbb{Z}$, then $(k - l)m + r_1 - r_2 = gm$, and then $r_1 = r_2$, since $r_1 - r_2$ cannot be g_1 times of m with $|g_1| > 0$.

Theorem 3.6.20. *The relation $\equiv (m)$ is an equivalence relation in $\mathbb{Z}$, which is compatible with addition and multiplication in $\mathbb{Z}$, i. e., from $a \equiv b(m)$ and $c \equiv d(m)$ there follows $a + c \equiv b + d(m)$ and $a \cdot c \equiv b \cdot d(m)$.*

Proof. First, $a \equiv a(m)$, since $a - a = 0 = 0 \cdot m$.

Second, $a \equiv b(m) \Rightarrow b \equiv a(m)$, since from $a - b = gm$ there follows $b - a = -gm$.

Third, $a \equiv b(m)$ and $b \equiv c(m) \Rightarrow a \equiv c(m)$, since from $a - b = g_1 m$ and $b - c = g_2 m$ there follows by addition of the equations $(a - b) + (b - c) = g_1 m + g_2 m$ and hence $a - c = (g_1 + g_2)m$, i. e., $a \equiv c(m)$.

Let $a \equiv b(m)$ and $c \equiv d(m)$, i. e., $a - b = g_1 m$ and $c - d = g_2 m$. From these equations, we get:

$$(a - b) + (c - d) = (a + c) - (b + d) = (g_1 + g_2)m \quad \text{and}$$
$$a(c - d) + d(a - b) = (ag_2 + dg_1)m \Leftrightarrow ac - bd = (ag_2 + dg_1)m.$$

Therefore, $a + c \equiv b + d(m)$ and $ac \equiv bd(m)$. □

Definition 3.6.21. The remainder class a modulo m of an integer a is the set of all integers which are congruent to a modulo m,

$$[a]_m := \{b \mid b \in \mathbb{Z} \wedge a \equiv b(m)\}.$$

Also $\mathbb{Z}/(m) = \{[0]_m, [1]_m, \ldots, [m-1]_m\}$ is the set of all remainder classes modulo m.

The equivalence relation congruence modulo m defines a partition of $\mathbb{Z}$ into remainder classes. From the infinite set $\mathbb{Z}$, we get the finite set $\mathbb{Z}_m := \mathbb{Z}/(m)$.

Theorem 3.6.22. *Defining on $\mathbb{Z}_m$ two binary operations, addition and multiplication, by*

I. $[a]_m + [b]_m = [a + b]_m$;

II. $[a]_m \cdot [b]_m = [a \cdot b]_m$

gives a commutative ring

$$\underline{\mathbb{Z}_m} = \underline{\mathbb{Z}/(m)} = (\mathbb{Z}/(m); +, \cdot),$$

the remainder ring modulo m.

Proof. The operations + and · are defined correctly, since from

$$[a]_m = [a']_m \wedge [b]_m = [b']_m, \quad \text{i. e., } a \equiv a'(m) \wedge b \equiv b'(m),$$

there follows

$$a + b \equiv a' + b'(m) \wedge ab \equiv a'b'(m),$$

i. e.,

$$[a]_m + [b]_m = [a']_m + [b']_m \wedge [a]_m[b]_m = [ab]_m.$$

Since calculations with remainder classes are defined by calculations with their representatives, the ring properties are easy to show. □

As an example, we consider the ring $\underline{\mathbb{Z}_4} = \underline{\mathbb{Z}/(4)}$ with the elements $\mathbb{Z}_4 = \{[0]_4, [1]_4, [2]_4, [3]_4\}$. Operations are given by the following tables:

$+$	$[0]_4$	$[1]_4$	$[2]_4$	$[3]_4$
$[0]_4$	$[0]_4$	$[1]_4$	$[2]_4$	$[3]_4$
$[1]_4$	$[1]_4$	$[2]_4$	$[3]_4$	$[0]_4$
$[2]_4$	$[2]_4$	$[3]_4$	$[0]_4$	$[1]_4$
$[3]_4$	$[3]_4$	$[0]_4$	$[1]_4$	$[2]_4$

$\cdot$	$[0]_4$	$[1]_4$	$[2]_4$	$[3]_4$
$[0]_4$	$[0]_4$	$[0]_4$	$[0]_4$	$[0]_4$
$[1]_4$	$[0]_4$	$[1]_4$	$[2]_4$	$[3]_4$
$[2]_4$	$[0]_4$	$[2]_4$	$[0]_4$	$[2]_4$
$[3]_4$	$[0]_4$	$[3]_4$	$[2]_4$	$[1]_4$

We see that $[2]_4[2]_4 = [0]_4$, where $[0]_4$ is the zero element of the ring $\underline{\mathbb{Z}_4}$, which can be seen from the first table. The ring $\underline{\mathbb{Z}_4}$ has zero-divisors, while $\underline{\mathbb{Z}}$ is zero-divisor-free.

The existence of zero-divisors follows from the fact that in $\mathbb{Z}_4$ the cancellation rule is not satisfied, e. g.,

$$[2]_4[1]_4 = [2]_4[3]_4 = [2]_4, \quad \text{but} \quad [1]_4 \neq [3]_4.$$

The second table shows that the equation $[2]_4[x]_4 = [3]_4$ has no solution. This means that the multiplication is not invertible. There arises the question of whether there are integers m such that the multiplication in $\underline{\mathbb{Z}_m}$ is invertible or whether there is a subset of $\underline{\mathbb{Z}_m}$ in which the multiplication is invertible.

There are not only differences of the rings $\underline{\mathbb{Z}}$ and $\underline{\mathbb{Z}_m}$, but also several similarities. By $a \mapsto [a]_m$ for all $a \in \mathbb{Z}$, a function $\varphi : \mathbb{Z} \to \mathbb{Z}_m$ is defined. This φ is surjective and has the important property to be compatible with the operations in $\mathbb{Z}$, resp. in $\mathbb{Z}_m$:

$$\varphi(a+b) = [a+b]_m = [a]_m + [b]_m = \varphi(a) + \varphi(b),$$

and similarly for the multiplication. The map φ is an example of a *homomorphism* of $\mathbb{Z}$ onto $\mathbb{Z}_m$, and $\mathbb{Z}_m$ is said to be a homomorphic image of $\mathbb{Z}$.

Now we consider remainder classes $[a]_m$ with $\gcd(a,m) = 1$. We can assume that $a > 0$ since our considerations are independent of the representatives, so we can choose positive ones.

Definition 3.6.23. A class $[a]_m$ is said to be a prime remainder class modulo m if $\gcd(a,m) = 1$. Let $P(m)$ be the collection of all prime remainder classes modulo m.

$$P(m) := \{[a]_m \mid \gcd(a,m) = 1\}.$$

Theorem 3.6.24. *The set $(P(m);\cdot)$ is a commutative semigroup in which the multiplication $\cdot$ is invertible.*

Proof. Let a and b be integers with $\gcd(a,m) = 1$ and $\gcd(b,m) = 1$. We show that $\gcd(ab,m) = 1$. Suppose $\gcd(ab,m) = d$ and $d \neq 1$. Then from $d|ab$ there follows $d|a$ or $d|b$. Together with $d|m$, this is a contradiction. Since $d \neq 1$, d has a prime divisor $p|d$. From $p|d$ and $d|ab$, there follows $p|ab$. From the existence and uniqueness of the prime factorization, there follows $p|a$ or $p|b$. Since $p|d$ and $d|m$ imply $p|m$, there follows $\gcd(a,m) = 1$ or $\gcd(b,m) = 1$, and thus $p|1$, a contradiction. Therefore, $\gcd(a,m) = 1$.

Associativity is satisfied for arbitrary remainder classes and therefore also for prime remainder classes.

Now $[1]_m$ belongs to $P(m)$ since $\gcd(1,m) = 1$. Therefore, $P(m)$ has an identity element.

We prove now that each prime remainder class modulo m has an inverse. Indeed, if $\gcd(a,m) = 1$, then 1 can be written as linear combination $ax + my = 1$ of a and m. Here $ax + my = 1$ means $ax \equiv 1(m)$ and x can be determined by the Euclidean algorithm. Instead of $ax \equiv 1(m)$, we can write $[ax]_m = [1]_m$ or $[a]_m[x]_m = [1]_m$. This means $[x]_m = [a]_m^{-1}$. □

As an example, we consider $P(8)$ with $[1]_8$, $[3]_8$, $[5]_8$, $[7]_8$, and the following multiplication table:

$\cdot$	$[1]_8$	$[3]_8$	$[5]_8$	$[7]_8$
$[1]_8$	$[1]_8$	$[3]_8$	$[5]_8$	$[7]_8$
$[3]_8$	$[3]_8$	$[1]_8$	$[7]_8$	$[5]_8$
$[5]_8$	$[5]_8$	$[7]_8$	$[1]_8$	$[3]_8$
$[7]_8$	$[7]_8$	$[5]_8$	$[3]_8$	$[1]_8$

Now with the prime remainder classes modulo m, we know an example of an algebraic structure with one binary operation which is associative and invertible. More generally, we define:

Definition 3.6.25. Let A be a nonempty set. Then $\mathcal{A} = (A; \cdot)$ is said to be a *group* if
1. $\cdot : A \times A \to A$ is a binary operation.
2. $\cdot$ is associative.
3. $\cdot$ is invertible, i. e.,

$$\forall a, b \in A \, \exists x, y \in A \, (a \cdot x = y \cdot a = b).$$

The group axiom 3 can be equivalently replaced by:
3.1. There exists an identity element $e \in A$ with $e \cdot a = a \cdot e = a$ for all $a \in A$, and
3.2. For each $a \in A$, there exists an inverse a^{-1} with $a \cdot a^{-1} = a^{-1} \cdot a = e$.

If the binary operation is written as addition +, then, instead of the identity element, one speaks of a zero element $0 \in A$ with $a + 0 = 0 + a = a$ and, instead of an inverse, each $a \in A$ has an opposite element $-a \in A$ which satisfies $a + (-a) = (-a) + a = 0$.

The number of the prime remainder classes modulo m can be determined by Euler's φ-function which is defined as follows:

Definition 3.6.26. The function $\varphi : \mathbb{N} \to \mathbb{N}$ *defined by*

$$m \mapsto \varphi(m) := \operatorname{card} P(m), \quad m \in \mathbb{N}, \, m \geq 1$$

is said to be *Euler's φ-function*.

The following table gives $\varphi(m)$ for $m = 1, \ldots, 15$:

m	1	2	3	4	5	6	7	8	9	10	11	12	13	14	15
$\varphi(m)$	1	1	2	2	4	2	6	4	6	4	10	4	12	6	8

Finally, we mention two formulas to calculate $\varphi(m)$ if for m the prime power factorization,

$$m = p_1^{\mu_1} \cdots p_r^{\mu_r},$$

is known:

$$\varphi(m) = \varphi(p_1^{\mu_1} \cdots p_r^{\mu_r}) = \varphi(p_1^{\mu_1})\varphi(p_2^{\mu_2}) \cdots \varphi(p_r^{\mu_r}),$$
$$\varphi(m) = m\left(1 - \frac{1}{p_1}\right) \cdots \left(1 - \frac{1}{p_r}\right) = p_1^{\mu_1 - 1} \cdots p_r^{\mu_r - 1}(p_1 - 1) \cdots (p_r - 1).$$

Depending on m, the remainder class ring $\underline{\mathbb{Z}_m}$ can be a *field*.

Definition 3.6.27. A nontrivial ring $\mathcal{A} = (A; +, \cdot)$ (i. e., A consisting of more than one element) is said to be a field if $A \setminus \{0\}$ is a group.

Corollary 3.6.28. *The remainder class ring modulo m,* $(\mathbb{Z}_m; +, \cdot)$, *is a field if and only if m is a prime number.*

Proof. Assume $m > 1$ is not a prime. Then there is a nontrivial decomposition of m into $m = ab$ with $1 < a, b < m$. Then in $\mathbb{Z}_m$ one has:

$$[a]_m[b]_m = [m]_m = [0]_m \quad \text{with } [a]_m \neq [0]_m \text{ and } [b]_m \neq [0]_m,$$

a contradiction, since a field has no zero divisors.

If, conversely, m is a prime, then the set of all remainder classes different from $[0]_m$ from $\mathbb{Z}_m$ is $P(m)$. Since $(P(m); \cdot)$ is a group, $\mathbb{Z}_m$ is a field. □

A consequence of the group property of $P(m)$ and the fact that $P(m)$ consists of $\varphi(m)$ elements is the so-called *small Fermat's theorem.*

Corollary 3.6.29. *Let a and m be integers,* $m > 1$, *and* $\gcd(a, m) = 1$. *Then* $a^{\varphi(m)} \equiv 1(m)$.

Proof. Let

$$P(m) := \{[a_1]_m, \ldots, [a_l]_m\} \quad \text{with } l = \varphi(m)$$

be the set of all prime remainder classes modulo m. Let $[a]_m$ be an arbitrary element from $P(m)$. Since $(P(m); \cdot)$ is a group, the set of all products $[a_1]_m[a]_m \cdots [a_l]_m[a]_m$ gives $P(m)$, since all these products belong to $P(m)$ and since the elements are pairwise different, which follows from the cancellation rule in $(P(m); \cdot)$. From

$$P(m) = \{[a_1]_m[a]_m, \ldots, [a_l]_m[a]_m\} = \{[a_1]_m, [a_2]_m, \ldots, [a_l]_m\}$$

there follow

$$[a_1]_m[a]_m \cdots [a_l]_m[a]_m = [a_1]_m \cdots [a_l]_m$$

and

$$[a_1]_m[a_2]_m \cdots [a_l]_m([a]_m)^l = [a_1]_m \cdots [a_l]_m[1]_m.$$

Using the cancellation rule, one gets

$$([a]_m)^l = [1]_m \quad \text{or, equivalently,} \quad a^{\varphi(m)} \equiv 1(m).$$ □

Corollary 3.6.29 is quite useful to solve problems of elementary number theory, for instance:

Show that $2^{256} - 1$ *is not a prime!*

Indeed, with $m = 257$, we have $\varphi(m) = 256$ (since 257 is a prime). From $\gcd(2, 257) = 1$, there follows $2^{256} \equiv 1(257)$ and $257|(2^{256} - 1)$, i. e., $2^{256} - 1$ has a nontrivial divisor and so cannot be a prime number.

In the previous section, we asked already for solutions $[x]_m$ of the equation $[a]_m[x]_m = [b]_m$. This question corresponds to the solution of a *linear congruence* $ax \equiv b(m)$ or to the solution of a *diophantine equation* $ax + my = b$ (*Diophant of Alexandria, around* 250 *AC*).

Definition 3.6.30. An equation of the form

$$a_1x_1 + \cdots + a_nx_n = b \quad \text{with } a_1, \ldots, a_n, b \in \mathbb{Z}$$

is said to be *linear diophantine equation* and $(x_1, \ldots, x_n)$ is said to be its solution if $x_1, \ldots, x_n$ are integers which satisfy this equation.

Example 3.6.31. A customer will spend 200 euros to buy wine glasses worth 16, 10, or 25 euro a piece. Which possibilities does the customer have? The diophantine equation which has to be solved is

$$16x_1 + 10x_2 + 25x_3 = 200.$$

Theorem 3.6.32. *The linear diophantine equation $a_1x_1 + \cdots + a_nx_n = b$ is solvable if and only if the greatest common divisor of $a_1, \ldots, a_n$ divides b. In this case the reduced diophantine equation $a_1'x_1 + \cdots + a_n'x_n = b'$ with $a_i = da_i'$, $b = db'$, and $d = \gcd(a_1, \ldots, a_n)$, $\gcd(a_1', \ldots, a_n') = 1$ has the same set of solutions.*

Proof. If $d = \gcd(a_1, \ldots, a_n)$, then d can be written as a linear combination of $a_1, \ldots, a_n$. Therefore, there are integers $c_1, \ldots, c_n$ with $d = c_1a_1 + \cdots + c_na_n$. If $d \mid b$, then there exists an integer b' with $b = db'$ and, multiplying the equation $d = c_1a_1 + \cdots + c_na_n$ with b', we obtain

$$(b'c_1)a_1 + \cdots + (b'c_n)a_n = b'd = b \quad \text{and} \quad x_1 = c_1b', \ldots, x_n = c_nb'.$$

If, conversely, the equation $a_1x_1 + \cdots + a_nx_n = b$ is solvable, then $d|a_1, \ldots, d|a_n$ and also $d|b$ by multiplication, resp. division, with d. □

Definition 3.6.33. A congruence $ax \equiv b(m)$ with $a, b, m \in \mathbb{Z}$, $m > 1$, is said to be a *linear congruence*, and each integer x which satisfies $ax \equiv b(m)$ is said to be a solution.

The linear congruence $ax \equiv b(m)$ corresponds to the linear diophantine equation $ax + my = b$. But conversely, to the linear diophantine equation $ax + my = b$ corresponds not only the linear congruence $ax \equiv b(m)$ but also the linear congruence $my \equiv b(a)$. This gives

Theorem 3.6.34. *A linear congruence $ax \equiv b(m)$ is solvable if and only if the greatest common divisor of a and m divides b. In this case, the reduced linear congruence $a'x \equiv$*

$b'(m')$ with $a = da'$, $b = db'$, $m = dm'$, and $\gcd(a', m') = 1$ has the same set of solutions as $ax \equiv b(m)$.

Example 3.6.35. We consider the linear congruence $9x \equiv 6(12)$. Since $\gcd(9, 12) = 3$ and $3|6$ this linear congruence is solvable and the reduced linear congruence $3x \equiv 2(4)$ has the same set of solutions.

Having answered to the question under which conditions there exist solutions, we want now to determine the set of solutions.

Theorem 3.6.36. *The set of solutions of $a'x \equiv b'(m')$ with $\gcd(a', m') = 1$ is exactly the set of all elements of $\mathbb{Z}$ which are congruent to x modulo m'. If a special solution x_0 is known, one obtains the general solution as $x \equiv x_0(m)$. This is the general solution of each linear congruence $ax \equiv b(m)$ with $d = \gcd(a, m)$ and $d|b$, from which $a'x \equiv b'(m')$ can be obtained by division with b.*

Proof. Element x_0 is a solution of $ax \equiv b(m)$ if and only if x_0 is a solution of $a'x \equiv b'(m')$, where $\gcd(a', m') = 1$. The latter congruence corresponds to the equation $[a']_{m'}[x]_{m'} = [b']_{m'}$. Thus, x_0 is a solution of $a'x \equiv b'(m')$ if and only if $[x_0]_{m'}$ is a solution of $[a']_{m'}[x]_{m'} = [b']_{m'}$. This solution can be given by $[x_0]_{m'} = [a']_{m'}^{-1}[b']_{m'}$. Therefore, x is a solution if and only if $x \in [x_0]_{m'}$. But now we have to present the set of solutions as a remainder class modulo m corresponding to the given problem:

$$\begin{aligned} x \in [x_0']_{m'} &\Leftrightarrow \exists g \in \mathbb{Z}\,(x = x_0 + gm') \\ &\Leftrightarrow \exists g \in \mathbb{Z}\, \exists 0 \le r < d\,(x = x_0 + (dq + r)m' = x_0 + rm' + dqm) \\ &\Leftrightarrow x \in [x_0]_m \cup [x_0 + m']_m \cup \cdots \cup [x_0 + (d-1)m']_m. \end{aligned}$$

Here $g = dq + r$ with $0 \le r < d$ follows by division with remainder. Moreover, we used $dm' = m$. □

Example 3.6.37. We consider again $9x \equiv 6(12)$. By division, we obtain the reduced linear congruence $3x \equiv 2(4)$. A special solution can be obtained by trial and error, $x_0 = 2$ is such a solution and then $L = \{x \mid x \in \mathbb{Z} \wedge x \equiv 2(4)\}$. To give the solutions of the first linear congruence modulo 12, we write $[2]_4$ as $[2]_4 = [2]_{12} \cup [6]_{12} \cup [10]_{12}$.

The special solution can also be obtained as a solution of $[3]_4[x]_4 = [2]_4$, i. e., $[x]_4 = [3]_4^{-1}[2]_4$. Because $3 \cdot 3 \equiv 1(4)$, we get $[3]_4^{-1} = [3]_4$ and then $[x]_4 = [3]_4[2]_4 = [2]_4$.

The relationship between linear congruences and linear diophantine equations allows determining the set of solutions of $ax + my = b$.

Corollary 3.6.38. *If (x_0, y_0) is a special solution of the diophantine equation $ax + my = b$, then the set of solutions L is given by*

$$L = \{(x, y) \mid x = x_0 + gm' \wedge y = y_0 - ga' \wedge g \in \mathbb{Z}\}$$

with $m = dm'$, $a = da'$, and $d = \gcd(a, m)$.

Proof. If $(x_0, y_0) \in L$, then $ax_0 \equiv b(m)$. Substituting into $ax + my = b$ and isolating y_0 gives $y_0 = \frac{b-ax_0}{m}$.

Similarly, from $(x,y) \in L$ we get $x \in [x_0]_{m'}$, and $y = \frac{b-ax}{m}$. Then $x = x_0 + gm'$ and

$$y = \frac{b - a(x_0 + gm')}{m} = \frac{b - ax_0}{m} - \frac{gam'}{m} = \frac{b - ax_0}{m} - \frac{ga}{d} = y_0 - ga', \quad g \in \mathbb{Z}.$$

Conversely, each pair (x,y) of integers of this form is a solution of the linear diophantine equation, since substitution of $x = x_0 + gm', y = y_0 - ga'$ into the linear diophantine equation gives

$$\begin{aligned} a(x_0 + gm') + m(y_0 - ga') &= ax_0 + agm' + my_0 - a'gm \\ &= (ax_0 + my_0) + a'dgm' - dm'ga' = b; \quad a = a'd,\ m = m'd. \end{aligned}$$

□

Example 3.6.39. We determine the set of solutions for $9x + 12y = 6$. Since $\gcd(9, 12) = 3$ and $3 \mid 6$, the equation is solvable. The reduced linear diophantine equation is $3x + 4y = 2$, which corresponds to the linear congruence $3x \equiv 2(4)$.

Since $3 \cdot 2 \equiv 2(4)$, we have the special solution $x_0 = 2$. Substituting $x_0 = 2$ into $3x + 4y = 2$ and isolating y gives $y_0 = -1$. Then

$$L = \{(x,y) \mid x = 2 + g \wedge y = -1 - 3g \wedge g \in \mathbb{Z}\}.$$

3.7 Rational numbers

In the semifield $\underline{\mathbb{Q}^+}$, the addition is in general not invertible. Since $\underline{\mathbb{Q}^+}$ is a semiring, we may apply Theorem 3.5.9 and can construct the difference ring of $\underline{\mathbb{Q}^+}$.

We prove that $D(\mathbb{Q}^+)$ forms a field.

Lemma 3.7.1. *The difference ring of the semifield of all nonnegative rational numbers is a field.*

Proof. The difference ring $D(\mathbb{Q}^+)$ has an identity element and this is $1 \in \mathbb{Q}^+$ since $1 \cdot (a - b) = 1 \cdot a - 1 \cdot b = a - b$. (We notice that $D(\mathbb{Q}^+)$ is commutative since $\mathbb{Q}^+$ is commutative.) Each element from $D(\mathbb{Q})$, which is different from the zero element, has an inverse element. For elements from $\mathbb{Q}^+$, i. e., for differences $a - b$ with $a, b \in \mathbb{Q}^+$, $a > b$, this is clear, since $\mathbb{Q}^+$ is a semifield. Let $a - b \in D(\mathbb{Q}^+)$ and $a, b \in \mathbb{Q}^+$ with $a < b$. Then $\frac{1}{b+a} - \frac{2b}{b^2-a^2} \in D(\mathbb{Q}^+)$ is the inverse element of $a - b$, since

$$(a - b)\left(\frac{1}{b+a} - \frac{2b}{b^2 - a^2}\right) = \left(\frac{a}{b+a} + \frac{2b^2}{b^2 - a^2}\right) - \left(\frac{b}{b+a} + \frac{2ab}{b^2 - a^2}\right) \sim (1, 0).$$

□

Definition 3.7.2. The minimal upper ring $D(\underline{\mathbb{Q}^+})$ of the semifield $\underline{\mathbb{Q}^+} = (\mathbb{Q}^+, +, \cdot)$ is said to be the field $\mathbb{Q}$ of rational numbers.

The field $\underline{\mathbb{Q}}$ consists of all differences $a - b$ with $a, b \in \underline{\mathbb{Q}^+}$. Calculation with the differences follows the rules

$$\begin{aligned}
&\text{I.} && a - b = c - d \Longleftrightarrow a + d = b + c, \quad a, b, c, d \in \mathbb{Q}^+,\\
&\text{II.} && (a - b) + (c - d) = (a + c) - (b + d),\\
&\text{III.} && (a - b) \cdot (c - d) = (ac + bd) - (ad + bc).
\end{aligned}$$

The field $\underline{\mathbb{Q}}$ is a minimal upper ring of $\underline{\mathbb{Q}^+}$, i. e., for each upper ring $\mathcal{R}$ of $\underline{\mathbb{Q}^+}$ with $\underline{\mathbb{Q}^+} \subset \mathcal{R} \subseteq \underline{\mathbb{Q}}$, one has $\mathcal{R} = \underline{\mathbb{Q}}$. Conversely, each minimal upper ring of $\underline{\mathbb{Q}^+}$ is a difference ring of $\underline{\mathbb{Q}^+}$ and therefore isomorphic to $\underline{\mathbb{Q}}$.

On the set $\mathbb{Q}$, a binary relation can be defined similar to the order on $\mathbb{Z}$.

Definition 3.7.3. Let $a - b, a' - b' \in \mathbb{Q}$. Then we define $a - b < a' - b'$ if and only if $a + b' < b + a'$, where $a, b', a', b \in \mathbb{Q}^+$ and where $a + b' < b + a'$ is the order in $\mathbb{Q}^+$.

Theorem 3.7.4. *The binary relation $<$ defined in Definition 3.7.3 is an irreflexive order relation in $\underline{\mathbb{Q}}$, which is compatible (monotone) with respect to addition and multiplication.*

The proof is similar to that of Theorem 3.5.15 (see [19]) and is left as an exercise to the reader.

We mention that the order relation on $\mathbb{Q}$ extends the order in $\mathbb{Q}^+$ and is the only possible order which extends the order in $\mathbb{Q}^+$. This means that if we apply the order defined on $\mathbb{Q}$ on elements of $\mathbb{Q}^+ \subseteq \mathbb{Q}$, then this is the order on $\mathbb{Q}^+$.

We recall the definition of an ordered ring $\mathcal{A} = (A; +, \cdot, <)$ as a ring $(A; +, \cdot)$ with an irreflexive order relation $<$ on A which is compatible with addition $+$ and multiplication $\cdot$. Then we obtain for rational numbers a similar algebraic characterization as for the semiring of natural numbers (Theorem 3.1.18), the ring of integers (Theorem 3.5.16), and the semifield of nonnegative rational numbers (Theorem 3.3.12).

Theorem 3.7.5. *Let $\mathcal{A} = (A, +, \cdot, <)$ be an ordered ring. Then the following statements are equivalent:*

(i) *$\mathcal{A}$ is isomorphic to $\underline{\mathbb{Q}} = (\underline{\mathbb{Q}}, +, \cdot, <)$.*

(ii) *$\mathcal{A}$ is isomorphic to the (ordered) difference ring of $\underline{\mathbb{Q}^+}$: $\underline{A} \overset{\sim}{\longleftrightarrow} D(\underline{\mathbb{Q}^+})$.*

(iii) *$\mathcal{A}$ is isomorphic to a minimal ordered upper ring of $\underline{\mathbb{Q}^+}$.*

(iv) *$\mathcal{A}$ is a minimal ordered field, i. e., $\underline{A}$ is an ordered field and contains no proper subfield.*

Proof. (i) $\Longrightarrow$ (ii) By Definition 3.7.2, $\underline{\mathbb{Q}} = D(\underline{\mathbb{Q}^+})$.

(ii) $\Longrightarrow$ (iii) Since $D(\underline{\mathbb{Q}^+})$ is, as an ordered difference ring, a minimal ordered upper ring of $\underline{\mathbb{Q}^+}$, (ii) implies (iii).

(iii) $\Longrightarrow$ (iv) By (iii), $\mathcal{A}$ is isomorphic to a minimal ordered upper ring of $\underline{\mathbb{Q}^+}$, which is isomorphic to the ordered difference ring $\mathcal{D}(\mathbb{Q}^+)$. Now $\mathcal{D}(\mathbb{Q}^+)$ is a field (Lemma 3.7.1) and therefore an ordered field. Assume there is a subfield $\mathcal{R} \subseteq \underline{\mathbb{Q}^+}$. From $1 \in R$, there follows $\underline{\mathbb{Q}^+} \subseteq \mathcal{R}$, i. e., $\mathcal{R}$ is an upper ring of $\underline{\mathbb{Q}^+}$. Since $\underline{\mathbb{Q}}$ is a minimal ordered upper ring of $\underline{\mathbb{Q}^+}$, we have $\mathcal{R} = \underline{\mathbb{Q}}$. Since $\mathcal{A}$ is isomorphic to $\underline{\mathbb{Q}}$, $\mathcal{A}$ is also a minimal ordered field.

(iv) $\Longrightarrow$ (*i*) If $\mathcal{A}$ is a minimal ordered field, then the intersection $\mathcal{U}$ of all subsemifields of $\mathcal{A}$ is a minimal ordered subsemifield of $\mathcal{A}$. Let $e \in A$, then $e \in U$ and $ne \in U$ for all $n \in \mathbb{N}$. Now $ne = me$ implies $n = m$ since, if $n < m$ $(n > m)$, then there would be an $x \neq 0$, $x \in \mathbb{Q}$, with $n+x = m$ (a $y \neq 0, y \in \mathbb{Q}$ with $m+y = n$). Then $ne+xe = me$ $(me+ye = ne)$ and, since $xe > 0$ $(ye > 0)$, one gets $ne < me$ $(me < ne)$. Therefore, $\mathcal{U}$ contains a subsemifield which is isomorphic to $\underline{\mathbb{N}}$ and then also $\underline{\mathbb{Q}}^+$ is contained in $\mathcal{U}$. Because of its minimality, $\mathcal{U}$ is isomorphic to $\underline{\mathbb{Q}}^+$. The isomorphism between $\mathcal{U}$ and $\underline{\mathbb{Q}}^+$ can be extended to an isomorphism between $\mathcal{A}$ and a minimal ordered upper ring of $\underline{\mathbb{Q}}^+$, i. e., to $D(\underline{\mathbb{Q}}^+)$. But this means that (i) is satisfied. □

There is another way to construct $\mathbb{Q}$ from $\mathbb{Q}^+$ (or $\mathbb{Z}$ from $\mathbb{N}$). Define $Q'_+ = \{-a \mid a \in \mathbb{Q}^+\}$ and consider the set $Q' = Q'_+ \cup \mathbb{Q}^+$ with $Q'_+ = \{-a \mid a \in \mathbb{Q}^+\}$. Then

Theorem 3.7.6. *Extending order, addition, and multiplication of* $\underline{\mathbb{Q}}^+$ *to* $Q' = Q'_+ \cup \mathbb{Q}^+$ *by*

(i)

$$-a > -b \quad \textit{if and only if } a < b,\ -b < 0,$$

(ii)

$$a + (-b) := \begin{cases} a - b & \textit{if } a > b,\ \textit{i. e.}, a = b + x \textit{ with } x \in \mathbb{Q}^+ \setminus \{0\}, \\ 0 & \textit{if } a = b, \\ -(b - a) & \textit{if } a < b,\ \textit{i. e.}, b = a + y \textit{ with } y \in \mathbb{Q}^+ \setminus \{0\}, \end{cases}$$

$$(-a) + (-b) := -(a + b).$$

(iii)

$$a \cdot (-b) = (-a) \cdot b := -ab, \quad (-a) \cdot (-b) := ab,$$

one obtains an ordered upper field $\mathcal{Q}' = (Q', +, \cdot, <)$*, namely* $\underline{\mathbb{Q}}$.

Proof. Ring $\underline{\mathbb{Q}}$ is a minimal ordered upper ring of $\underline{\mathbb{Q}}^+$, $\mathbb{Q}^+ \cup Q'_+ \subseteq \mathbb{Q}$ and $\mathbb{Q}^+ \cap Q'_+ = \emptyset$. Then one proves (i)–(iii). Thus $\mathcal{Q}'$ is an ordered subring of $\underline{\mathbb{Q}}$ which contains $\underline{\mathbb{Q}}^+$. Since $\underline{\mathbb{Q}}$ is minimal, one obtains $\mathcal{Q}' = \underline{\mathbb{Q}}$. Moreover, $\underline{\mathbb{Q}}$ is an ordered field containing $\underline{\mathbb{Q}}^+$. □

3.8 Historical remarks

Numbers play a crucial role in our day-to-day life, in science, technology, economy, and culture. During the nineteenth century, mathematicians developed many different abstractions which have some properties in common with numbers. Hypercomplex numbers consist of extensions and modifications of complex numbers. Number systems are examples of more general algebraic structures as semirings, semifields, rings, and fields. Arithmetic and number theory are the mathematical disciplines with numbers as their main topic. The German mathematician *Leopold Kronecker* (1823–1891) requested that all mathematics has to be founded on numbers and all numbers should be based on natural numbers. “God made the integers, all else is the work of man,” is probably his most famous quote. Indeed, the approach to number systems of this book started with an axiomatic foundation of natural numbers as cardinalities of finite sets and by Peano axioms. All other number systems are constructed from the natural numbers. At first glance, the algebraic characterizations of the natural numbers, the integers, the non-negative rational numbers, and the rational numbers, given in Theorems 3.1.18, 3.5.16, 3.3.12, and 3.7.5 as minimal algebraic structures with two binary operations, one binary relation, and satisfying certain properties, look like some principles of minimum energy in nature, as if made by the gods. Having a deeper insight, we will recognize the reason.

Leopold Kronecker worked from 1855 on as a private docent at the University of Berlin, which means without a regular salary. Nevertheless, Kronecker gave the mathematical department (I. Mathematisches Institut) life and color by his ideas. Only in 1882, after Ernst Eduard Kummer's (1810–1893) retirement, he accepted a call to take over his chair. Both mathematicians knew each other for a long time since Kummer was Kronecker's highschool teacher in Liegnitz, today Legnica (Silesia). The number system of real numbers, which was not considered in this chapter, consists of rational and irrational numbers. While rational numbers can be written as fractions of integers $\frac{p}{q}, p, q \in \mathbb{Z}, q \neq 0$, or as finite or infinite periodic decimal fractions, irrational real numbers can be represented as infinite nonperiodic decimal fractions. Richard Dedekind (1831–1916) was a professor of mathematics at the Technical University of Braunschweig (Germany) and developed the theory of irrationality, especially in his books "Stetigkeit und irrationale Zahlen" [26] and "Was sind und was sollen die Zahlen" [25]. He is especially known by the "Dedekind cut," a method to construct the real numbers from the rational ones. Dedekind's friend, Georg Cantor (1845–1918), a former student of Kronecker, was a professor at University of Halle. He introduced set theory as a new research object. In his paper [16], he introduced transfinite cardinal numbers as cardinalities of infinite sets and its arithmetic. The smallest infinite cardinal number is the cardinality of $\mathbb{N}$ which is called countably infinite and which is equal to the cardinality of $\mathbb{Z}$ and $\mathbb{Q}^+$, resp. $\mathbb{Q}$. This infinite cardinal number is called $\aleph_0$. The next one is the cardinality of the set of real numbers, called $\aleph_1$. Then follows the cardinality of the power set, $\aleph_2 := \operatorname{card}(\mathcal{P}(\mathbb{R}))$, etc. Cantor introduced operations for infinite cardinalities. Infinite cardinalities can be ordered such that $\aleph_0 \ll \aleph_1 \ll \aleph_2 \ll \cdots$. It is not clear and an unsolved problem whether there is an infinite cardinality between $\aleph_0$ and $\aleph_1$. The *continuum hypothesis* states that there is no transfinite cardinality between $\aleph_0$ and $\aleph_1$.

4 Linear diophantine equations and consumer theory

In this chapter we continue our study of commodities which we began in Chapter 2. We ask for a consumer's possibilities to spend a given budget for its preferred n-tuple of commodities (we will speak of a bundle of commodities), supposing that the prices of the commodities are fixed. We consider two typical examples and show how a customer's preferences can be expressed by a utility function. Similar to Chapter 2, Definition 2.3.2, where we defined the binary relation $\sim_{\mathrm{SIR}}$ by the kernel of the labor function lab, in this chapter we define the *indifference relation* $\sim$ as the kernel of a utility function u. This allows us once more to apply Chapter 1, Theorems 1.4.9 and 1.7.13. For more details on consumer theory see e. g. [6].

https://doi.org/10.1515/9783112234433-004

4.1 The consumer problem

Consumer theory is concerned with the question how a rational consumer will make consumption decisions. Actually, the consumer problem belongs to the choice theory, but becomes a particular instance since any decision depends on the consumer's income, his budget, and his (or her) preferences are determined by the usefulness of the bundle of goods he (or she) prefers. Clearly, a majority of consumers has only a restricted freedom to choose between different bundles of commodities or services, restricted by the budget. The idea is that the consumer chooses a vector $(x_1, \ldots, x_n)$ of commodities or services with fixed prices $(p_1, \ldots, p_n)$ and wants to maximize its usefulness, subject to a budget constraint, which means the consumer, as a rational consumer, will not spend more money as he has or is willing to pay. Utility maximization means purchasing those bundles of commodities that bring the greatest benefit.

Consumer theory needs several assumptions, for instance, *nonsatiation*. Roughly speaking, consumers always want to consume more. Another assumption is *decreasing marginal utility*, meaning that consumers lose satisfaction with a product the more they have consumed it. Consumers are assumed to be able to decide rationally. Decisions can be difficult if people are not familiar with all properties of the product. There could be an emotional component involved in the decision making process that cannot be captured economically. So, consumers are sometimes indifferent to the choices available. We will restrict ourselves to a period of time where the prices are fixed.

4.2 The budget equation

We assume that the consumer wants to spend m units of money. If the prices $p_1, \ldots, p_n$ are fixed and she (he) wants to buy x_i units of commodity i, $1 \leq i \leq n$, then the overall expenditures are $p_1x_1 + \cdots + p_nx_n$ and are equal to m. Moreover, we assume that the consumer wants to spend the whole amount of money with the aim to maximize the utility. We will start with the following example:

Example 4.2.1. John wants to allocate 200 euros to video games and pizza. The price for a video game is 50 euros and the price for a pizza is 10 euros. John asks for all pairs (x_1, x_2) of nonnegative integers for which the diophantine equation

$$50x_1 + 10x_2 = 200$$

is satisfied. This is a linear diophantine equation in x_1 and x_2. If John does not like pizza, he can spend all of his money on video games. Then $x_2 = 0$ and $x_1 = 4$. If he is very hungry and not interested in video games, then $x_1 = 0$ and $x_2 = 20$, and he can buy and eat 20 pizzas. This leads to the additional conditions

$$0 \leq x_1 \leq 4, \; 0 \leq x_2 \leq 20.$$

If, in the general case, $p_1, \ldots, p_n$ are not natural numbers, but nonnegative rational numbers, i. e., fractions $\frac{p_1}{p_1'} \ldots, \frac{p_n}{p_n'}, \frac{m}{m'}$ with $p_1', \ldots, p_n', m' \neq 0$ of natural numbers, then instead of

$$p_1x_1 + \cdots + p_nx_n = m$$

we have the equation

$$\frac{p_1}{p_1'}x_1 + \cdots + \frac{p_n}{p_n'}x_n = \frac{m}{m'}. \tag{*}$$

Multiplying the latter equation by the least common multiple

$$\operatorname{lcm}(p_1{}', \ldots, p_n{}', m'),$$

we obtain

$$p_1\frac{\operatorname{lcm}(p_1{}', \ldots, p_n{}', m')}{p_1'}x_1 + \cdots + p_n\frac{\operatorname{lcm}(p_1{}', \ldots, p_n{}', m')}{p_n'}x_n = m\frac{\operatorname{lcm}(p_1{}', \ldots, p_n{}', m')}{m'}. \tag{**}$$

Equations (*) and (**) have the same set of solutions and the last one is a linear diophantine equation (indeed, it is an equation with nonnegative natural numbers as coefficients and we are interested in nonnegative natural numbers as solutions since if we assume that there are nonnegative rational, not natural solutions, we can repeat the previous procedure). This shows that prices of commodities and the amount of budget can be assumed to be given by nonnegative natural numbers, and therefore we restrict our considerations to linear diophantine equations.

4.3 The cases $n = 2$ and $n = 3$

Now we will come back to John and will calculate all his possibilities to spend 200 euros on food.

Example 4.3.1. Dividing both sides of the linear diophantine equation

$$50x_1 + 10x_2 = 200 \tag{†}$$

by 10, we obtain the shortened linear diophantine equation

$$5x_1 + x_2 = 20. \tag{††}$$

This equation is equivalent to (†), i. e., it has the same set of solutions, supposing it is solvable. By Theorem 3.6.32 from Chapter 3, it is solvable if and only if the greatest common divisor of 5 and 1, $\gcd(5, 1) = 1$, divides 20. Therefore (††) is solvable and (†) is also solvable, and both equations have the same set of solutions.

Equation (††) corresponds to the linear congruence $x_2 \equiv 20(5) \equiv 0(5)$ or $[x_2]_5 = [0]_5$ where $[x_2]_5 = \{5g \mid g \in \mathbb{Z}\}$ is the remainder class modulo 5. Since $x_2 \geq 0$, we may assume that $g \in \mathbb{N}$. Substituting $x_2 = 5g$ into (††) gives $5x_1 + 5g = 20$, and then $x_1 = 4 - g$. Because of $0 \leq x_1 \leq 4$, we get $0 \leq g \leq 4$ and obtain exactly the following possibilities:

g	x_1	x_2	(x_1, x_2)
0	4	0	(4, 0)
1	3	5	(3, 5)
2	2	10	(2, 10)
3	1	15	(1, 15)
4	0	20	(0, 20)

John has exactly 5 possibilities to spend his 200 euros.

Now we come to the general case

$$p_1x_1 + \cdots + p_nx_n = m, \quad p_1, \ldots, p_n, m \in \mathbb{N}. \tag{*}$$

Problem. Determine all n-tuples of natural numbers which satisfy this equation!

For $x_1 = x_2 = \cdots = x_{i-1} = x_{i+1} = \cdots = x_n = 0, 1 \leq i \leq n$, we have $p_ix_i = m$ and then

$$0 \leq x_i \leq \frac{m}{p_i}, \ 1 \leq i \leq n$$

as additional conditions.

At first, we check whether (*) is solvable, i. e., whether $d = \gcd(p_1, \ldots, p_n)$ divides m, i. e., $d|m$. If (*) is solvable, we go to the reduced linear diophantine equation dividing both sides of (*) by d. This gives

$$\begin{aligned} &p_1'x_1 + \cdots + p_n'x_n = m' \quad \text{with } p_1 = dp_1', \ldots, p_n = dp_n', m = dm', \\ &\gcd(p_1', \ldots, p_n') = 1. \end{aligned} \tag{**}$$

Equation (**) has the same set of solutions as (*).

By the Euclidean algorithm, d is a linear combination of $p_1, \ldots, p_n$. Therefore, there are natural numbers $c_1, \ldots c_n$ with

$$c_1p_1 + \cdots + c_np_n = d.$$

Multiplying with m', we get

$$p_1(c_1m') + \cdots + p_n(c_nm') = m,$$

and then

$$(x_1, \ldots, x_n) = (x_1m', \ldots, c_nm')$$

is a special solution.

Now we consider an example for $n = 3$ and show a method to obtain the set of all solutions from a special solution.

Example 4.3.2. John will spend his 200 euros on food for himself and his friends. Today the restaurant offers 3 meals: fish and ships for 15 euros a portion, fried chicken with french fries for 20 euros, and a superpizza for 25 euros. He wants to know how many friends he can invite. Each of them, included himself, should be given a meal and should have the choice between 3 meals. All three meals should be ordered.

This time the linear diophantine equation with additional conditions

$$15x_1 + 20x_2 + 25x_3 = 200, \quad x_1, x_2, x_3 \in \mathbb{N},\ 0 \le x_1 \le 13,\ 0 \le x_2 \le 10,\ 0 \le x_3 \le 8,$$

has to be solved. Since $d = \gcd(10, 20, 25) = 5$ divides 200, the equation is solvable and $3x_1 + 4x_2 + 5x_3 = 40$ is the reduced diophantine equation. For the moment, we fix x_3 and consider the linear diophantine equation with 2 unknowns

$$3x_1 + 4x_2 = 40 - 5x_3$$

which corresponds to the linear congruence

$$4x_2 \equiv (40 - 5x_3)(3),$$

which is equivalent to $x_2 \equiv (1 + x_3)(3)$ since $40 \equiv 1(3)$ and $-5 \equiv 1(3)$. We consider all possibilities for x_3, $0 \le x_3 \le 8$, and calculate the corresponding values for x_1 and x_3:

$$x_3 = 0 \Rightarrow x_2 \equiv 1(3) \Rightarrow x_2 = 1 + 3g$$

$$\overset{\text{by substitution and reduction}}{\Longrightarrow} \quad x_1 = 12 - 4g \Rightarrow 0 \le g \le 3$$

g	x_1	x_2	x_3
0	12	1	0
1	8	4	0
2	4	7	0
3	0	10	0

$$x_3 = 1 \Rightarrow x_2 \equiv 2(3) \Rightarrow x_2 = 2 + 3g,$$

$$\overset{\text{by substitution and reduction}}{\Longrightarrow} \quad x_1 = 9 - 4g \Rightarrow 0 \le g \le 2$$

g	x_1	x_2	x_3
0	9	2	1
1	5	5	1
2	1	8	1

$$x_3 = 2 \Rightarrow x_2 \equiv 0(3) \Rightarrow x_2 = 3g,$$

$$\overset{\text{by substitution and reduction}}{\Longrightarrow} \quad x_1 = 10 - 4g \Rightarrow 0 \le g \le 2$$

g	x_1	x_2	x_3
0	10	0	2
1	6	3	2
2	2	6	2

$$x_3 = 3 \Rightarrow x_2 \equiv 1(3) \Rightarrow x_2 = 1 + 3g, \overset{\text{by substitution and reduction}}{\Longrightarrow} x_1 = 7 - 4g \Rightarrow 0 \le g \le 1$$

g	x_1	x_2	x_3
0	7	1	3
1	3	4	3

$$x_3 = 4 \Rightarrow x_2 \equiv 2(3) \Rightarrow x_2 = 2 + 3g, \overset{\text{by substitution and reduction}}{\Longrightarrow} x_1 = 4 - 4g \Rightarrow 0 \le g \le 1$$

g	x_1	x_2	x_3
0	4	2	4
1	0	5	4

$$x_3 = 5 \Rightarrow x_2 \equiv 0(3) \Rightarrow x_2 = 3g, \overset{\text{by substitution and reduction}}{\Longrightarrow} x_1 = 5 - 4g \Rightarrow 0 \le g \le 1$$

g	x_1	x_2	x_3
0	5	0	5
1	1	3	5

$$x_3 = 6 \Rightarrow x_2 \equiv 1(3) \Rightarrow x_2 = 1 + 3g, \overset{\text{by substitution and reduction}}{\Longrightarrow} x_1 = 2 - 4g \Rightarrow g = 0$$

g	x_1	x_2	x_3
0	2	1	6

$$x_3 = 7 \Rightarrow x_2 \equiv 2(3) \Rightarrow x_2 = 2 + 3g, \overset{\text{by substitution and reduction}}{\Longrightarrow} x_1 = -1 - 4g \Rightarrow \nexists g \ge 0.$$

If $x_3 = 8$, then $x_1 = x_2 = 0$ since $5 \cdot 8 = 40$.

These are all the solutions, and so the set of all solutions of the linear diophantine equation with additional conditions is

$$L = \{(12,1,0),(8,4,0),(4,7,0),(0,10,0),(9,2,1),(5,5,1),(1,8,1),(10,0,2),(6,3,2),$$
$$(2,6,2),(7,1,3),(3,4,3),(4,2,4),(0,5,4),(5,0,5),(1,3,5),(2,1,6),(0,0,8)\}.$$

Not all of these 18 solutions solve John's problem. Since each meal should be ordered, we can cancel the solutions

$$(12,1,0),(8,4,0),(4,7,0),(0,10,0),(10,0,2),(0,5,4),(5,0,5)(0,0,8),$$

i. e.,

$$L' = \{(9,2,1),(5,5,1),(1,8,1),(6,3,2),(2,6,2),(7,1,3),(3,4,3),(4,2,4),(1,3,5),(2,1,6)\}.$$

If John, for instance, were to invite 8 friends, then his preference could be expressed by $x_1 + x_2 + x_3 = 9$ (including himself) and the set of solutions would be $L_9 = \{(1,3,5),(2,1,6)\}$. John's preference is to invite as many people as possible, i. e., he wants to maximize the sum $u(x_1, x_2, x_3) = x_1 + x_2 + x_3$. Since L' is the set of all the solutions of the problem, we have $u(x_1, x_2, x_3) = x_1 + x_2 + x_3 \in \{9, 10, 11, 12\}$ and John can invite at most 11 friends since the maximum of $x_1 + x_2 + x_3$ is 12. There is exactly one solution where the sum of the entries is 12 and this is $(9, 2, 1)$. John can invite at most 11 friends and then they have to order 9 times fish and ships, 2 times fried chicken with french fries, and one superpizza.

4.4 Consumer's preferences

Let $\underline{x} = (x_1, \ldots, x_n)$ and $\underline{y} = (y_1, \ldots, y_n)$ be two bundles of goods, where $x_i, y_j \in \mathbb{Q}^+$. We define order relations $\succ$ and $\succeq$, but not componentwise as we did in Chapter 1, Section 1.6, when considering the partial order relations $\prec$ and $\preceq$, respectively.

Definition 4.4.1. The relation $\underline{x} \succ \underline{y}$ means that the consumer strictly prefers $\underline{x}$ to $\underline{y}$. The relation $\underline{x} \succeq \underline{y}$ means that the consumer weakly prefers $\underline{x}$ to $\underline{y}$. This includes the cases where the consumer strictly prefers $\underline{x}$ to $\underline{y}$ or finds $\underline{x}$ as good as $\underline{y}$. The relation $\underline{x} \sim \underline{y}$ means that the consumer is indifferent between the goods.

Let C be the set of all bundles of goods under consideration; $\underline{C} = (C; \succeq)$ will be assumed to be a totally (linearly) ordered set.

Indeed, the binary relation $\succeq$ is

- transitive: if the consumer prefers $\underline{x}$ to $\underline{y}$ and $\underline{y}$ to $\underline{z}$, then he prefers $\underline{x}$ to $\underline{z}$;
- antisymmetric: if $\underline{x} \succeq \underline{y}$ and $\underline{y} \succeq \underline{x}$, then $x \sim y$ (because this means that the consumer cannot decide which one he (she) likes better);
- reflexive: $\underline{x} \succeq \underline{x}$ for all $\underline{x} \in C$.

If we assume that any two bundles are comparable, i. e., $\underline{x} \succeq \underline{y}$ or $\underline{y} \succeq \underline{x}$, then we exclude the cases where the consumer is indifferent. Therefore, only in this case $\succeq$ is a total order on C.

The relation $\succ$ will be assumed to be an irreflexive order relation since $\succ$ is

- transitive;
- asymmetric: if $\underline{x} \succ \underline{y}$ then $\neg(\underline{y} \succ \underline{x})$;
- irreflexive: $\neg(\underline{x} \succ \underline{x})$.

We assume that $\sim$ is an equivalence relation, i. e., we assume that $\underline{x} \sim \underline{x}, \underline{x} \sim \underline{y} \Rightarrow \underline{y} \sim \underline{x}, \underline{x} \sim \underline{y} \wedge \underline{y} \sim \underline{z} \Rightarrow \underline{x} \sim \underline{z}$. (This makes sense if preferences are expressed by a utility function.) The equivalence relation $\sim$ defines a partition of C into equivalence classes where the equivalence class $[\underline{x}]_\sim$ is defined to be the set of all bundles $\underline{x}'$ with $\underline{x}' \sim \underline{x}$. This is the set of all bundles for which the consumer is indifferent with respect to $\underline{x}$.

Remark 4.4.2. If we interpolate the discrete set $[x]_\sim$ of pairs to a set of pairs consisting of real numbers, then $[x]_\sim$ can be interpreted as a set of all points of some space. Especially for $n = 2$, one obtains the set of all points of a curve which is said to be *curve of* *indifferences*. From the properties of a partition that $[x]_\sim \cap [y]_\sim = \emptyset$ for $\underline{x} \nsim \underline{y}$ (see Definition 1.4.8), we obtain that two different curves of indifferences have no point in common and, because $x \in [x]_\sim$ for all $\underline{x} \in C$, for any $\underline{x} \in C$ there is a curve which goes through the point corresponding to $\underline{x}$.

As we mentioned at the beginning of this section, another way to compare two bundles $\underline{x}$ and $\underline{y}$ of commodities consists of comparing the entries x_i, y_i of the n-tuples, i. e., the amounts of the goods

$$\underline{x} \geq \underline{y} :\Leftrightarrow \forall 1 \leq i \leq n\,(x_i \geq y_i) \wedge \exists j \in \{1,\ldots,n\}\,(x_j > y_j).$$

4.5 Perfect substitutes and perfect complements

In the second part of our example, John's preference is that each person he has invited, say all 8 friends and himself, will be given a meal and that all three different dishes will be ordered. Both conditions are satisfied for the two solutions $(1,3,5)$ and $(2,1,6)$. John is indifferent with respect to both bundles. The second has one superpizza more and one times fish and ships more than the first choice. The first also has two times fried chicken with french fries more than the second. The sum of the ordered meals is constant and equal to 9. In the economic literature, one speaks of *perfect substitutes* which will be considered mainly for $n = 2$, but can be defined for arbitrary n.

Perfect complements are commodities or services which will be consumed in a certain relationship to each other. When the consumption of one good increases, also does the consumption of the complementary good. The situation of the consumer will not be improved if the consumption of only one of two goods increases. A typical example is that of left and right shoes, regarded as different goods. It becomes no more useful for the consumer if the consumption of right shoes increases from 1 to 100 while the consumption of left shoes is constantly 1. If we assume that the consumer can throw away the right shoes, which he cannot use, for free, then he is indifferent to the bundles $(1,1), (1,2), \ldots, (1,n)$ (and correspondingly to $(1,1), (2,1), \ldots, (n,1)$ for the left shoes). This means that for indifferent bundles the minimum of the entries is 1. Of course, this can also be considered if $n > 2$. Then for all n goods, the consumption of one of the goods is related to the consumption of the other goods.

4.6 Preferences and utility functions

In the second part of our example, we considered not only the budget-restriction function $f(x_1,x_2,x_3) = 3x_1 + 4x_2 + 5x_3$ and the resulting equation $f(x_1,x_2,x_3) - 40 = 0$, but also the function $u(x_1,x_2,x_3) = x_1 + x_2 + x_3$ and the equation $u(x_1,x_2,x_3) - 9 = 0$ if John wants to invite 8 friends or, more generally, one of the equations $u(x_1,x_2,x_3) - k = 0$,

$k \in \{9, 10, 11, 12\}$ if John wants to invite $k-1$ friends. The functions f, u are ternary functions defined on the set $\mathbb{N}$,

$$f, u : \mathbb{N}^3 \to \mathbb{N}.$$

The function u can be used to express the consumer's preferences. Therefore we define

$$(x_1, x_2, x_3) \succeq (y_1, y_2, y_3) :\Leftrightarrow u(x_1, x_2, x_3) \geq u(y_1, y_2, y_3)$$

and

$$(x_1, x_2, x_3) \sim (y_1, y_2, y_3) :\Leftrightarrow u(x_1, x_2, x_3) = u(y_1, y_2, y_3).$$

Here $\geq$ is the order relation on the set $\mathbb{N}$ of natural numbers.

Functions like u are said to be *utility functions*. For arbitrary $n \geq 1$, utility functions are n-ary functions

$$u : \mathbb{Q}^n \to \mathbb{Q}.$$

The consumer wants to maximize the utility function. We mention that utility functions for perfect substitutes have the form $u(x_1, \ldots, x_n) = a_1 x_1 + \cdots + a_n x_n$, $a_i \in \mathbb{Q}$. (See Example 4.3.2 for $n = 3$ and $a_1 = a_2 = a_3 = 1$.) Utility functions for perfect complements and their generalizations to arbitrary $n \geq 1$ have the form $u(x_1, \ldots, x_n) = \min(a_1 x_1, \ldots, a_n x_n)$, $a_i \in \mathbb{Q}$.

Since in our example $u(1, 3, 5) = u(2, 1, 6) = 9$, John is indifferent to ordering meals corresponding to $(1, 3, 5)$ or $(2, 1, 6)$.

If the preference relation $\succeq$ is defined by a utility function u, then the equivalence relation $\sim$ is the kernel of the utility function,

$$\sim = \succeq \cap \preceq = \ker u.$$

This is a general situation, also if the utility function is considered to be a real function $f : \mathbb{R}^n \to \mathbb{R}$. Indeed, indifference curves for $n = 2$ or indifference spaces for arbitrary natural n, where n is the arity of the utility function, are equivalence classes with respect to the kernel of the utility function or, equivalently, sets of the partition defined by the kernel of the utility function. This knowledge makes clear, without any proof, why different indifference curves cannot intersect and why there is an indifference curve through any point of C.

In our example, the kernel of $u(x_1, x_2, x_3) = x_1 + x_2 + x_3$ defines the following partition of the 18-element set of bundles:

$$\begin{aligned}\{(0,0,8)\} &\cup \{(0,5,4),(1,3,5),(2,1,6)\} \\ &\cup \{(0,10,0),(1,8,1),(2,6,2),(3,4,3),(4,2,4),(5,0,5)\} \\ &\cup \{(4,7,0),(5,5,1),(6,3,2),(7,1,3)\} \\ &\cup \{(8,4,0),(9,2,1),(10,0,2)\} \\ &\cup \{(12,1,0)\}.\end{aligned}$$

The quotient set

$$L/\sim = \{[(0,0,8)]_\sim, [(0,5,4)]_\sim, [(0,10,0)]_\sim, [(4,7,0)]_\sim, [(8,4,0)]_\sim, [(12,1,0)]_\sim\}$$

consists of 6 elements corresponding to the image of function u,

$$\mathrm{Im}(u) = \{8,9,10,11,12,13\}.$$

For the solution of John's problem, we have to restrict the utility function u to the domain L' and to maximize the new utility function $u' := u|_{L'}$.

4.7 Historical remarks

Diophantus (Diophant) of Alexandria was a Greek mathematician. He was born in the 3rd century AD in Alexandria. His main opus is entitled "Arithmetica" and consists of more than 10 volumes. Unfortunately, only six of them survived until our days. "Arithmetica" is a collection of problems from arithmetic and their solutions. Diophant tried to solve algebraic equations of the form

$$Ax^2 + Bx + C = y^2 \quad \text{and} \quad Ax^3 + Bx^2 + Cx + D = y^2$$

or systems of such equations. He was interested only in positive rational solutions. This leads to rational coefficients A, B, C, D. Diophant gives no general solutions, but only numerical examples. Nevertheless, his work influenced the development of algebra and number theory. "Arithmetica" was several times edited and annotated, for instance, in 1621 in Paris by *Caspar Claudius Bachet* [4]. The young French lawyer *Pierre Fermat* (later "de Fermat"), born in 1601 in Beaumont de Lomagne, near to Toulouse, had a chance to read the book. He had studied civil law at the University of Orleans and kept an office of a "counselor" at the Parliament of Toulouse. In his free time, Fermat studied Diophant's work, gave new solutions, and formulated new problems. He expressed his mathematical ideas, mostly without complete proofs, in letters to friends or as short notices, written into Diophant's book. His most famous sidenote concerns the equation $x^n + y^n = z^n$. He maintained to have found a beautiful proof that this equation has for any $n \geq 3$ no positive integer as solution. The world most famous mathematicians and many nonmathematicians looked for a proof, since the problem is easy to formulate and to understand. It is not difficult to see that it is enough to solve the problem for odd prime numbers and for $n = 4$. *Ernst Eduard Kummer* (1810–1893) proved in 1847 that the conjecture is true for all so-called "regular" prime numbers. The biggest sensation of the years 1993–1994 was that the British mathematician *Andrew Wiles* found the proof of "Fermat's last theorem" using methods of modern mathematics.

In this book we consider linear diophantine equations where the unknowns $x_1, \ldots, x_n$ have no exponent greater than 1 and their applications in consumer theory.

Pierre de Fermat published in 1670 a small booklet entitled "Remarks to Diophant" [30]. For other editions of Diophant's work, see [58, 63].

5 Graphs

Many problems can be understood easier if one uses graphs for illustration. But also the solution can be easier as the following “classical” example shows:

Example 5.0.1 (The bridges of Königsberg). The problem is to devise a walk through the city of Königsberg (today Kaliningrad, Russia) that crosses each of the seven bridges leading over the *Pregel* river exactly once. The starting and ending points of the walk need not be the same. Does such a route exist?

The problem was published by *Leonhard Euler* (1707–1783) in 1736. The map of Königsberg can be replaced by the following symbolic picture showing the river, the land areas I–IV, and the seven bridges (Figure 5.1).

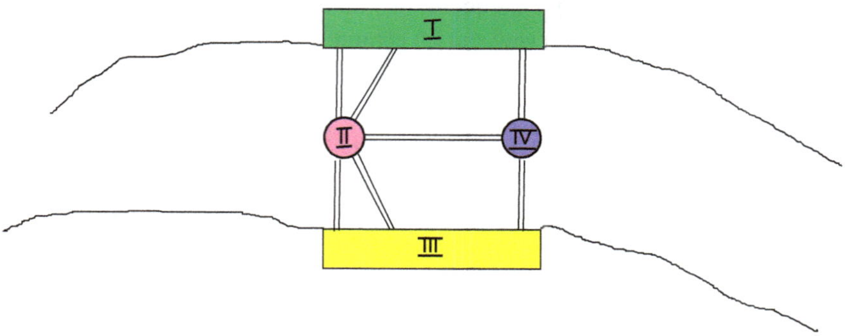

Figure 5.1: The bridges of Königsberg.

Clearly, such a walk has to begin or end in I, since from the three bridges going to I two bridges lead in the same direction and one in the opposite. But for III and IV, it is the same and there are five bridges leading to II. Since such a route has only one starting point and only one ending point, it is impossible to fulfil the requested condition. There can be only two land areas to which an odd number of bridges lead. All others must be connected by an even number of bridges with another area.

The four land areas can be regarded as four vertices of a graph, and the bridges as seven edges connecting the vertices. The graph can be pictured as in Figure 5.2.

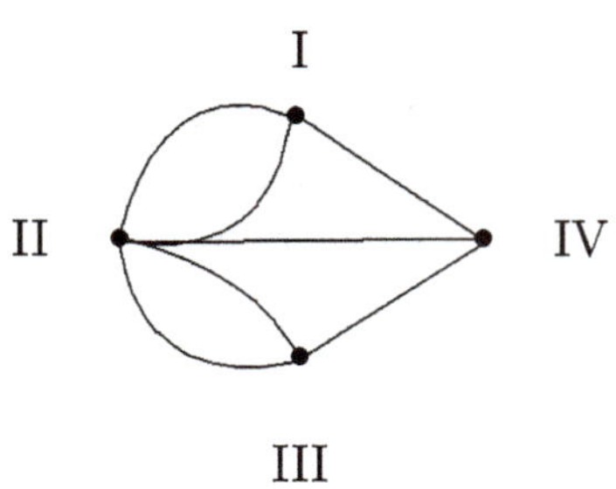

Figure 5.2: Graph of the bridges of Königsberg problem.

https://doi.org/10.1515/9783112234433-005

5.1 Basic concepts

Definition 5.1.1. A graph is a pair $G = (V, E)$, where V is a set and E is a set of two-element subsets of V, i. e.,

$$E \subseteq \{\{v, w\} \mid v, w \in V, v \neq w\}.$$

The elements of V are said to be the vertices, and the elements of E are said to be the edges of the graph.

We introduce some notation: if $e = \{v, w\}$ is an edge of a graph, we write also $e = vw$ and say "e connects v with w." The vertices v and w are said to be *adjacent* with vw. The graph $G = (V, E)$ is said to be finite if V is a finite set.

The *adjacency matrix* of a graph is a quadratic number table whose entries are 0 or 1; $a_j \in \{0, 1\}$ is the element in the ith row and jth column which is defined by

$$a_{ij} = \begin{cases} 1 & \text{for } \{i, j\} \in E, \\ 0 & \text{for } \{i, j\} \notin E. \end{cases}$$

Example 5.1.2. Determine a graph with $V = \{1, 2, 3, 4, 5\}$ whose adjacency matrix is

$$\begin{pmatrix} 0 & 1 & 1 & 0 & 0 \\ 1 & 0 & 1 & 1 & 0 \\ 1 & 1 & 0 & 1 & 1 \\ 0 & 1 & 1 & 0 & 0 \\ 0 & 0 & 1 & 0 & 0 \end{pmatrix}.$$

Clearly, only the adjacency is important for the graph, not the notation of the vertices. This can be expressed by the concept of an *isomorphism*. Instead, we define the concept of a *graph homomorphism* which is a bit more general.

Definition 5.1.3. The graph $G_2 = (V_2, E_2)$ is a homomorphic image of $G_1 = (V_1, E_1)$ if there is a surjective mapping $\varphi : V_1 \to V_2$ which satisfies

$$\forall x, y \in V_1(\{x, y\} \in E_1 \Leftrightarrow \{\varphi(x), \varphi(y)\} \in E_2).$$

The mapping φ is said to be a homomorphism from G_1 onto G_2. If φ is one-to-one, φ is said to be an isomorphism from G_1 onto G_2 and we write $G_1 \cong G_2$. If both graphs are equal, i. e., $G_1 = G_2 = G$, then the isomorphism φ is said to be an automorphism.

In principle, this means that G_1 and G_2 are isomorphic if and only if only the notation of the vertices has changed.

Definition 5.1.4. Let $G = (V, E)$ be a graph. A graph $H = (W, F)$ with $W \subseteq V$ and $F \subseteq E$ is said to be subgraph of G, written as $H \subseteq G$. If additionally

$$F = E \cap \{\{v, w\} \mid v, w \in W\},$$

then H is said to be a subgraph of G induced by W, written as $H \leq_W G$.

If G is a subgraph of H and $V = W$, then G is said to be a spanning subgraph of H (Figure 5.4).

Example 5.1.5. Are the graphs of Figure 5.3 isomorphic? It is easy to check that an isomorphism is given by the mapping φ which maps

$$a \mapsto p, b \mapsto q, c \mapsto r, d \mapsto s, e \mapsto t, f \mapsto u.$$

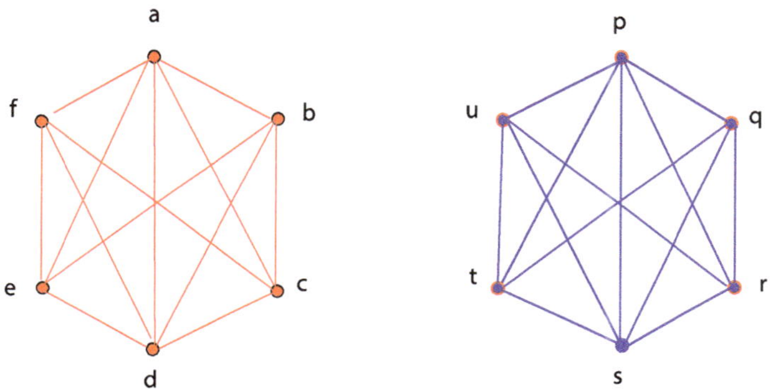

Figure 5.3: Isomorphism of graphs.

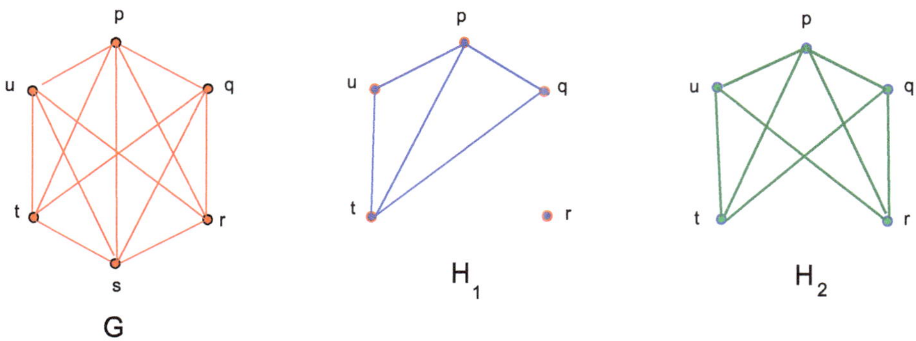

Figure 5.4: Subgraphs.

If we ask questions about graphs, the answer is mostly related to classes of isomorphic graphs. For instance, if we ask for all graphs with three vertices, we mean the number of isomorphism classes of three-element graphs.

Example 5.1.6. Graph H_2 of Figure 5.4 is the subgraph of G induced by $W = \{p, q, r, t, u\}$ since H_2 contains all edges between these vertices. Graph H_1 is a subgraph of G.

When comparing graphs, it makes sense to count the edges going to a vertex.

Definition 5.1.7. For each vertex v of a graph, the natural number

$$\deg v := |\{e \in E \mid v \in e\}|$$

is said to be *the degree* of v. A graph G is said to be regular, more precisely, r-regular, if all vertices have the same degree r.

The following statement follows directly from the definition of an isomorphism: If φ is an isomorphism of graph $G = (V, E)$ to graph H, then

$$\forall v \in V(\deg v = \deg \varphi(v)).$$

The next statement is sometimes called the "handshaking lemma." We think of a group V of people greeting each other with a hand shake. Let E be the set of pairs of the group of people shaking hands with each other.

Lemma 5.1.8. *For any finite graph $G = (V, E)$,*

$$\sum_{v \in V} \deg v = 2|E|.$$

Proof. Since an edge connects two vertices, it will be counted twice when we add up the degrees of vertices of G, $\sum \deg v$. So, the sum of the degrees is twice the number of edges (see Figure 5.5). □

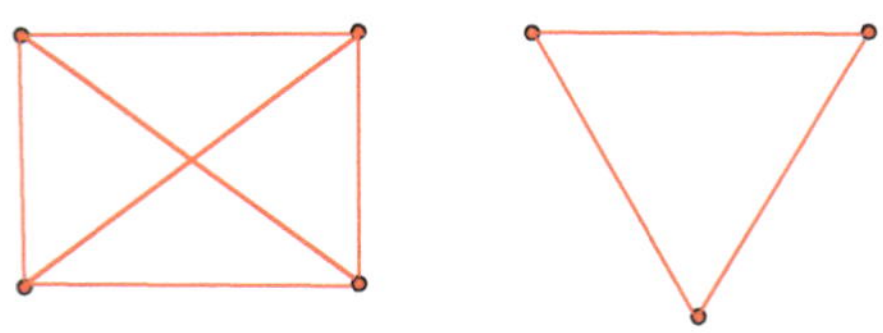

Figure 5.5: $|E| = 3$, $\sum \deg v = 6$, $|E| = 6$, $\sum \deg v = 12$.

The handshaking lemma allows counting the number of edges in a graph. It is half the sum of the degrees of the vertices. We notice that if G is r-regular and has n vertices, then the number of edges in G must be $\frac{nr}{2}$, by the handshaking lemma. As another consequence of this lemma, the number of vertices of odd degree in any graph must be even.

Definition 5.1.9. Let $G = (V, E)$ be a graph. A sequence $(v_0, \ldots, v_n)$ of vertices of G is said to be a walk, if $e_i := \{v_i, v_{i+1}\} \in E$ for $i = 0, \ldots, n-1$, i. e., if two successive vertices are connected by an edge. If $v_0 = v_n$, then the walk is closed, otherwise open. If the v_i are all distinct and the walk is open, then it is said to be a path. If the v_i are all different and the walk is closed, then it is a cycle. The length of a walk is the number of the edges contained in the walk, i. e., the walk $(v_0, \dots, v_n)$ has length n.

Example 5.1.10. For the graph given by Figure 5.6, (a, b, c, d, e, c, b, f) is a walk of length 7, (a, b, d) is not a walk, $(a, b, c, d, e, c, b, f, a)$ is a closed walk, (a, b, c, d, e, c, g) is a path, and (a, b, c, d, e, f, a) is a circle.

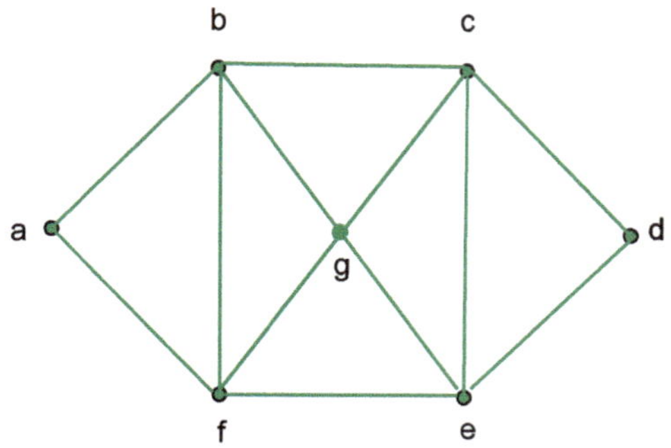

Figure 5.6: Walk, path, and cycle.

Lemma 5.1.11. *Let $G = (V, E)$ be a graph and $u, v \in V$ be two distinct vertices. Let W be a shortest walk from u to v in G. Then W is a path.*

Proof. Suppose $W = v_0v_1 \dots v_n$, $v_0 = u$, $v_n = v$. If W is not a path, then $v_i = v_j$ for some $i < j$ with $\{v_i, v_j\} \neq \{u, v\}$. We now define a new walk

$$W' = v_0v_1 \dots v_iv_{j+1} \dots v_k.$$

Then the length of W' is less than the length of W, a contradiction. So, W is a path. □

Definition 5.1.12. Two vertices v, w of a graph $G = (V, E)$ are said to be connected if there is a path from v to w. The graph G is said to be connected if any two of its vertices are linked by a path in G.

Connectedness is a binary relation on the set V, which is defined by

$$v_1 \sim v_2 :\Leftrightarrow \exists \text{ path } (v_1, \dots, v_2)$$

Clearly, $\sim$ is an equivalence relation and therefore it defines a partition of V. The equivalence classes of V with respect to the relation $\sim$ are said to be *connectedness components* of the graph.

Example 5.1.13. The graph in Figure 5.7 is not connected (it is disconnected), since the vertices 1 and 5 cannot be linked. $\{1, 2, 3, 4\}$, $\{5, 6, 9, 10\}$, $\{7, 8\}$, $\{11\}$ are the connectedness components.

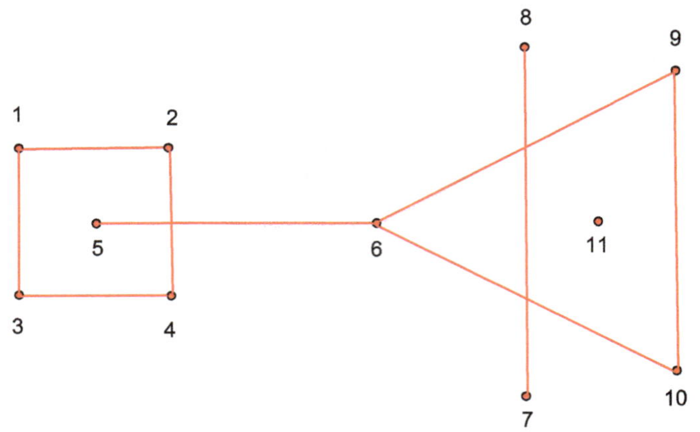

Figure 5.7: Connectedness components.

Lemma 5.1.14. *A graph G is connected if and only if, for each (disjoint) partition $V = V_1 \cup V_2$, $V_1 \cap V_2 = \emptyset$ of V with $V_1, V_2 \neq \emptyset$, there exists an edge $e = \{v, w\}$ with $v \in V_1$, $w \in V_2$.*

Proof. ($\Rightarrow$) Let $V = V_1 \cup V_2$, $V_1 \cap V_2 = \emptyset$ be any disjoint partition of V with $a \in V_1$, $b \in V_2$. Since G is connected, there exists a path $(a = v_0, v_1, \dots, v_n = b)$. Then there is an i with $v_i \in V_1$, $v_{i+1} \in V_2$ and $e := \{v_i, v_{i+1}\}$ is the edge we are looking for.

($\Leftarrow$) Assume G is not connected. Let V_1 be a connectedness component of G and $V_2 = V \setminus V_1$. By assumption, there is an edge $e = \{v, w\}$ with $v \in V_1$, $w \in V_2$. Then v and w belong to the same connectedness component of G. This contradiction shows that G is connected. □

The concept of a graph can be generalized by properties of the edges. The first generalization which we want to consider concerns directed edges.

Definition 5.1.15. A directed graph (or digraph) is a pair $G = (V, E)$, where V is a set and E is a set of ordered pairs (v, w), $v, w \in V$ of elements from V. The pair (v, w) is said to be a directed edge. Also in this case we will write vw; $v = w$ is possible. Then we will speak of a loop. A sequence $(v_0, \dots, v_n)$ of elements from V is said to be a walk of the directed graph G if this is a walk of the corresponding undirected graph. Paths and cycles will be replaced by directed paths and directed cycles. A directed graph is said to be connected if the corresponding undirected graph is connected, and strongly connected if for any two vertices v, w there is a directed path from v to w.

Definition 5.1.16. Let $G = (V, E)$ be a directed graph and let $\nu : E \to \mathbb{Q}^+$ be a mapping which maps each directed edge from E to a nonnegative rational number. Then $G' = (V, E, \nu)$ is said to be an edge-valued directed graph.

Remark 5.1.17. In a similar way, one can also define edge-valued undirected graphs. Moreover, we notice that a directed graph $G = (V, E)$ is a special case of a relational system which is a pair consisting of a set and a set of relations defined on this set. Here G' is an example of a many-sorted algebra, a class of algebraic structures which will be considered in the following chapters.

The following binary operation maps each pair of directed (or undirected) graphs to a new directed (or undirected) graph.

Definition 5.1.18. Let $G_1 = (V_1, E_1)$, $G_2 = (V_2, E_2)$ be any two directed (or undirected) graphs. Then $G_1 \cup G_2 := (V_1 \cup V_2, E_1 \cup E_2)$ is a directed (undirected) graph which is said to be the union of G_1, G_2.

5.2 Eulerian and Hamiltonian graphs

We will come back to the starting example of this chapter, the bridges of Königsberg problem and its negative solution.

Definition 5.2.1. An *Eulerian walk* (resp. a *closed Eulerian walk* of a graph) is a walk (resp., a closed walk) which contains each edge from G exactly once. A graph which contains a closed Eulerian walk will be called Eulerian.

Obviously, the negative solution of the bridges of Königsberg problem had something to do with the degrees of the vertices.

Theorem 5.2.2. *A finite connected graph is Eulerian if and only if all of its vertices have even degree.*

Proof. ($\Rightarrow$) Let $G = (E, V)$ be a finite connected graph with a closed Eulerian walk $(v_0, v_1, \ldots, v_{m-1}, v_0)$. If the vertex v occurs in the sequence $v_0, \ldots, v_{m-1}$ exactly l-times, then $\deg v = 2l$, i. e., v has even degree.

($\Leftarrow$) Suppose that all vertices of G have even degree. Let $t = (v_1, \ldots, v_k)$ be the longest walk in G. If $v_k \neq v_1$ then one can check that an odd number of edges from t contain both v_1 and v_2. Therefore, there is an edge $\{v_k, v_{k+1}\}$ of G which does not occur in t. Now $(v_1, v_2, \ldots, v_k, v_{k+1})$ is a longer walk than t, a contradiction, and $v_k = v_1$, i. e., t is closed. Since G is connected, there is an edge $e = \{v_i, v\} \in E$ which does not occur in t. Now there are two cases:

1. v is not a vertex of t. Then

$$(v_i, v_{i+1}, \ldots, v_1, v_2, \ldots, v_{i-1}, v_i)$$

is a closed walk which has the same length as t and, if we add $\{v_i, v\}$, we get the walk

$$(v_i, v_{i+1}, \ldots, v_k, v_1, v_2, \ldots, v_{i-1}, v_i, v)$$

which is longer than t, a contradiction.

2. v is a vertex of t. Let $v = v_j, j < i$. Then

$$(v_i, v_{i+1}, \dots, v_k, v_1, \dots, v_{j-1}, v_j, v_{i-1}, \dots, v_{j+1}, v_j)$$

is a walk that contains e and is longer than t, a contradiction. □

In a similar way, we can request that there is a walk or a closed walk that contains each vertex exactly once.

Definition 5.2.3. A *Hamiltonian walk* (resp. a *closed Hamiltonian walk* of a graph G) is a walk (resp. closed walk) which contains each vertex of the graph exactly once. A graph with a closed Hamiltonian walk is said to be Hamiltonian.

Example 5.2.4. The graph given by Figure 5.8 is Hamiltonian. A closed Hamiltonian walk is (a, b, c, d, e, f).

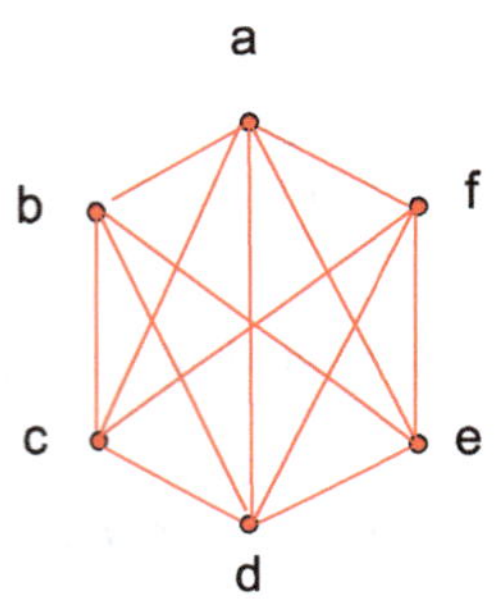

Figure 5.8: A Hamiltonian graph.

For Hamiltonian graphs, there are several sufficient conditions, one of them is

Theorem 5.2.5. *Let G be a graph with $n \geq 3$ vertices. If all vertices of G have a degree which is greater than or equal to $\frac{n}{2}$, then G is Hamiltonian.*

Proof. From G, we construct a graph G' by adding k vertices, each of which is connected with all the vertices of G by a new edge. Let k be the least number of vertices which is needed to make G' Hamiltonian. We will find a contradiction to the assumption that $k > 0$, i. e., G is not Hamiltonian. Let $K = (a, x, b, \dots, a)$ be a closed Hamiltonian walk of G' with vertices a, b of G and a new vertex x. Clearly, a and b are not neighbors, since otherwise x is not needed for the closed Hamiltonian walk and thus (because of $n \geq 3$) for G'. In K, a vertex b' which is a neighbor of b cannot directly follow a vertex a' which is a neighbor of a, since otherwise one could replace $K = (a, x, b, \dots, a', b', \dots, a)$ by $K' = (a, a', \dots, b, b', \dots, a)$ and x would be not needed (see Figure 5.9).

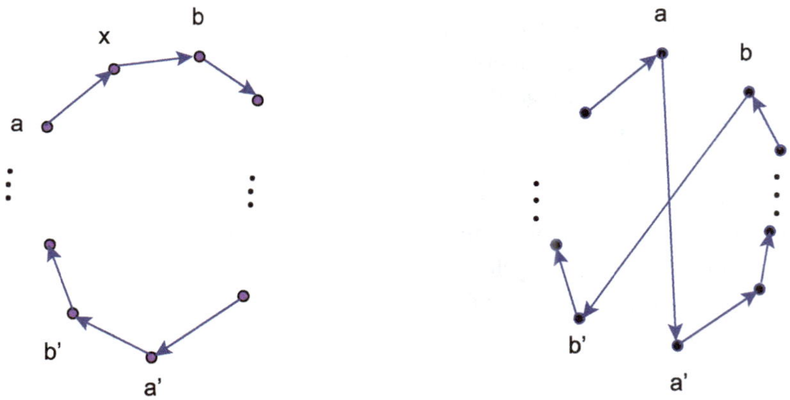

Figure 5.9: A sufficient condition.

We set
- $M :=$ set of all neighbors of a in G',
- $N :=$ set of all neighbors of b in G',
- $O :=$ set of all vertices of G', which follow in K directly a vertex from M.

Then,
- $N \cap O = \emptyset$ (since a and b are not neighbors).
- $|N| \geq \frac{n}{2}+k$, $|O| = |M| \geq \frac{n}{2}+k$. This implies $|N \cup O| = |N|+|O| \geq n+2k$, a contradiction, since G' has only $n + k$ vertices. □

5.3 Trees

Definition 5.3.1. A tree is a connected graph without cycles (with more than one vertex).

Definition 5.3.2. A forest is a graph, which contains no cycles (with more than one vertex).

Therefore, a graph is a forest if and only if all its connectedness components are trees. Trees can be characterized as follows:

Theorem 5.3.3. *The following conditions are equivalent:*
(i) *G is a tree with n vertices.*
(ii) *Any two vertices of the graph are connected by exactly one path.*
(iii) *G is connected, but for any edge e of G, $G \setminus e$ is not connected.*
(iv) *G is connected and has exactly $n-1$ edges.*
(v) *G has no cycle and exactly $n-1$ edges.*
(vi) *G has no cycle, but for any two not adjacent vertices u, v of G, $G \cup \{u, v\}$ has exactly one cycle.*

(Note that the notations $G \setminus e$ and $G \cup \{u, v\}$ are short, but incorrect. Improve!)

Proof. (i) $\Rightarrow$ (ii) Since a tree is connected, any two vertices are connected by a path. If there were more than one path, then there would be cycles.

(ii) $\Rightarrow$ (iii) Let $e = \{v, w\}$ be an edge of G. Since $\{v, w\}$ is the only path of G, connecting v and w, $G \setminus \{v, w\}$ is not connected.

(iii) $\Rightarrow$ (iv) Deleting one edge from G, the number of connectedness components increases by 1. Since G consists of one connectedness component and since after deleting all edges the number of connectedness components is n, graph G has exactly $n-1$ edges.

(iv) $\Rightarrow$ (v) Assume G contains a cycle K which has exactly k vertices and then also k edges. Since G is connected, the edges of G can be chosen in such a way that they connect all $n-k$ vertices, which are left, with K. To do this, one needs $n-k$ edges, since any additional edge connects at most one more vertex with K. Altogether G has then $k + (n-k) = n$ edges, a contradiction.

(v) $\Rightarrow$ (vi) Graph G is connected, since adding successively the edges to G, the number of connectedness components increases by 1 for each edge. Let $\{v, w\}$ be not an edge of G. Each cycle of $G \cup \{v, w\}$ which contains $\{v, w\}$ consists of $\{v, w\}$ and a path from w to v. Since G has no cycles, there is only one such path, and thus only one cycle of $G \cup \{v.w\}$ containing $\{v, w\}$.

(vi) $\Rightarrow$ (i) We show that G is connected. Assume v and w belong to different connectedness components of G, then $G \cup \{v, w\}$ cannot contain a cycle. □

Since the connectedness components of a forest are trees, we get

Corollary 5.3.4. *A forest with n vertices and k connectedness components has $n-k$ edges.*

Proposition 5.3.5. *Any finite tree has at least one vertex of degree ≤ 1 (such vertices are called leaves).*

Proof. Suppose a tree G has n vertices. Then, by Lemma 5.1.8, the sum of the degrees of all the vertices is twice the number of edges which is $n-1$. Therefore, $\sum \deg v = 2n-2$. Assume that each of the n vertices has a degree ≥ 2. Then $\sum \deg v \geq 2n$, a contradiction. □

Note that for $n \geq 2$ there are no vertices with degree 0. Then all leaves have degree 1. Therefore, there are at least two leaves. Sometimes there is one vertex of a tree which is considered to be special. Such a vertex is then called the root of the tree. A tree G with a fixed root is said to be a *rooted* tree. Let for the tree G and any vertex y of G, the uniquely determined (since a tree has no cycle) path from r to y be denoted by rGy. We define a partial order relation on the set E of all vertices of the tree G as follows:

Definition 5.3.6. Let $G = (V, E)$ be a rooted tree with root r. For any $x, y \in V$, let

$$x \leq y :\Leftrightarrow x \in rGy.$$

It is easy to check and left as an exercise to the reader that $\leq$ is a partial order on the set V of vertices of the tree G. Clearly, $\leq$ is not a total (linear) order since there are incomparable elements in V, for instance, any two different leaves. The partial order $\leq$ is also said to be the *tree order of G associated with the root r*. Note that the root r is the least element and the leaves are the maximal elements. Also $(V; \leq)$ is a partially ordered set. The ends of any edge of G are comparable. The length of the path from r to $y \in V$ is also called the height of y.

Trees can be used to visualize terms. Then the vertices, except for leaves, are labeled by operation symbols and the leaves are labeled by variables. The root is labeled by the outermost operation symbol of the term. Figure 5.10 shows the tree which visualizes the term

$$f_2(f_1(f_2(f_2(x_1, x_2), f_1(x_3))), f_1(f_2(f_1(x_1), f_1(f_1(x_2))))).$$

The operation symbol f_2 is binary, i. e., it has two inputs and labels the root and all vertices of degree 3, while f_1 is unary as it has one input and labels vertices of degree 2. The variables x_1, x_2 label the five leaves of the tree. The first input of f_2 corresponds to the edge on the left- and the second to the edge on the right-hand side.

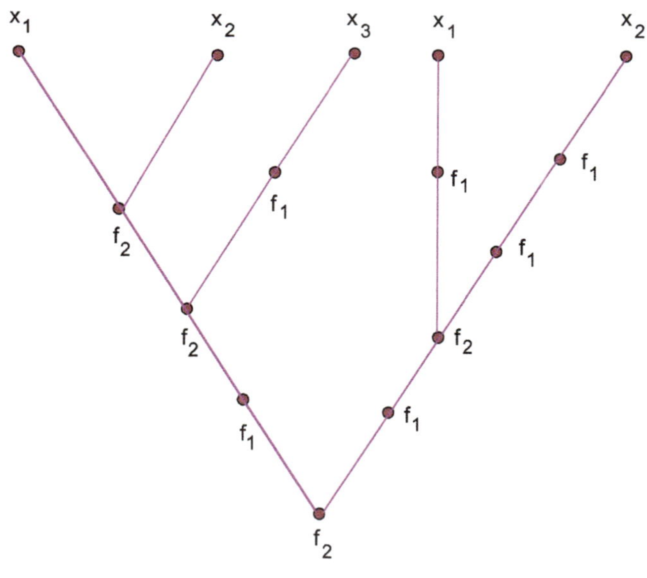

Figure 5.10: Term and tree.

5.4 Bipartite graphs

A tree which consists only of a root and of leaves is a special case of a bipartite graph. Graph $G = (V, E)$ is bipartite if the set V of vertices admits a partition into 2 classes such that every edge has its end in the other class and the vertices in the same class of the partition are not adjacent. The partition classes can be one-element. There are no loops. This can be generalized to the concept of an r-partite graph.

Definition 5.4.1. Let $r \geq 2$ be a natural number. A graph $G = (V, E)$ is called r-partite if V admits a partition into r classes such that every edge v has its end in a class different from the class of v, and if v and v' belong to the same class of the partition then they are not adjacent.

Example 5.4.2. The classical transportation problem asks for minimal costs when commodities shall be transferred from several producers (sources) to several consumers (destinations) and some supply constraints of the sources and demand constraints of the consumers are satisfied. Let $S = \{s_1, s_2, \ldots, s_m\}$ be the set of sources and let $D = \{d_1, d_2, \ldots, d_n\}$ be the set of destinations. Then we consider a graph $G = (V, E)$ with $V = S \cup D$ and

$$E \subseteq \big\{\{s_i, d_j\} \mid i \in \{1, \ldots, m\}, j \in \{1, \ldots, n\}\big\}.$$

Here $\{s_i, d_j\} \in E$ means that commodities from producer s_i will be delivered to consumer d_j. Clearly, G is a bipartite graph. In the special case, when $E = \{\{s_i, d_j\} \mid i \in \{1, \ldots, m\}, j \in \{1, \ldots, n\}\}$, each producer sends commodities to each consumer. The corresponding bipartite graph is called *complete* (Figure 5.11). For more details on the transportation problem see e.g. [43].

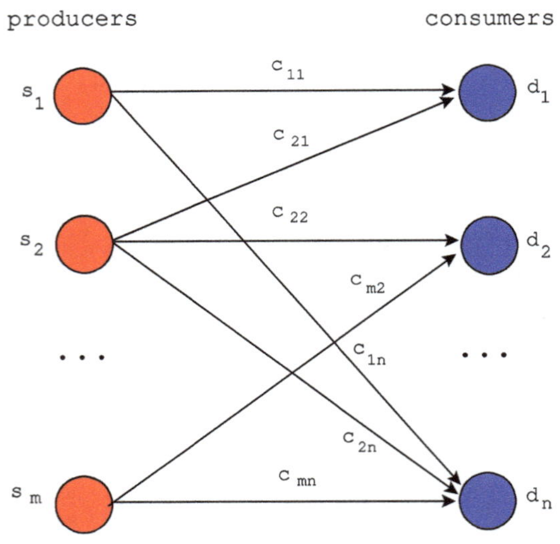

Figure 5.11: Transportation problem.

Definition 5.4.3. Let $G = (V, E)$ be an r-partite graph such that any two vertices from different classes of the partition are adjacent. Then G is said to be a complete r-partite graph. Let K_t^r be the complete r-partite graph where each class of the partition of V contains t vertices.

Since commodities are sent from the producer to the consumer, one should better consider the directed bipartite graph $G = (V, E')$ with $E' \subseteq S \times D$. The graph-theoretical model can be improved applying labeling of vertices and edges. The vertices from S are labeled by the production capacities and the vertices from D are labeled by the consumer's demands. The edges are labeled by the costs.

The following theorem characterizes bipartite graphs (for the proof, see [27]).

Theorem 5.4.4 ([27]). *A graph is bipartite if and only if it contains no odd cycle.*

5.5 Planar graphs

To obtain nice diagrams of graphs, one tries to draw the diagram in the plane without any two of its edges crossing. We say that such graphs can be embedded into the plane. To define such an embedding, one needs the concept of a Jordan curve $f(t)$. An *embedding* of a graph $G = (V, E)$ is a function $f : V \cup E \to \mathbb{R}^2 \cup C$ where C is the set of continuous curves in $\mathbb{R}^2$ such that f is one-to-one, $f(v)$ is a point in $\mathbb{R}^2$ for each $v \in V$, and $f(\{u, v\})$ is a continuous curve in the two-dimensional Euclidean space $\mathbb{R}^2$ with end points u and v if $\{u, v\} \in E$.

Definition 5.5.1. A graph is said to be planar if it can be embedded into the plane. The corresponding diagrams are called plane (or planar) diagrams.

The planar diagram of a graph G partitions $\mathbb{R}^2 \setminus G$ into a union of disjoint connected plane regions, the *faces* of G. Each graph with a plane diagram has a face which is infinite. Let f be the number of faces of a planar graph. Between the numbers of vertices, edges, and faces of a planar graph, there is the following relationship, known as *Euler's formula*:

Theorem 5.5.2 (Euler's formula). *Let f be the number of faces of the plane diagram of a planar connected graph with n vertices and m edges. Then*

$$n + f = m + 2.$$

Proof. The proof is by induction on the number of edges. If $m = 0$, then $n = 1$, since G is connected. Then $f = 1$ (since a single point cannot partition the plane). Thus,

$$n + f = 1 + 1 = 0 + 2 = m + 2.$$

(For $m = 1$, one has $n = 2, f = 1$, and $n + f = 2 + 1 = 1 + 2 = m + 2$.)

Suppose the formula is satisfied for a connected planar graph with $m - 1$ edges. We have to show that this implies the corresponding formula for connected planar graphs with m edges. Now consider the following cases:

- Case 1: G has no cycles (i. e., G is a tree since G is connected). Then $m = n - 1$ and $f = 1$, since G has no cycles, leading to

$$n + f = n + 1 = n - 1 + 2 = m + 2.$$

– Case 2: G is not a tree. Then G has at least one cycle. Delete an edge e of G, which is contained in that cycle, i. e., form $G' := G \setminus e$. This is a connected graph, otherwise, by Theorem 5.3.3, G must be a tree. For G', the inductive hypothesis is satisfied, i. e., if the numbers of faces and edges of G' are denoted by f' and m', respectively, then $n + f' = m' + 2$. Each plane diagram of G arises by adjoining e to a plane diagram of G'. This divides the face into two parts since a cycle arises. The number of faces increases by 1 and the number of edges increases also by 1. Adding 1 to the equation $n + f' = m' + 2$ gives $n + f' + 1 = m' + 1 + 2$, and therefore $n + f = m + 2$. (An example can be seen in Figure 5.12). □

Remark 5.5.3. The following statements which we give here without proof are useful for showing that a graph is planar:
1. For each planar graph with n vertices and $m \geq 2$ edges, there holds $m \leq 3n - 6$.
2. For each planar graph with n vertices and $m \geq 2$ edges without cycles of length 3, there holds $m \leq 2n - 4$.

Example 5.5.4. We consider the following example to Theorem 5.5.2:

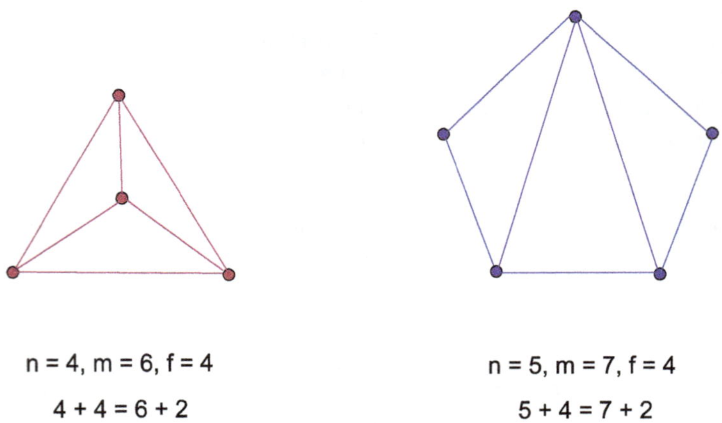

Figure 5.12: Euler's formula.

Example 5.5.5. We consider the following examples to Remarks 5.5.3 (Figure 5.13):
For graph K_5, we have $n = 5$, $m = 10$, $10 > 3 \cdot 5 - 6 = 9$. Therefore, K_5 is not planar.
For graph $K_{3,3}$, we have $n = 6$, $m = 9$, $9 > 2 \cdot 6 - 4 = 8$. Therefore, $K_{3,3}$ is not planar.

Definition 5.5.6. Let $G = (V, E)$, $G' = (V', E')$ be graphs with $\{a, b\} \in E$, $x \notin V$, $V' = V \cup \{x\}$, $E' = (E \setminus \{a, b\}) \cup \{\{a, x\}, \{x, b\}\}$ (i. e., the edge $\{a, b\}$ will be deleted and the edges

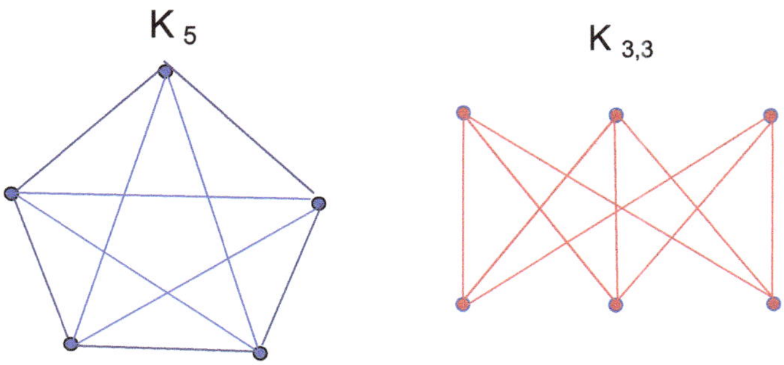

Figure 5.13: Nonplanar graphs.

$\{a,x\}, \{x,b\}$ will be adjoined). Then we say that G' arises from G by adjoining the vertex x to the edge $\{a,b\}$. A graph H is said to be a subdivision of a graph G if it arises from G by successive adjoining of finitely many new vertices.

Then one can prove the following characterization of planar graphs.

Theorem 5.5.7. *A finite graph is planar if and only if it contains no subdivision of K_5 or $K_{3,3}$ as subgraphs.*

Planar graphs, which can be represented by plane diagrams with symmetries as many as possible, are said to be *Platonian graphs*.

Definition 5.5.8. A finite planar graph is said to be Platonian if there are natural numbers $r, s \geq 3$ such that:
(1) G is r-regular,
(2) G has a plane diagram in which all faces are bordered by cycles of length s.

Theorem 5.5.9. *Up to isomorphism, there are exactly* 5 *Platonian graphs, represented by the following plane diagrams* (i)–(v) *(Figure* 5.14*).*

Assume s and r are given by the following table:

	r	s
(i)	3	3
(ii)	4	3
(iii)	3	4
(iv)	5	3
(v)	3	5

We mention that the Platonian graphs correspond to the 5 *Platonian solids*, i. e., convex regular polyhedra in the three-dimensional Euclidean space: *tetrahedron, hexahedron (cube), octahedron, icosahedron*, and *dodecahedron*. All faces of the polyhedra are regular and congruent.

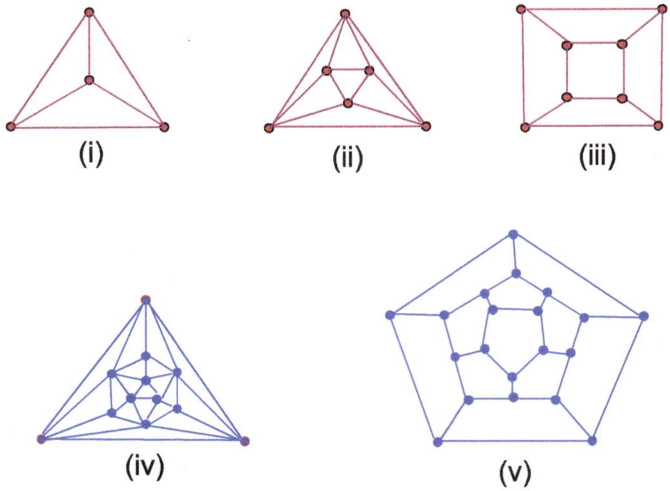

Figure 5.14: The Platonian graphs.

5.6 Coloring of graphs

When maps are colored, there arises the problem of how many colors are needed to color adjacent countries differently. At the first glance, one could think that the solution depends on the number of countries or on the number of borders between the countries. But this is wrong. We will prove that for any map five colors are enough. For many years, it was an open problem whether even only four colors are enough. This "four color problem" was solved in 1976 when *Appel, Haken, and Koch* [1] proved that four colors are enough. Since their proof is based on complex calculations and not on logical arguments, it is not fully accepted. Each map can be considered a diagram of a finite planar graph. The borders correspond to the edges of a graph and the intersections of borders are the vertices. One can assume that there are no multiple edges between two vertices. The faces of the diagram of this graph are colored differently. The colors are understood as elements of some sets. Instead of faces, one can also color vertices or edges.

Definition 5.6.1. A finite graph is said to be *k(-vertices) colorable* if to each vertex of the graph one of k colors can be assigned in such a way that different vertices are colored differently. The least natural number n for which the graph G is n-colorable is called *the chromatic number* $\chi(G)$ of graph G.

The problem to color a map can be translated into a problem of graph coloring.

We will now prove that in any case 5 colors are enough. This was first proved by *Heawood* in 1890.

Theorem 5.6.2. *Each finite planar graph is* 5*-colorable.*

Proof. Let n be the number of vertices of a planar graph. For $n \leq 5$, the statement is true. We will prove the theorem by induction on n. Assume that the statement is satisfied for graphs with n vertices. Let G be a planar graph with $n+1$ vertices. At first, we show that G has a vertex v of degree ≤ 5. Assume, $\deg w \geq 6$ for all vertices w. Then by the handshaking lemma, $2m = \sum_{w \in V} \deg w \geq 6n$, where m is the number of vertices. This contradicts Remark 5.5.3(1) since G is planar. Consider the subgraph G' of G arising by the deletion of vertex v. By the inductive hypothesis, G' is 5-colorable with the colors $1, 2, 3, 4, 5$. If one of the five colors is not used to color one of the vertices in the neighborhood of v, this color will be used for v and the proof is finished. Let $\deg v = 5$ and assume that all colors are used for the vertices in the neighborhood of v.

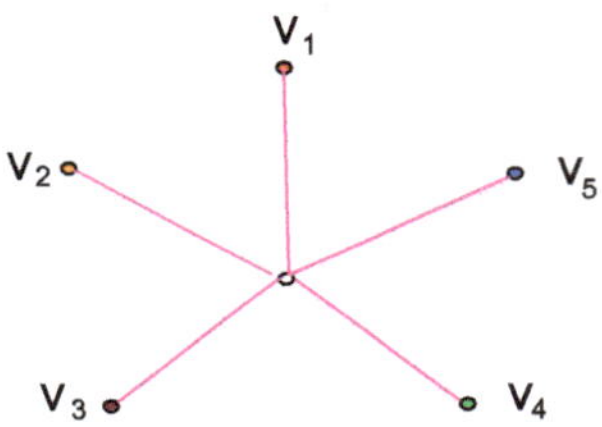

In the diagram of G, let v_i be the vertex in the neighborhood of v which is colored by i. Let $G_{1,3}$ be the subgraph of G spanned by the vertices of G' which are colored with 1 or 3. If there is no path between v_1 and v_3 in G' containing only vertices colored by 1 or 3, then in the connectedness component $G_{1,3}$, to which v_1 belongs, the colors 1 and 3 can be exchanged everywhere. In this way, we obtain a 5-coloring of G', in which no neighbor of v is colored with 1. Now we color v with 1 and the proof is finished.

If v_1 and v_3 belong in G' to the same connectedness component of $G_{1,3}$, then there is a path connecting v_1 and v_3 whose vertices are colored only with 1 or 3. This path forms, together with (v_3, v, v_1), a cycle which either contains v_2 or the vertices v_4 and v_5.

Since G is planar, in both cases v_2 and v_4 belong to different connectedness components of the subgraph $G_{2,4}$ of G' spanned by the vertices of G which are colored by 2 or 4. Now in the connectedness component of $G_{2,4}$ containing v_2 we exchange the colors 2 and 4. In this way, we obtain a 5-coloring of G', in which the neighbors of v are not colored with 2. Now we color v with 2 and finish the proof. □

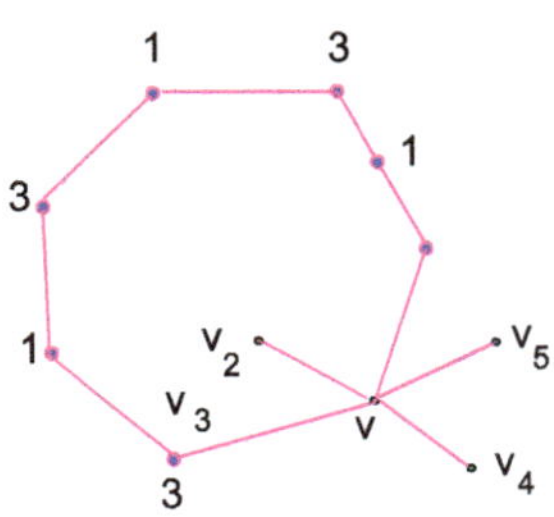

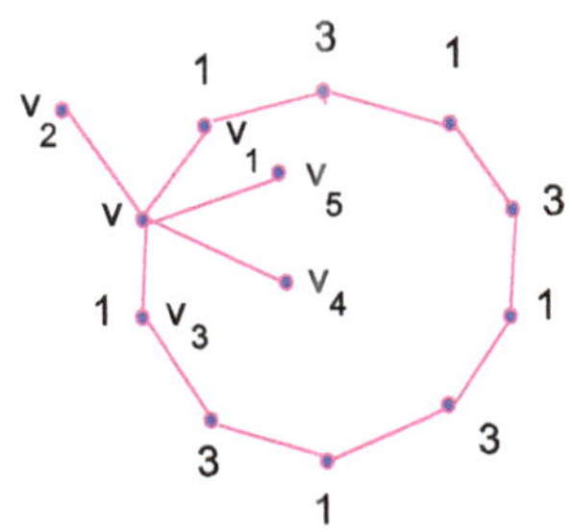

5.7 Historical remarks

Leonhard Euler's paper on "the bridges of Königsberg problem," entitled "Solutio problematis ad geometriam situs pertinentis," published in 1741 at the Petersburg Academy of Sciences, can be regarded as the hour of birth for graph theory. In 1857, A. Cayley wrote a paper on trees [14]. The notion of a "graph" goes probably back to J. J. Sylvester (1814–1897). Dénes Kőnig (1884–1944) published in 1936 the first textbook on graph theory [39]. The "four color problem" is one of the big problems in graph theory. Its solution had a great influence on the development of the subject. The problem of how many colors are needed to color adjacent countries of a map differently was asked first by F. Guthrie (1831–1899), a professor of mathematics at the South African College in Cape Town. His brother Frederick (1833–1886) was at this time a student of mathematics in Cambridge and asked his academic teacher A. de Morgan (1806–1871), the first president of the London Mathematical Society, for the solution. Arthur Cayley explained in 1878 at a session of the London Mathematical Society the difficulties of the problem. Nevertheless, in 1878 the lawyer A. B. Kempe, a member of the London Mathematical Society, believed to have found a solution. His proof was wrong, as P. J. Heawood (1861–1955) in 1890 showed, but he could use Kempe's idea to solve the five color problem. When on July 22, 1976, a computer in Illinois after 1200 hours of work and after 10 billions of decisions finally stopped, K. Appel and W. Haken could inform the world that the "four colors are enough." In [2], the reader can check the proof, written on 741 pages. Later on, Robertson, Sanders, Seymour, and Thomas found a shorter and simpler proof [57] – the reader should better read this one.

6 Matrices

A matrix is a rectangular table of elements from a field $\mathcal{K}$. In economical applications, the field $\mathcal{K}$ is the field of real numbers or the field of rational numbers. But we have also to consider matrices with entries from $\mathbb{Q}^+$, $\mathbb{Z}$, or $\mathbb{N}$. The definition of a matrix becomes useful if matrices are considered as elements of algebraic structures. This makes it possible to calculate with matrices. We introduce the concept of a vector space and study its properties. We define and calculate determinants and formulate and prove the main theorems on equation systems (for additional information see e. g. [64]).

https://doi.org/10.1515/9783112234433-006

6.1 Calculation with matrices

Matrices can be written as

$$A = \begin{pmatrix} a_{11} & \dots & a_{1n} \\ \vdots & \ddots & \vdots \\ a_{m1} & \dots & a_{mn} \end{pmatrix} = (a_{ik}), \quad i = 1, \dots, m, k = 1, \dots, n, a_{ik} \in K.$$

The first index i of a_{ik} is called the *row index* and k is called the *column index*. The ordered pair (m, n) is called the *type* of matrix A. A matrix is said to be quadratic if $m = n$. Let $K_{m,n}$ be the set of all matrices over $\mathcal{K}$, i. e., with entries from $\mathcal{K}$ which is an algebraic structure with two binary operations (in most cases, $\mathcal{K}$ is a field, semifield, ring, or semiring).

Definition 6.1.1. Two matrices are equal if both have the same type and if for all $i = 1, 2, \dots, m$, $k = 1, 2, \dots, n$, the equations $a_{ik} = b_{ik}$ are satisfied.

On the set $K_{m,n}$ of all matrices $A = (a_{ik})$, $B = (b_{ik})$ of type (m, n), we introduce an addition by $C = A + B$, where $C = (c_{ik})$ is a matrix of type (m, n) which satisfies for all $i = 1, \dots, m, k = 1, \dots, n, c_{ik} = a_{ik} + b_{ik}$. In the latter equation, by $+$ we mean the addition in the field $\mathcal{K}$. By this definition, the properties of the addition in $\mathcal{K}$ are transferred to the properties of the addition in $K_{m,n}$. We assume that $\mathcal{K}$ is a field. Other cases will be considered later.

Theorem 6.1.2. *Let $\mathcal{K}$ be a field and let $K_{m,n}$ be the set of all matrices of type (m, n). Then $\mathcal{K}_{m,n} = (K_{m,n}; +)$ is a commutative group.*

Clearly, the sum of two matrices of type (m, n) is a matrix of type (m, n). We have to show that the matrix addition is associative and commutative. But this follows directly from the associative and commutative laws for the addition in the field $\mathcal{K}$. Since the addition in the field $\mathcal{K}$ is invertible, or equivalently, since there exist a zero element $0 \in K$ and to each $a \in K$ also the opposite $-a \in K$, the addition in $\mathcal{K}_{m,n}$ is also invertible. The details of the proof are left as an exercise to the reader.

The second operation defined on $K_{m,n}$ is a multiplication of elements $\lambda \in K$ with matrices from $K_{m,n}$:

$$\lambda \circ (a_{ik}) := \lambda(a_{ik}) := (\lambda a_{ik}).$$

The multiplication $\circ$ is called *scalar multiplication.*

Theorem 6.1.3. *Let $\mathcal{K}$ be a field and let $K_{m,n}$ be the set of all matrices of type (m, n) over $\mathcal{K}$. Then for all $\lambda, \mu \in K, A, B \in K_{m,n}$ there holds:*

(i) $\lambda(\mu A) = (\lambda\mu)A$.

(ii) $(\lambda + \mu)A = \lambda A + \mu A$.

(iii) $\lambda(A + B) = \lambda A + \mu A$.
(iv) $eA = A$, *here e is the identity element of the field $\mathcal{K}$.*

The proof uses the properties of the field $\mathcal{K}$ and is left as an exercise to the reader.

The advantages of the matrix calculus become clear after introducing the second binary operation on $K_{m,n}$, the matrix multiplication. The multiplication of two matrices A, B is defined if the number of columns of A is equal to the number of rows of B. This means that if the type of A is (m, n) and the type of B is (p, q), then $n = p$ has to be satisfied. We consider the following example:

Example 6.1.4. Let A and B be the matrices

$$\begin{pmatrix} 2 & 5 & -1 \\ 3 & 4 & 1 \\ 7 & 1 & -1 \\ 1 & 0 & 2 \end{pmatrix} \quad \text{and} \quad \begin{pmatrix} 2 & 1 \\ 7 & -4 \\ 2 & 3 \end{pmatrix}.$$

Since the type of A is $(4, 3)$ and the type of B is $(3, 2)$, the product $A \cdot B$ is defined and we ask for the result.

Definition 6.1.5. Assume that a matrix $A = (a_{ik})$ has type (m, n) and matrix $B = (b_{ik})$ has type (n, p). Then $C = A \cdot B$ is the matrix of type (m, p) whose elements c_{ik} are defined by

$$c_{ik} = \sum_{j=1}^{n} a_{ij} b_{jk} = a_{i1} b_{1k} + a_{i2} b_{2k} + \cdots + a_{in} b_{nk}.$$

Example 6.1.6. The matrices

$$\begin{pmatrix} 2 & 5 & -1 \\ 3 & 4 & 1 \\ 7 & 1 & -1 \\ 1 & 0 & 2 \end{pmatrix} \quad \text{and} \quad B = \begin{pmatrix} 2 & 1 \\ 7 & -4 \\ 2 & 3 \end{pmatrix}$$

can be multiplied as we have already seen and the product is a matrix of type $(4, 2)$. The element in the first row, first column is $2 \cdot 2 + 5 \cdot 7 + (-1) \cdot 2$. The element in position $(1, 2)$ is $2 \cdot 1 + 5 \cdot (-4) + (-1) \cdot 3$. Finally, we get the product matrix as

$$\begin{pmatrix} 37 & -21 \\ 36 & -10 \\ 19 & 0 \\ 6 & 7 \end{pmatrix}.$$

The matrix product can also be written as

$$(a_{ik}) \cdot (b_{ik}) = \left(\sum_{j=1}^{n} a_{ij} b_{jk} \right).$$

We ask which laws this multiplication satisfies. If the type of A is (m, n) and the type of B is (n, q), then the product $A \cdot B$ exists, but if $q \neq m$, then $B \cdot A$ is not defined. This shows that the commutative law is not satisfied in general. One could suggest that at least the multiplication of quadratic matrices is commutative. The following counterexample shows that in general this is also false.

Example 6.1.7.

$$\begin{pmatrix} 3 & 4 \\ 0 & 2 \end{pmatrix} \cdot \begin{pmatrix} -1 & 2 \\ 1 & 1 \end{pmatrix} = \begin{pmatrix} 1 & 10 \\ 2 & 2 \end{pmatrix}, \quad \text{but} \quad \begin{pmatrix} -1 & 2 \\ 1 & 1 \end{pmatrix} \cdot \begin{pmatrix} 3 & 4 \\ 0 & 2 \end{pmatrix} = \begin{pmatrix} -3 & 0 \\ 3 & 6 \end{pmatrix}.$$

It is not difficult to prove and left as an exercise to the reader that the multiplication of matrices has the following properties:

Theorem 6.1.8.
(i) $(A + B) \cdot C = A \cdot C + B \cdot C$, $A \cdot (B + C) = A \cdot B + A \cdot C$ *(distributive laws).*
(ii) $A \cdot (B \cdot C) = (A \cdot B) \cdot C$ *(associative law).*
(iii) $\lambda(A \cdot B) = (\lambda A) \cdot B = A \cdot (\lambda B)$.

Another important operation on $K_{m,n}$ is the unary operation of forming the *transpose*. For a given matrix, another uniquely determined matrix is defined by switching the row and column indices of the given matrix.

Definition 6.1.9. The transpose of matrix A is the matrix A^T which arises from A by exchanging the rows and columns of A,

$$(a_{ik})^T := (a_{ki}).$$

If matrix A has type (m, n), then its transpose A^T has type (n, m). Matrix A is called *symmetric* if $A = A^T$. Clearly, each symmetric matrix is quadratic, i. e., $m = n$. For the transpose the following rules are satisfied and easy to prove:

Theorem 6.1.10.
(i) *If A and B are matrices of the same type, then $(A + B)^T = A^T + B^T$.*
(ii) *If $A \in K_{m,n}$ and $\lambda \in K$, then $(\lambda A)^T = \lambda A^T$.*
(iii) *If $A \in K_{m,n}$ and $B \in K_{n,q}$, then $(A \cdot B)^T = B^T \cdot A^T$.*

On the set $K_{n,n}$ of all quadratic matrices of type (n, n), the multiplication is always defined. Since the multiplication is associative, as $K_{n,n}$ forms a commutative group with respect to addition and the distributive laws are satisfied, we have:

Corollary 6.1.11. *$\mathcal{K}_{n,n} = (K_{n,n}; +, \cdot)$ is a ring.*

6.2 Vector spaces

The algebraic concept of a ring summarizes not all possibilities to calculate with matrices. The multiplication of $A \in K_{m,n}$ with elements $\lambda \in K$ and its properties cannot be described by the ring properties. We need another algebraic structure.

Definition 6.2.1. Let $\mathcal{K}$ be a field with elements $\lambda, \mu, \alpha, \beta, \ldots$ The triple $\mathcal{V} := (V; +, \circ)$ is said to be a vector space over $\mathcal{K}$ (also a $\mathcal{K}$-vector space) if V is a nonempty set with elements $a, b, c, \ldots,$ called vectors, together with an addition on V and a scalar multiplication (operator application) of the elements from K with the elements from V, if the following properties are satisfied:
(i) $(V; +)$ is an abelian (commutative) group;
(ii) $\lambda(\mu a) = (\lambda\mu)a$;
(iii) $(\lambda + \mu)a = \lambda a + \mu a$;
(iv) $\lambda(a + b) = \lambda a + \lambda b$;
(v) $ea = a$ where e is the identity element of the field $\mathcal{K}$. (Instead of $\alpha \circ a$, we simply write αa.) If $\mathcal{K}$ is the field of real numbers, we speak of a real vector space.

Corollary 6.2.2. *If 0 is the zero element of the $\mathcal{K}$-vector space V and if $-a$ is the opposite of a, then for all $\alpha \in K$ and for all $a \in V$:*
(i) $\alpha 0 = 0$;
(ii) $0a = 0$;
(iii) $(-e)a = -a$;
(iv) $\alpha a = 0 \Rightarrow \alpha = 0 \vee a = 0$.
(Note that here the zero element of the field and the zero vector have the same notation.)

Proof. (i) Clearly, $\alpha 0 = \alpha(0 + 0) = \alpha 0 + \alpha 0$. Moreover, $\alpha 0 + 0 = \alpha 0$. Thus $\alpha 0 + 0 = \alpha 0 + \alpha 0$. An application of the group properties of $(V; +)$ gives $\alpha 0 = 0$.

(ii) From $0a = (0 + 0)a = 0a + 0a$ and $0a + 0 = 0a$ there follows as in the proof of (i), $0a = 0$.

(iii) Now we have $a + (-a) = 0 = 0a = (e + (-e))a$. Applying the distributive law and the properties of addition, we get (iii).

(iv) Let $\alpha \neq 0$. Then in the field $\mathcal{K}$ the inverse α^{-1} exists and we have $a = ea = (\alpha^{-1}\alpha)a = (\alpha)^{-1}(\alpha a) = \alpha^{-1}0 = 0$. □

As a consequence of Definition 6.2.1, we have

Corollary 6.2.3. *For any field $\mathcal{K}$, the set $K_{m,n}$ of all matrices of type (m, n) forms a $\mathcal{K}$-vector space $\mathcal{K}_{m,n} = (K_{m,n}; +, \circ)$.*

Especially for the types $(1, n)$ and $(m, 1)$, the matrices are rows $(a_1 \ldots a_n)$ or columns $(a_1 \ldots a_m)^T$. Addition and multiplication are simply

$$(a_1 \ldots a_n) + (b_1 \ldots b_n) = (a_1 + b_1 \ldots a_n + b_n)$$

and

$$a(a_1 \dots a_n) = (aa_1 \dots aa_n)$$

for rows and

$$(a_1 \dots a_m)^T + (b_1 \dots b_m)^T = (a_1 + b_1 \dots a_m + b_m)^T$$

and

$$a(a_1 \dots a_m)^T = (aa_1 \dots aa_m)^T$$

for columns. For a special class of vector spaces, both $\mathcal{K}$-vector spaces play an important role as we will see later.

The following examples show that the concept of a $\mathcal{K}$-vector space occurs in different branches of mathematics.

Example 6.2.4.

1. The set of all polynomials of degree n, i. e., the set
$$\{f(x) | f(x) = a_0 + a_1x + \cdots + a_nx^n, a_i \in \mathbb{R}\}$$
forms, together with addition and scalar multiplication, given by
$$(f+g)(x) := f(x) + g(x), \quad (af)(x) := af(x), a \in \mathbb{R},$$
a real vector space.
2. The set of all functions defined on the real interval $[0,1]$ forms, together with addition and scalar multiplication, defined similar to example 1, a real vector space.
3. Let $\mathcal{K}$ be a field and M be a set. Then the set
$$V = \{f | f : M \to K\}$$
of all functions from M to K, together with
$$(f+g)(x) := f(x) + g(x), \quad (af)(x) := af(x), a \in K,$$
forms a $\mathcal{K}$-vector space.

Definition 6.2.5. A vector space $\mathcal{U}$ is said to be a subspace of a $\mathcal{K}$-vector space $\mathcal{V}$ if $\mathcal{U}$ is a $\mathcal{K}$-vector space with $U \subseteq V$ such that addition and scalar multiplication in $\mathcal{V}$ and $\mathcal{U}$ agree.

Theorem 6.2.6 (Subspace criterion). *A subset U of V forms a subspace of the $\mathcal{K}$-vector space $\mathcal{V}$ if and only if the following conditions are satisfied:*

(i) $\emptyset \neq U \subseteq V$;

(ii) $a, b \in U \Rightarrow a + b \in U$;

(iii) $\alpha \in K, a \in U \Rightarrow \alpha a \in U$.

Proof. If $\mathcal{U}$ is a subspace of $\mathcal{V}$, then (i)–(iii) are satisfied.

If, conversely, these conditions are satisfied, then the sum of any two elements from U is again an element from U. Therefore, the restriction of the addition in V onto the elements of U is a binary operation in U. The associative and commutative laws are satisfied in $\mathcal{V}$, consequently also in $\mathcal{U}$. From $0 = 0a \in U$ and $-a = (-e)a \in U$, we conclude that $\mathcal{U}$ has a zero element and to each $a \in U$ also the opposite element $-a$ belongs to U. Therefore, $(U; +)$ is an abelian group. Since $a \in U$ implies $\alpha a \in U$ and the axioms of the scalar multiplication are valid in $\mathcal{V}$, they are also valid in $\mathcal{U}$. □

The subspace criterion can be used to prove the following statement:

Theorem 6.2.7. *Let $U_1, \dots, U_k$ be universes (carrier sets) of subspaces $\mathcal{U}_1, \dots, \mathcal{U}_k$ of a $\mathcal{K}$-vector space $\mathcal{V}$, then the intersection $U_1 \cap \dots \cap U_k$ is the universe of a subspace of $\mathcal{V}$.*

Proof. Since the zero vector belongs to each subspace, it belongs to the intersection. Thus the intersection is nonempty. Also $a, b \in U_i$ for all $i = 1, \dots, k$ and $\alpha \in K$ implies $a + b, \alpha a \in U_i$ for all $i = 1, \dots, k$ and then $a + b, \alpha a \in U_1 \cap \dots \cap U_k$. This shows that all the three conditions of the subspace criterion are satisfied. □

The following example shows that the union of the universes of subspaces of a $\mathcal{K}$-vector space $\mathcal{V}$ is in general not the universe of a subspace of $\mathcal{V}$.

Example 6.2.8. The sets

$$V_1 = \left\{ \alpha \begin{pmatrix} 0 \\ 1 \end{pmatrix} \middle| \alpha \in \mathbb{R} \right\}, \quad V_2 = \left\{ \alpha \begin{pmatrix} 1 \\ 0 \end{pmatrix} \middle| \alpha \in \mathbb{R} \right\}$$

form, with respect to the operations defined by

$$\alpha \begin{pmatrix} 0 \\ 1 \end{pmatrix} + \beta \begin{pmatrix} 0 \\ 1 \end{pmatrix} = (\alpha + \beta) \begin{pmatrix} 0 \\ 1 \end{pmatrix}, \quad \beta \left(\alpha \begin{pmatrix} 0 \\ 1 \end{pmatrix} \right) = \beta\alpha \begin{pmatrix} 0 \\ 1 \end{pmatrix}$$

for V_1 and in the corresponding way for V_2, subspaces of the real vector space $\mathcal{V}$ with

$$V = \left\{ \alpha \begin{pmatrix} 0 \\ 1 \end{pmatrix} + \beta \begin{pmatrix} 1 \\ 0 \end{pmatrix} \middle| \alpha, \beta \in \mathbb{R} \right\}.$$

But $V_1 \cup V_2$ is not the universe of a subspace of $\mathcal{V}$ since, for instance, the vector $\begin{pmatrix} 1 \\ 1 \end{pmatrix}$ as a sum of two vectors from $V_1 \cup V_2$ does not belong to $V_1 \cup V_2$.

In this case we may define the subspace generated by the union of the universes of subspaces. For this definition, we need the concept of a *linear span* of a vector space.

Definition 6.2.9. Let $\mathcal{V}$ be a $\mathcal{K}$-vector space, and let $M \subseteq V$ be a set of vectors. Then the set of all linear combinations of vectors from M, i. e., the set of all vectors b from V, for which there exist a natural number n, vectors $a_1, \dots, a_n \in M$, and elements $\alpha_1, \dots, \alpha_n \in K$ with $b = \alpha_1 a_1 + \dots + \alpha_n a_n$, is called *the linear span* $\langle M \rangle$ of M.

We give some examples.

Example 6.2.10.
1. Let $V = \mathbb{R}^2$ be the set of all pairs of real numbers. Together with the componentwise defined addition and the scalar multiplication with real numbers, $\mathcal{V}$ becomes a real vector space. We choose $M = \{(0,1),(1,0)\}$. Then $\langle M\rangle = \{\alpha(0,1)+\beta(1,0) \mid \alpha,\beta \in \mathbb{R}\} = \{(\alpha,\beta) \mid \alpha,\beta \in \mathbb{R}\} = \mathbb{R}^2$. Therefore, the linear span of M is all of $V = \mathbb{R}^2$.
2. Let $V = \mathbb{R}^2$ and $M = \{(0,1)\}$. Then $\langle M\rangle = \{(0,\alpha) \mid \alpha \in \mathbb{R}\}$.

The first example shows that the whole vector space can be generated from only two vectors. In some sense, all information contained in the vector space is contained in a two-element subset. Step by step, this will open the way to one of the most important concepts of linear algebra. But we need at first some more results and concepts.

Theorem 6.2.11. *Let $\mathcal{V}$ be a $\mathcal{K}$-vector space and $M \subseteq V$ a nonempty set of vectors of V. Then $\langle \mathcal{M}\rangle := (\langle M\rangle; +, \circ)$ is a subspace of the $\mathcal{K}$-vector space $\mathcal{V}$.*

Proof. To prove that $\langle \mathcal{M}\rangle$ is a subspace of $\mathcal{V}$, we apply the subspace criterion. Since $M \neq \emptyset$, there is a vector $a \in M$, and the vector $a = ea$, where e is the identity element of $\mathcal{K}$, belongs to $\langle \mathcal{M}\rangle$. Thus $\langle \mathcal{M}\rangle \neq \emptyset$. Clearly, the sum of two linear combinations of elements from M is a linear combination since

$$(\alpha_1 a_1 + \cdots + \alpha_n a_n) + (\beta_1 a_1 + \cdots + \beta_n a_n) = (\alpha_1+\beta_1)a_1 + \cdots + (\alpha_n+\beta_n)a_n.$$

The scalar product of $\alpha \in K$ with a linear combination is again a linear combination,

$$\beta(\alpha_1 a_1 + \cdots + \alpha_n a_n) = \beta\alpha_1 a_1 + \cdots + \beta\alpha_n a_n. \qquad \square$$

Definition 6.2.12. The subspace $\langle \mathcal{M}\rangle := (\langle M\rangle; +, \circ)$ of $\mathcal{V}$ is said to be generated by M, and M is said to be a generating system of $\langle \mathcal{M}\rangle$.

If we map each subset M of V to its linear span $\langle M\rangle$, then this defines a function, we say an *operator*, $\langle\cdot\rangle : \mathcal{P}(V) \to \mathcal{P}(V)$, which has the following properties:

Theorem 6.2.13. *Let $\mathcal{V}$ be a $\mathcal{K}$-vector space and $M \subseteq V$. The operator $\langle\cdot\rangle$ has the following properties:*
(H1) $M \subseteq \langle M\rangle$,
(H2) $M_1 \subseteq M_2 \Rightarrow \langle M_1\rangle \subseteq \langle M_2\rangle$,
(H3) $\langle\langle M\rangle\rangle = \langle M\rangle$.

Proof. (H1) From $a \in M$, there follows with the identity element e of the field $\mathcal{K}$, $a = ea \in \langle M\rangle$, since ea is a linear combination.

(H2) From $b = \alpha_1 a_1 + \cdots + \alpha_n a_n \in \langle M_1\rangle$ with $a_1,\ldots,a_n \in M_1$, there follows $a_1,\ldots,a_n \in M_2$ because of $M_1 \subseteq M_2$ and thus $b = \alpha_1 a_1 + \cdots + \alpha_n a_n \in \langle M_2\rangle$.

(H3) $\langle M\rangle \subseteq \langle\langle M\rangle\rangle$ follows from (H1). To prove the opposite inclusion, let $b \in \langle\langle M\rangle\rangle$, i. e., $b = \alpha_1 a_1 + \cdots + \alpha_n a_n$ with $a_i \in \langle M\rangle$. Since $\langle M\rangle$, as a subspace of $\mathcal{V}$, is closed with respect to addition and scalar multiplication, we obtain $b \in \langle M\rangle$. □

A generating system should be minimal. It should not contain vectors which can be obtained as linear combinations of other vectors of the generating system. The following definition will realize this idea.

Definition 6.2.14. Let $\mathcal{V} = (V; +, \circ)$ be a $\mathcal{K}$-vector space. A set $\{a_1, \ldots, a_r\}$ of pairwise distinct vectors is said to be linearly independent if each nontrivial linear combination of these vectors is different from the zero vector, i. e., if from $x_1 a_1 + \cdots + x_r a_r = 0$ there follows $x_1 = \cdots = x_r = 0$. The set $\{a_1, \ldots, a_r\}$ is said to be linearly dependent if it is not linearly independent, i. e., if there exists a system $x_1, \ldots, x_r$ of elements of the field $\mathcal{K}$ with $x_i \neq 0$ for at least one $i \in \{1, \ldots, r\}$ but $x_1 a_1 + \cdots + x_r a_r = 0$.

In the latter case, the vector a_i can be written as a linear combination of the vectors $\{a_1, \ldots, a_r\}$.

Example 6.2.15.

1. Let $\mathcal{V} = \mathbb{R}^2$. Then $M = \{(1,0),(0,1)\}$ is linearly independent, since from

$$x \cdot \begin{pmatrix} 0 \\ 1 \end{pmatrix} + y \cdot \begin{pmatrix} 1 \\ 0 \end{pmatrix} = \begin{pmatrix} 0 \\ 0 \end{pmatrix},$$

i. e., from $0 \cdot x + 1 \cdot y = 0, 1 \cdot x + 0 \cdot y = 0$ there follows $x = y = 0$.

2. Are the vectors

$$\begin{pmatrix} 0 \\ 1 \\ 2 \end{pmatrix}, \begin{pmatrix} -1 \\ 2 \\ 1 \end{pmatrix}, \begin{pmatrix} 3 \\ 1 \\ 1 \end{pmatrix}$$

linearly independent? We obtain

$$x_1 \cdot \begin{pmatrix} 0 \\ 1 \\ 2 \end{pmatrix} + x_2 \cdot \begin{pmatrix} -1 \\ 2 \\ 1 \end{pmatrix} + x_3 \cdot \begin{pmatrix} 3 \\ 1 \\ 1 \end{pmatrix} = \begin{pmatrix} 0 \\ 0 \\ 0 \end{pmatrix}.$$

This corresponds to the linear equational system.

$$\begin{aligned} -x_2 + 3x_3 &= 0, \\ x_1 + 2x_2 + x_3 &= 0, \\ 2x_1 + x_2 + x_3 &= 0. \end{aligned}$$

In Section 6.4 we will solve systems of linear equations. From the previous system, we obtain the equivalent system

$$\begin{aligned} x_1 + 2x_2 + x_3 &= 0, \\ -x_2 + 3x_3 &= 0, \\ 10x_3 &= 0, \end{aligned}$$

which has the uniquely determined solution $x_1 = x_2 = x_3 = 0$.

The following statement is a simple exercise for the reader.

Corollary 6.2.16. *A system of vectors consisting of only one vector is linearly dependent if and only if this is the zero vector.*

Which properties do linearly independent systems of vectors have?

Theorem 6.2.17. *Let $\mathcal{V} = (V; +, \circ)$ be a $\mathcal{K}$-vector space. Then:*
(i) *Each subsystem of a linearly independent system is linearly independent.*
(ii) *If a subsystem of a vector system is linearly dependent, then the whole system is linearly dependent.*
(iii) *A linearly independent system does not contain the zero vector. Conversely, each system of vectors which contains the zero vector is linearly dependent.*
(iv) *A system $\{a_1, \dots, a_n\}$ is linearly dependent if and only if at least one of the vectors is a linear combination of the others.*

Proof. (i) Let $\{a_1, \dots, a_r\}$ be linearly independent and let $\{a_{i_1}, \dots, a_{i_s}\} \subseteq \{a_1, \dots, a_r\}$. If $\alpha_{i_1} a_{i_1} + \dots + \alpha_{i_s} a_{i_s} = 0$, then $0a_1 + \dots + 0a_{i_1-1} + \alpha_{i_1} a_{i_1} + \dots + \alpha_{i_s} a_{i_s} + 0a_{i_s+1} + \dots + 0a_r = 0$. Since the given system is linearly independent, all coefficients are 0, consequently the system $\{a_{i_1}, \dots, a_{i_s}\}$ is linearly independent.

(ii) is the contraposition of (i).

(iii) If $\{a_1, \dots, a_n\}$ is linearly independent and $a_i = 0$, then $\lambda a_i = 0$ for $\lambda \neq 0$, i. e., $\{a_i\}, i \in \{1, \dots, n\}$, is a linearly dependent subsystem. This contradicts (ii).

(iv) If $a_i = \lambda_1 a_1 + \dots + \lambda_{i-1} a_{i-1} + \lambda_{i+1} a_{i+1} + \dots + \lambda_n a_n$, then $\lambda_1 a_1 + \dots + \lambda_{i-1} a_{i-1} - 1 \cdot a_i + \lambda_{i+1} a_{i+1} + \dots + \lambda_n a_n = 0$, but not all coefficients are 0. Thus the system is linearly dependent. If, conversely, $\{a_1, \dots, a_n\}$ is linearly dependent, then $\lambda_1 a_1 + \dots + \lambda_n a_n = 0$ with $\lambda_i \neq 0$ for an i. Thus the equation can be solved for a_i. □

Our leading question now is: Does there for the vector space $\mathcal{V} = (V; +, \circ)$ exist a (possibly finite) subset of vectors such that each vector of the given vector space is a linear combination of this subset and such that no vector of the subset can be removed?

Definition 6.2.18. A set $B = \{a_1, \dots, a_r\} \subseteq M \subseteq V$ is called a basis of M if and only if the following conditions are satisfied:
(i) B is linearly independent.
(ii) Each element from M is a linear combination of elements of B.

Example 6.2.19. Consider $\mathcal{V} = \mathbb{R}^n$, $e_1 := (1, 0, \dots, 0)$, $e_2 := (0, 1, 0, \dots, 0)$, …, $e_n := (0, \dots, 0, 1)$. Then each vector $b = (b_1, \dots, b_n)$ is a linear combination of the vectors $e_1, \dots, e_n$ since $b = b_1 e_1 + \dots + b_n e_n$, which means that (ii) is satisfied. Because of

$$x_1 e_1 + \dots + x_n e_n = 0 \Rightarrow (x_1, \dots, x_n) = (0, \dots, 0) \Rightarrow \forall i \in \{1, \dots, n\}\ (x_i = 0),$$

the first condition is also satisfied. Therefore, $\{e_1, \dots, e_n\}$ is a basis of the vector space $\mathbb{R}^n$ of all n-tuples of real numbers.

If a vector space $\mathcal{V}$ has a finite basis, it is said to be *finitely based*, otherwise, i. e., if $\mathcal{V}$ has a basis which is not finite, it is said to be *infinitely based.*

An important observation is that the representation of a vector as a linear combination of a given basis is uniquely determined. As an interesting fact we note that the uniqueness of the representation in a given basis is equivalent to the linear independence.

Theorem 6.2.20. *Let $\mathcal{U} = (U;+,\circ)$ be a finitely generated $\mathcal{K}$-vector space. Then the m-element set $\{a_1,\dots a_m\}, m \geq 1$, is a basis of $\mathcal{U}$ if and only if each vector from U is a uniquely determined linear combination of these vectors.*

Proof. 1. Let $\{a_1,\dots,a_m\}$ be an m-element basis of $\mathcal{U}$. Since each basis is a generating system, each vector of U has at least one representation as a linear combination of the basis vectors. We assume that there are two representations of x, $x = x_1a_1+\dots+x_ma_m = y_1a_1+\dots+y_ma_m$. Then $(x_1-y_1)a_1+\dots+(x_m-y_m)a_m = 0$. From the linear independence, there follow $x_i = y_i$ for all $i = 1,\dots,m$ and so the representation is unique.

2. Let now the representation as a linear combination be uniquely determined. Assume the generating system $\{a_1,\dots,a_m\}$ is not a basis. Then it is linearly dependent. Therefore from $\alpha_1a_1+\dots\alpha_ma_m = 0$ there follows that there is at least one $\alpha_i \neq 0$. From $x = x_1a_1+\dots+x_ma_m$ there follows the equation $x = x+0 = (x_1+\alpha_1)a_1+\dots+(x_m+\alpha_m)a_m$. This gives two representations of x as linear combinations, a contradiction to the uniqueness. Thus $\{a_1,\dots,a_m\}$ is linearly independent. □

If the vector space is finitely generated, then a basis is a minimal set of generators.

Theorem 6.2.21. *Let $\mathcal{V}$ be a $\mathcal{K}$-vector space and $M \subseteq V$. Let $B \subseteq M$ be a set consisting of r vectors. Then B is a basis of M if and only if*
(i) *B is linearly independent.*
(ii) *Any $r+1$ different elements of M are linearly dependent.*

Proof. 1. Let B be a basis of M. Then by the definition of a basis, (i) is satisfied. Let $b_1,\dots,b_{r+1}$ be different elements of M and $B = \{a_1,\dots,a_r\}$. Then

$$\begin{aligned} b_1 &= \alpha_{11}a_1 + \dots + \alpha_{1r}a_r, \\ &\vdots \\ b_{r+1} &= \alpha_{(r+1)1}a_1 + \dots + \alpha_{(r+1)r}a_r. \end{aligned}$$

We have to show that the equation $x_1b_1+\dots+x_{r+1}b_{r+1} = 0$ has nontrivial solutions. By substitution of $b_1,\dots,b_{r+1}$ into the left-hand side of the last equation, we get:

$$\begin{aligned} &x_1(\alpha_{11}a_1+\dots+\alpha_{1r}a_r)+\dots+x_{r+1}(\alpha_{(r+1)1}a_1+\dots+\alpha_{r+1r}a_r) = 0, \\ \Longrightarrow\ &(x_1\alpha_{11}+\dots+x_{r+1}\alpha_{(r+1)1})a_1+\dots+(x_1\alpha_{1r}+\dots+x_{r+1}\alpha_{r+1r})a_r = 0. \end{aligned}$$

Linear independence of $\{a_1,\dots a_r\}$ implies

$$x_1a_{11} + \cdots + x_{r+1}a_{r+11} = 0,$$
$$\vdots$$
$$x_1a_{1r} + \cdots + x_{r+1}a_{r+1r} = 0.$$

We mention that this linear system of r equations in $r + 1$ unknowns $x_1, \ldots, x_{r+1}$ has nontrivial solutions $\{x_1, \ldots, x_{r+1}\}$, which are not all equal to 0 (for the details, see Sections 6.4/ 6.5). Therefore, $\{b_1, \ldots, b_{r+1}\}$ is linearly dependent.

2. Now we assume that (i) and (ii) are satisfied. We consider the following cases:

2.1. Let $b \in B$. Then b is one of the elements $a_1, \ldots, a_r$. Without loss of generality, $b = a_j$. Then b is a linear combination of $a_1, \ldots, a_r$,

$$b = 0a_1 + \cdots + 0a_{j-1} + 1a_j + 0a_{j+1} + \cdots + 0a_r.$$

2.2. Assume now that $b \notin B$. Then by (ii), $a_1, \ldots, a_r, b$ are linearly dependent. Then there is a linear combination $x_1a_1 + \cdots + x_ra_r + xb = 0$ with $a_i \neq 0$ for at least one coefficient. Then a_i must be x, since if $x = 0$, then $x_1a_1 + \cdots + x_ra_r = 0$. From (i), there follows $x_1 = \cdots = x_r = 0$ and all the coefficients have to be 0. This contradicts the linear dependence of $a_1, \ldots, a_r, b$. Then $b = (-\frac{x_1}{x})a_1 + \cdots + (-\frac{x_r}{x_1})a_r$, and b is a linear combination of elements from B. Therefore the second condition of the definition of a basis is satisfied. □

Not all vector spaces are finitely generated. We consider the set $V = \{f(x) \mid f(x) = a_0 + a_1x + \cdots\}$ of all polynomials of arbitrary degree with coefficients from a field $\mathcal{K}$ with addition of polynomials and scalar multiplication with elements from the field. To write any polynomial as a linear combination, one needs all polynomials $\{e, x, x^2, \ldots\}$, where e is the identity element of $\mathcal{K}$. This is an infinite generating system which is even a basis since from $a_0e + a_1x + \cdots + a_nx^n + \cdots = 0$ there follows $a_i = 0$ for all $i \in \mathbb{N}$.

Clearly, from each finite generating system, one can make a basis by deleting all vectors which are not needed. This means that each finitely generated vector space has a finite basis. Moreover, the following statements are also clear:

Remark 6.2.22.
1. If a $\mathcal{K}$-vector space $\mathcal{V}$ has a finite basis then each basis of $\mathcal{V}$ is finite.
2. Any two finite bases of $\mathcal{V}$ have the same length.

Definition 6.2.23. Let $\mathcal{V}$ be a $\mathcal{K}$-vector space. If $M \subseteq V$ has a basis $\{a_1, \ldots, a_r\}$, then the number r is called *the rank* of M, denoted rank (M). Moreover, we define rank$(\{0\}) = 0$. If $M = V$, then the rank is called dimension of $\mathcal{V}, r = \dim(\mathcal{V})$. If the dimension is r, the vector space is called an r-dimensional $\mathcal{K}$-vector space. If there is no finite basis, the vector space is *infinitely-dimensional.*

Example 6.2.24. We have that $\dim \mathbb{R}^n = n$, since the n vectors $(1, 0, \dots, 0), \dots, (0, \dots, 0, 1)$ form a basis of $\mathbb{R}^n$.

Without proofs, we mention the following facts:

Corollary 6.2.25.
(i) *If $\dim(V) = n$, then each system of n linearly independent vectors from V is a basis of V.*
(ii) *If $\mathcal{V}$ is a finitely generated vector space and $\mathcal{V} = \langle\{a_1, \dots, a_n\}\rangle$, then each system $\{b_1, \dots, b_s\}$ of $s \geq 0$ linearly independent vectors of $\mathcal{V}$ can be completed to a basis of $\mathcal{V}$.*

Remark 6.2.26.
1. If $\{a_1, \dots, a_n\}$ is a basis of $\mathcal{V}$ and if b, c are vectors from V, then $\{b, c, a_2, \dots, a_n\}$ is linearly dependent.
2. If $\{b_1, \dots, b_n\}$ is a basis and if $\{a_1, \dots, a_r\}$ is a generating system, then $n \leq r$. Therefore, a basis is a minimal generating system.
3. If $\mathcal{U}$ is a subspace of $\mathcal{V}$ and if $\dim(\mathcal{V})$ is finite, then $\dim(\mathcal{U}) \leq \dim(\mathcal{V})$.

Using the concept of the rank of a system of vectors, we define the rank of a matrix.

Definition 6.2.27. The rank of a matrix A is the rank of the system of its column vectors:

$$\operatorname{rank}\begin{pmatrix} a_{11} & \dots & a_{1n} \\ \vdots & \ddots & \vdots \\ a_{m1} & \dots & a_{mn} \end{pmatrix} = \operatorname{rank}(\{\underline{a_1}, \dots, \underline{a_n}\}).$$

We consider the following example:

Example 6.2.28. Let the matrix

$$A = \begin{pmatrix} 6 & 1 & 2 & -3 \\ 1 & -2 & 1 & 1 \\ 3 & 7 & -1 & -6 \\ 1 & 1 & 0 & 0 \end{pmatrix}$$

$$\text{with } \underline{a_1} = \begin{pmatrix} 6 \\ 1 \\ 3 \\ 1 \end{pmatrix}, \underline{a_2} = \begin{pmatrix} 1 \\ -2 \\ 7 \\ 1 \end{pmatrix}, \underline{a_3} = \begin{pmatrix} 2 \\ 1 \\ -1 \\ 0 \end{pmatrix}, \underline{a_4} = \begin{pmatrix} -3 \\ 1 \\ -6 \\ 0 \end{pmatrix}$$

be given. We have to determine a basis for the system $\{\underline{a_1}, \underline{a_2}, \underline{a_3}, \underline{a_4}\}$ of vectors. This means that we have to check which vector is a linear combination of the others. We see that

$$\begin{pmatrix} -3 \\ 1 \\ -6 \\ 0 \end{pmatrix} = x \begin{pmatrix} 6 \\ 1 \\ 3 \\ 1 \end{pmatrix} + y \begin{pmatrix} 1 \\ -2 \\ 7 \\ 1 \end{pmatrix} + z \begin{pmatrix} 2 \\ 1 \\ -1 \\ 0 \end{pmatrix}$$

is solvable with $x = 5, y = -5, z = -14$, since

$$\begin{pmatrix} -3 \\ 1 \\ -6 \\ 0 \end{pmatrix} = 5 \begin{pmatrix} 6 \\ 1 \\ 3 \\ 1 \end{pmatrix} - 5 \begin{pmatrix} 1 \\ -2 \\ 7 \\ 1 \end{pmatrix} - 14 \begin{pmatrix} 2 \\ 1 \\ -1 \\ 0 \end{pmatrix}.$$

Therefore, we cancel $\underline{a_4}$. The system $\underline{a_1}, \underline{a_2}, \underline{a_3}$ is linearly independent, since

$$x \begin{pmatrix} -6 \\ 1 \\ 3 \\ 1 \end{pmatrix} + y \begin{pmatrix} 1 \\ -2 \\ 7 \\ 1 \end{pmatrix} + z \begin{pmatrix} 2 \\ 1 \\ -1 \\ 0 \end{pmatrix} = \begin{pmatrix} 0 \\ 0 \\ 0 \\ 0 \end{pmatrix}.$$

is only trivially solvable with $x = y = z = 0$. Thus, $\{\underline{a_1}, \underline{a_2}, \underline{a_3}\}$ is basis of $\{\underline{a_1}, \underline{a_2}, \underline{a_3}, \underline{a_4}\}$ and $\operatorname{rank}(A) = 3$.

One could ask why we have preferred the columns to define the rank of a matrix. The reason is that the rank of the system of the row vectors is equal to the rank of the system of the column vectors of a matrix A.

Theorem 6.2.29. *Let A be a matrix. Then*

$$\operatorname{rank}(A) = \operatorname{rank}(A^T).$$

The proof will be given in Section 6.5 as an application of the main theorems on the solution of linear equation systems (see, e. g., [33] or [12]).

We are looking for an easier method to determine the rank of a matrix. Clearly, $\operatorname{rank} A \leq n$, since the rank of a matrix cannot be greater than the number of columns.

Theorem 6.2.30. *If a matrix B can be produced from matrix A by the following steps:*
I. *exchanging of two columns,*
II. *multiplying one column with $c \in \mathbb{R}^* (:= \mathbb{R} \setminus \{0\})$,*
III. *adding to one column the scalar product of c with another column,*

then $\operatorname{rank}(B) = \operatorname{rank}(A)$.

Proof. Let $\{\underline{a_1}, \ldots, \underline{a_n}\}$ be the system of columns of matrix A. If $\operatorname{rank}(A) = \operatorname{rank}(\{\underline{a_1}, \ldots, \underline{a_n}\}) = r \leq n$, then there is a basis $\{\underline{a_{i_1}}, \ldots, \underline{a_{i_r}}\} \subseteq \{\underline{a_1}, \ldots, \underline{a_n}\}$ of $\{\underline{a_1}, \ldots, \underline{a_n}\}$ consisting of r vectors. This is independent of the order of the vectors and shows the first statement.

If $\{\underline{a_{i_1}}, \ldots, \underline{a_{i_r}}\}$ is a basis of $\{\underline{a_1}, \ldots, \underline{a_n}\}$, then $\{\underline{a_{i_1}}, \ldots, c\underline{a_{ij}}, \ldots, \underline{a_{i_r}}\}$ is also a basis, since these vectors form again a linearly independent system of vectors. This proves claim II.

If $\{\underline{a_{i_1}}, \ldots, \underline{a_{i_r}}\}$ is a basis, then $\{\underline{a_{i_1}}, \ldots, \underline{a_{i_l}} + c\underline{a_{i_j}}, \ldots, \underline{a_{i_r}}\}$ is also a basis, thus claim III is also true. □

The aim of applying operations I–III is to use the Gaussian algorithm, which will be introduced in Section 6.4. By this algorithm, we produce a matrix B which has the same rank as matrix A from which we started.

$$\operatorname{rank}(A)=\operatorname{rank}(B)=\operatorname{rank}\begin{pmatrix} a_{11} & a_{12} & \dots & a_{1r} & a_{1r+1} & \dots & a_{1n}\\ 0 & a_{22} & \dots & a_{2r} & a_{2r+1} & \dots & a_{2n}\\ \vdots & \vdots & \ddots & \vdots & \vdots & \ddots & \vdots\\ 0 & 0 & \dots & a_{rr} & a_{rr+1} & \dots & a_{rn}\end{pmatrix} \tag{*}$$

with $a_{ik}=0$ for all $i>k$, $a_{\text{ii}}\neq 0$ for all i.

Theorem 6.2.31. *If a matrix A has the form (*), then $r=\operatorname{rank}(A)$.*

Proof. We have to show that the vectors

$$\underline{a_1}=\begin{pmatrix} a_{11}\\ 0\\ \vdots\\ 0\end{pmatrix},\quad \underline{a_2}=\begin{pmatrix} a_{12}\\ a_{22}\\ \vdots\\ 0\end{pmatrix},\quad \dots,\quad \underline{a_n}=\begin{pmatrix} a_{1n}\\ a_{2n}\\ \vdots\\ a_{rn}\end{pmatrix},$$

form a linear independent system, i. e., from

$$\lambda_1\underline{a_1}+\lambda_2\underline{a_2}+\dots+\lambda_n\underline{a_n}=\underline{0},$$

i. e., from

$$\lambda_1\begin{pmatrix} a_{11}\\ 0\\ \vdots\\ 0\end{pmatrix}+\lambda_2\begin{pmatrix} a_{12}\\ a_{22}\\ \vdots\\ 0\end{pmatrix}+\dots+\lambda_n\begin{pmatrix} a_{1n}\\ a_{2n}\\ \vdots\\ a_{rn}\\ 0\end{pmatrix}=\begin{pmatrix} 0\\ 0\\ \vdots\\ 0\end{pmatrix}$$

and thus from

$$\begin{aligned}
\lambda_1 a_{11}+\lambda_2 a_{12}+\dots+\lambda_r a_{1r}+\dots+\lambda_n a_{1n}&=0,\\
\lambda_2 a_{22}+\dots+\lambda_r a_{2r}+\dots+\lambda_n a_{2n}&=0,\\
\ddots\qquad&\\
\lambda_r a_{rr}+\dots+\lambda_n a_{rn}&=0,\\
0&=0,\\
&\vdots\\
0&=0,
\end{aligned}$$

there follows that all coefficients $\lambda_1,\dots,\lambda_n$ are equal to 0. At first, we show $\operatorname{rank}(A)=\operatorname{rank}(A^T)$ and consider the transpose

$$A^T = \begin{pmatrix} a_{11} & 0 & 0 & 0 & \dots & 0 \\ a_{12} & a_{22} & 0 & 0 & \dots & 0 \\ a_{13} & a_{23} & a_{33} & 0 & \dots & 0 \\ \vdots & \vdots & \vdots & \vdots & \ddots & \vdots \\ a_{1n} & a_{2n} & a_{3n} & a_{4n} & \dots & a_{rn} \end{pmatrix}.$$

From

$$\mu_1 \begin{pmatrix} a_{11} \\ a_{12} \\ \vdots \\ a_{1n} \end{pmatrix} + \mu_2 \begin{pmatrix} 0 \\ a_{22} \\ \vdots \\ a_{2n} \end{pmatrix} + \dots + \mu_r \begin{pmatrix} 0 \\ \vdots \\ a_{rr} \\ a_{rn} \end{pmatrix} = \begin{pmatrix} 0 \\ 0 \\ \vdots \\ 0 \end{pmatrix},$$

there follows $\mu_1 = \mu_2 = \dots = \mu_r = 0$. This means that the rank of the transpose A^T is r. Since $\operatorname{rank}(A) = \operatorname{rank}(A^T)$, the proof is finished. □

6.3 Regular matrices and determinants

One of the main problems of linear algebra is the solution of linear equation systems. In the next section, we will prove the main theorems on systems of linear equations. As a first example, we consider here the system $A\underline{x} = \underline{b}$ with

$$A = \begin{pmatrix} a_{11} & a_{12} \\ a_{21} & a_{22} \end{pmatrix}, \quad \underline{x} = \begin{pmatrix} x_1 \\ x_2 \end{pmatrix}, \quad \underline{b} = \begin{pmatrix} b_1 \\ b_2 \end{pmatrix}, \quad \text{i. e.,}$$

$$\begin{aligned} a_{11}x_1 + a_{12}x_2 &= b_1, \\ a_{21}x_1 + a_{22}x_2 &= b_2. \end{aligned}$$

The solution is $x_1 = \frac{b_1 a_{22} - b_2 a_{12}}{a_{11}a_{22} - a_{12}a_{21}}$, $x_2 = \frac{b_1 a_{21} - b_2 a_{11}}{a_{11}a_{22} - a_{12}a_{21}}$.
The real number which occurs in the fractions, namely

$$|A| = \begin{vmatrix} a_{11} & a_{12} \\ a_{21} & a_{22} \end{vmatrix} := a_{11}a_{22} - a_{12}a_{21},$$

is said to be a *determinant of order* 2.

With the permutations $S_2 = \{s_1 = (1), s_2 = (12)\}$, we can write

$$D = |A| = a_{1s_1(1)}a_{2s_1(2)} - a_{1s_2(1)}a_{2s_2(1)} = \sum_{s \in S_2} \operatorname{sgn}(s) a_{1s(1)} a_{2s(2)}.$$

If

$$A = \begin{pmatrix} a_{11} & a_{12} & a_{13} \\ a_{21} & a_{22} & a_{23} \\ a_{31} & a_{32} & a_{33} \end{pmatrix}$$

is a matrix of type (3, 3), we calculate its determinant by

$$\begin{aligned} D = |A| &= \begin{vmatrix} a_{11} & a_{12} & a_{13} \\ a_{21} & a_{22} & a_{23} \\ a_{31} & a_{32} & a_{33} \end{vmatrix} \\ &= a_{11}a_{22}a_{33} + a_{12}a_{23}a_{31} + a_{13}a_{21}a_{32} - a_{12}a_{21}a_{33} - a_{11}a_{23}a_{32} - a_{13}a_{22}a_{31} \\ &= \sum_{s \in S_3} \operatorname{sgn}(s) a_{1s(1)} a_{2s(2)} a_{3s(3)}. \end{aligned}$$

This equation is called the *rule of Sarrus*. We notice that *Sarrus'* rule helps in calculating the 6 products (Figure 6.1).

Indeed, we form the sum of 3 diagonal products from the top left to the bottom right and subtract the sum of the 3 diagonal products from the top right to the bottom left:

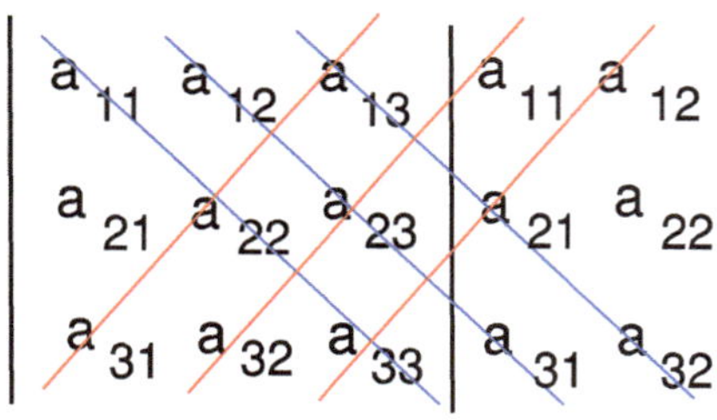

Figure 6.1: Rule of Sarrus.

In the general case, we define:

Definition 6.3.1 (G. W. Leibniz (1646–1716)). Let $A = (a_{ik})$, $a_{ik} \in K_{n,n}$, be a matrix of type (n, n) over the field $\mathcal{K}$, and let

$$s = \begin{pmatrix} 1 & 2 & \dots & n \\ s(1) & s(2) & \dots & s(n) \end{pmatrix}$$

be a permutation. Then $D = \sum_{s \in S_n} \operatorname{sgn}(s) a_{1s(1)} \cdots a_{ns(n)}$ is said to be a determinant of order n.

By this definition, a determinant is an element of K. For its calculation, one needs from each row and column precisely one element. Directly from the definition, we get some properties of determinants:

1. Let

$$A = \begin{pmatrix} a_{11} & a_{12} & \dots & a_{1n} \\ \vdots & \vdots & \ddots & \vdots \\ a_{n1} & a_{n2} & \dots & a_{nn} \end{pmatrix}$$

be a matrix of type (n, n) over the field $\mathcal{K}$. Let $D = |A|$ be its determinant. If there is a row in A consisting only of zeros, then $D = 0 \in K$.

Proof. Each summand of $D = \sum_{s \in S_n} \operatorname{sgn}(s) a_{1s(1)} \cdots a_{ns(n)}$ contains from each row and column precisely one element. Therefore, $D = 0$. □

2. If

$$A = \begin{pmatrix} a_{11} & a_{12} & \dots & a_{1n} \\ \vdots & \vdots & \ddots & \vdots \\ a_{i1} & a_{i2} & \dots & a_{in} \\ \vdots & \vdots & \ddots & \vdots \\ a_{n1} & a_{n2} & \dots & a_{nn} \end{pmatrix}$$

and

$$A' = \begin{pmatrix} a_{11} & a_{12} & \dots & a_{1n} \\ \vdots & \vdots & \ddots & \vdots \\ ca_{i1} & ca_{i2} & \dots & ca_{in} \\ \vdots & \vdots & \ddots & \vdots \\ a_{n1} & a_{n2} & \dots & a_{nn} \end{pmatrix},$$

then

$$D' = |A'| = cD = c|A|.$$

Proof. Clearly,

$$D' = |A'| = \sum_{s \in S_n} \operatorname{sgn}(s) c a_{1s(1)} \cdots a_{ns(n)} = cD,$$

by definition. □

3. If

$$A' = \begin{pmatrix} a_{11} & a_{12} & \dots & a_{1n} \\ \vdots & \vdots & \ddots & \vdots \\ a'_{i1} & a'_{i2} & \dots & a'_{in} \\ \vdots & \vdots & \ddots & \vdots \\ a_{n1} & a_{n2} & \dots & a_{nn} \end{pmatrix}, \quad A'' = \begin{pmatrix} a_{11} & a_{12} & \dots & a_{1n} \\ \vdots & \vdots & \ddots & \vdots \\ a''_{i1} & a''_{i2} & \dots & a''_{in} \\ \vdots & \vdots & \ddots & \vdots \\ a_{n1} & a_{n2} & \dots & a_{nn} \end{pmatrix},$$

$$A = \begin{pmatrix} a_{11} & a_{12} & \cdots & a_{1n} \\ \vdots & \vdots & \ddots & \vdots \\ a'_{i1} + a''_{i1} & a'_{i2} + a''_{i2} & \cdots & a'_{in} + a''_{in} \\ \vdots & \vdots & \ddots & \vdots \\ a_{n1} & a_{n2} & \cdots & a_{nn} \end{pmatrix},$$

then $D = |A| = |A'| + |A''|$.

Proof. We have

$$\begin{aligned} &\begin{vmatrix} a_{11} & a_{12} & \cdots & a_{1n} \\ \vdots & \vdots & \ddots & \vdots \\ a'_{i1} + a''_{i1} & a'_{i2} + a''_{i2} & \cdots & a'_{in} + a''_{in} \\ \vdots & \vdots & \ddots & \vdots \\ a_{n1} & a_{n2} & \cdots & a_{nn} \end{vmatrix} \\ &= \sum_{s \in S_n} \operatorname{sgn}(s) a_{1s(1)} \cdots (a'_{is(i)} + a''_{is(i)}) \cdots a_{ns(n)} \\ &= \sum_{s \in S_n} \operatorname{sgn}(s) a_{1s(1)} \cdots a'_{is(i)} \cdots a_{ns(n)} + \sum_{s \in S_n} \operatorname{sgn}(s) a_{1s(1)} \cdots a''_{is(i)} \cdots a_{ns(n)} \\ &= \begin{vmatrix} a_{11} & a_{12} & \cdots & a_{1n} \\ \vdots & \vdots & \ddots & \vdots \\ a'_{i1} & a'_{i2} & \cdots & a'_{in} \\ \vdots & \vdots & \ddots & \vdots \\ a_{n1} & a_{n2} & \cdots & a_{nn} \end{vmatrix} + \begin{vmatrix} a_{11} & a_{12} & \cdots & a_{1n} \\ \vdots & \vdots & \ddots & \vdots \\ a''_{i1} & a''_{i2} & \cdots & a''_{in} \\ \vdots & \vdots & \ddots & \vdots \\ a_{n1} & a_{n2} & \cdots & a_{nn} \end{vmatrix}. \end{aligned}$$

□

Altogether, we obtain the following properties:

I. If we obtain the quadratic matrix A' from a quadratic matrix A by exchanging two rows, then $D' = |A'| = -D = -|A|$. (This follows from $\operatorname{sgn}(s) = -\operatorname{sgn}(s')$ if $s' = (ij)s$ for two permutations s, s' of order n and the transposition (ij).)

II. Multiplying any row with $c \in K$, gives $D' = cD$, $c \neq 0$.

III. Multiplying a row with c and adding the new row to any other row yields $D' = cD$.

Then we obtain

Theorem 6.3.2. *For any quadratic matrix*

$$A = \begin{pmatrix} a_{11} & \cdots & a_{1n} \\ \vdots & & \\ a_{m1} & \cdots & a_{mn} \end{pmatrix},$$

there exist an element $c \in K$ and a quadratic matrix of the form

$$A' = \begin{pmatrix} a'_{11} & a'_{12} & \dots & a'_{1n} \\ 0 & a'_{22} & \dots & a'_{2n} \\ \vdots & \vdots & \ddots & \vdots \\ 0 & 0 & \dots & a'_{mn} \end{pmatrix}$$

such that

$$D = |A| = c \cdot |A'| = c \cdot D'.$$

The following example shows how we can use this result to calculate a determinant of order 4.

Example 6.3.3.

$$\begin{vmatrix} 3 & 2 & 1 & 4 \\ 3 & 1 & 5 & 1 \\ 2 & 0 & 2 & 1 \\ 1 & 2 & 1 & 2 \end{vmatrix} = -\begin{vmatrix} 2 & 3 & 1 & 4 \\ 1 & 3 & 5 & 1 \\ 0 & 2 & 2 & 1 \\ 2 & 1 & 1 & 2 \end{vmatrix} = \begin{vmatrix} 1 & 3 & 5 & 1 \\ 2 & 3 & 1 & 4 \\ 0 & 2 & 2 & 1 \\ 2 & 1 & 1 & 2 \end{vmatrix}$$

$$= \begin{vmatrix} 1 & 3 & 5 & 1 \\ 0 & -3 & -9 & 2 \\ 0 & 2 & 2 & 1 \\ 0 & -5 & -9 & 0 \end{vmatrix} = \frac{1}{3 \cdot 3} \begin{vmatrix} 1 & 3 & 5 & 1 \\ 0 & -3 & -9 & 2 \\ 0 & 0 & -12 & 7 \\ 0 & 0 & 18 & -10 \end{vmatrix}$$

$$= \frac{1}{3 \cdot 3 \cdot 2} \begin{vmatrix} 1 & 3 & 5 & 1 \\ 0 & -3 & -9 & 2 \\ 0 & 0 & -12 & 7 \\ 0 & 0 & 0 & 1 \end{vmatrix} = \frac{1}{3 \cdot 3 \cdot 2} 1 \cdot (-3) \cdot (-12) \cdot 1 = \frac{36}{18} = 2.$$

As a consequence of the following proposition, any rule for rows can be formulated for columns as well.

Proposition 6.3.4. *If A^T is the transpose of a quadratic matrix A, then $|A| = |A^T|$.*

Proof. From $D = |A| = \sum_{s \in S_n} \operatorname{sgn}(s) a_{1s(1)} \cdots a_{ns(n)}$, there follows

$$|A^T| = \sum_{s \in S_n} \operatorname{sgn}(s) a_{s(1)1} \cdots a_{s(n)n}.$$

If $s = \left(\begin{smallmatrix} 1 & \dots & n \\ s(1) & \dots & s(n) \end{smallmatrix} \right)$, then $s' := s^{-1} = \left(\begin{smallmatrix} s(1) & \dots & s(n) \\ 1 & \dots & n \end{smallmatrix} \right)$.

Obviously, $\operatorname{sgn}(s) = \operatorname{sgn}(s^{-1})$ and then

$$|A^T| = \sum_{s' \in S_n} \operatorname{sgn}(s') a_{1s'(1)} \cdots a_{ns'(n)} = D.$$ □

Now the determinant of the matrix A' from Theorem 6.3.2 can easily be calculated.

Theorem 6.3.5. *The determinant of the matrix A' from Theorem* 6.3.2 *is given by*

$$\begin{vmatrix} a_{11} & a_{12} & \dots & a_{1n} \\ 0 & a_{22} & \dots & a_{2n} \\ \vdots & \vdots & \ddots & \vdots \\ 0 & 0 & \dots & a_{nn} \end{vmatrix} = a_{11}a_{22}\cdots a_{nn}.$$

Proof. In each product $a_{1s(1)}a_{2s(2)}\cdots a_{ns(n)}$, there exists for $s \neq (1)$ a number i with $i > s(i)$. Indeed, from $1 \leq s(1), 2 \leq s(2), \dots, n \leq s(n)$, there follows $s(n) = n$, since in $\{1, \dots, n\}$ the number n is the greatest. Since s is bijective, from $n-1 \leq s(n-1)$ and $s(n) = n$, there follows $s(n-1) = n-1, \dots, s(2) = 2, s(1) = 1$. Thus, $s = (1)$. Because $a_{ik} = 0$ for $i > k$, the product $a_{11}a_{22}\cdots a_{nn}$ is the only product in the definition of a determinant which is different from 0 if all $a_{ii} \neq 0$. If $a_{jj} = 0$ for one $j \in \{1, \dots, n\}$, then the theorem is also true. □

Another possibility to calculate the determinant follows from the *cofactor expansion*. At first, we define the concept of a *minor*.

Definition 6.3.6. Omitting in the quadratic matrix A with $D := |A|$ the ith row and the jth column, one obtains a quadratic matrix whose determinant is said to be the minor D_{ij} of A,

$$D = \left|\begin{array}{cccc|c|cc} a_{11} & a_{12} & \dots & & a_{1j} & \dots & a_{1n} \\ \vdots & \vdots & & & \vdots & & \vdots \\ \hline a_{i1} & a_{i2} & \dots & & a_{ij} & \dots & a_{in} \\ \hline \vdots & \vdots & & & \vdots & & \vdots \\ a_{n1} & a_{n2} & \dots & & a_{nj} & \dots & a_{nn} \end{array}\right|.$$

Moreover, we define $A_{ij} = (-1)^{i+j}D_{ij}$.

Example 6.3.7. The determinant

$$D = \begin{vmatrix} a_{11} & a_{12} & a_{13} \\ a_{21} & a_{22} & a_{23} \\ a_{31} & a_{32} & a_{33} \end{vmatrix}$$

of a matrix of type (3, 3) has the minors

$$D_{11} = \begin{vmatrix} a_{22} & a_{23} \\ a_{32} & a_{33} \end{vmatrix}, \quad D_{12} = \begin{vmatrix} a_{21} & a_{23} \\ a_{31} & a_{33} \end{vmatrix}, \quad D_{13} = \begin{vmatrix} a_{21} & a_{22} \\ a_{31} & a_{32} \end{vmatrix},$$

$$D_{21} = \begin{vmatrix} a_{12} & a_{13} \\ a_{32} & a_{33} \end{vmatrix}, \quad D_{22} = \begin{vmatrix} a_{11} & a_{13} \\ a_{31} & a_{33} \end{vmatrix}, \quad D_{23} = \begin{vmatrix} a_{11} & a_{12} \\ a_{31} & a_{32} \end{vmatrix},$$

$$D_{31} = \begin{vmatrix} a_{12} & a_{13} \\ a_{22} & a_{23} \end{vmatrix}, \quad D_{32} = \begin{vmatrix} a_{11} & a_{13} \\ a_{21} & a_{23} \end{vmatrix}, \quad D_{33} = \begin{vmatrix} a_{11} & a_{12} \\ a_{21} & a_{22} \end{vmatrix}.$$

Then $A_{11} = D_{11}, A_{12} = -D_{12}, A_{13} = D_{13}, A_{21} = -D_{21}, A_{22} = D_{22}, A_{23} = -D_{23}, A_{31} = D_{31}, A_{32} = -D_{32}$, and $A_{33} = D_{33}$.

Theorem 6.3.8 (Cofactor expansion by the ith row). *For the determinant of a matrix $A \in K_{n,n}$, we have*

$$D = a_{i1}A_{i1} + \cdots + a_{in}A_{in} = \sum_{j=1}^{n} a_{ij}A_{ij}, \quad i = 1, \ldots, n.$$

Proof. From $D = \sum_{s \in S_n} \operatorname{sgn}(s) a_{1k_1} \cdots a_{nk_n}$, we factor out step by step $a_{11}, \ldots, a_{1n}$ and determine the cofactors since each summand contains an element from the first row. This can be done for each row i, and we obtain $D = a_{i1}b_{i1} + \cdots + a_{in}b_{in}$ for $i = 1, \ldots, n$. In this way, also b_{nn} can be calculated as the sum of all summands for which $k_n = n$ if we factor out a_{nn} from all the summands. Then we get

$$a_{nn}b_{nn} = a_{nn} \sum_{s \in S_{n-1}} a_{1k_1} \cdots a_{n-1k_{n-1}} = a_{nn} \begin{vmatrix} a_{11} & \cdots & a_{1n-1} \\ \vdots & \ddots & \vdots \\ a_{n-11} & \cdots & a_{n-1n-1} \end{vmatrix} = a_{nn}D_{nn}.$$

By permuting rows and columns, the determinant $\bar{D}$, where a_{ij} is in position (n, n), can be calculated. Altogether, there are $n-i+n-j = 2n-i-j$ changes of rows, resp. columns. Each time the sign changes. Therefore, $\bar{D} = (-1)^{2n-i-j}D = (-1)^{-i-j}D = \overline{a_{n1}}\,\overline{b_{n1}} + \cdots + \overline{a_{nn}}\,\overline{b_{nn}}$ with $\overline{a_{nn}} = a_{ij}$, $\overline{b_{nn}} = \overline{D_{nn}} = D_{ij}$. Thus $D = (-1)^{i+j}\bar{D} = (-1)^{i+j}(\overline{a_{n1}}\,\overline{b_{n1}} + \cdots + \overline{a_{nn}}\,\overline{b_{nn}}) = a_{ij}b_{ij} + R_{ij} = (-1)^{i+j}a_{ij}D_{ij} + R_{ij}$. Here R_{ij} is the sum of all the summands of D, in which a_{ij} does not occur. Then we have $a_{ij}b_{ij} = (-1)^{i+j}a_{ij}D_{ij}$ and then $b_{ij} = (-1)^{i+j}D_{ij}$, if $A_{ij} \neq 0$. □

The same can be done for columns and gives $D = \sum_{j=1}^{n} a_{ji}A_{ji}$, $i = 1, \ldots, n$ (expansion by the ith column).

Example 6.3.9. We calculate the following determinant:

$$D = \begin{vmatrix} 1 & 2 & 3 \\ 4 & 0 & 2 \\ 3 & 1 & 2 \end{vmatrix} = 1\begin{vmatrix} 0 & 2 \\ 1 & 2 \end{vmatrix} - 2\begin{vmatrix} 4 & 2 \\ 3 & 2 \end{vmatrix} + 3\begin{vmatrix} 4 & 0 \\ 3 & 1 \end{vmatrix}$$
$$= -2 - 2(8-6) + 3 \cdot 4 = -2 - 4 + 12 = 6.$$

If the rows of matrix A are denoted by $a_1, \ldots, a_n$, we write

$$D(a_1, \ldots, a_n) = \sum_{j=1}^{n} (-1)^{k+j} a_{jk}D_{jk}.$$

If we take the elements a_{ij} from another row (column) as the minors, then we obtain 0, as the following proposition states:

Proposition 6.3.10. *For $k \neq i$ and $k \neq j$,*

$$\sum_{j=1}^{n} a_{ij} A_{kj} = 0 \quad \textit{and} \quad \sum_{i=1}^{n} a_{ij} A_{ik} = 0.$$

Proof. In

$$D = \begin{vmatrix} a_{11} & \dots & a_{1n} \\ \vdots & & \vdots \\ a_{i1} & \dots & a_{in} \\ \vdots & & \vdots \\ a_{k1} & \dots & a_{kn} \\ \vdots & & \vdots \\ a_{n1} & \dots & a_{nn} \end{vmatrix}$$

we replace for $i < k$ the kth row by the ith row and obtain

$$D' = \begin{vmatrix} a_{11} & \dots & a_{1n} \\ \vdots & & \vdots \\ a_{i1} & \dots & a_{in} \\ \vdots & & \vdots \\ a_{i1} & \dots & a_{in} \\ \vdots & & \vdots \\ a_{n1} & \dots & a_{nn} \end{vmatrix}.$$

Expansion of D' by the kth row gives $D' = a_{i1}A_{k1} + \dots + a_{in}A_{kn} = 0, k \neq i$, since in D' two rows are equal. □

If we calculate the determinant of a quadratic matrix which is an element of the field $\mathcal{K}$, we have a mapping

$$\det : K_{n,n} \to K$$

and we can ask whether this mapping is compatible with the multiplications in $\mathcal{K}_{n,n}$ and $\mathcal{K}$, respectively. The answer gives the following theorem:

Theorem 6.3.11 (Multiplication theorem). *The determinant of the product of two matrices is the product of its determinants, $|A \cdot B| = |A||B|$ or, equivalently,*

$$\det(A \cdot B) = \det(A) \cdot \det(B).$$

(The multiplication symbols on both sides have different meanings.)

Proof. Let $A = (a_{ij})$, $B = (b_{jk})$ be matrices of type (n,n) and $C = A \cdot B = (c_{ik})$ with $c_{ik} = \sum_{j=1}^{n} a_{ij}b_{jk}$, $1 \le i, k \le n$, its product.

With $\beta_j = (b_{j1}, \ldots, b_{jn})$, $1 \le j \le n$, we obtain

$$\begin{aligned} |A \cdot B| &= D\left(\sum_{j_1=1}^{n} a_{1j_1}\beta_{j_1}, \sum_{j_2=1}^{n} a_{2j_2}\beta_{j_2}, \ldots, \sum_{j_n=1}^{n} a_{nj_n}\beta_{j_n}\right) \\ &= \sum_{j_1=1}^{n}\sum_{j_2=1}^{n} \cdots \sum_{j_n=1}^{n} a_{1j_1} \cdots a_{nj_n} D(\beta_{j_1}, \ldots, \beta_{j_n}) \\ &= \sum_{s \in S_n} a_{1s(1)} \cdots a_{ns(n)} \operatorname{sgn}(s) D(\beta_1, \ldots, \beta_n) = |A||B|. \end{aligned}$$

□

There is a close connection between the rank of a quadratic matrix and its determinant. Let $A \in K_{n,n}$ be a matrix.

Theorem 6.3.12. $\operatorname{rank}(A) = n \Leftrightarrow |A| \neq 0$, *i. e.*, $\operatorname{rank}(A) < n \Leftrightarrow |A| = 0$.

Proof. If A' is a matrix of the form

$$A' = \begin{pmatrix} a'_{11} & a'_{12} & \ldots & a'_{1n} \\ 0 & a'_{22} & \ldots & a'_{2n} \\ \vdots & \vdots & \ddots & \vdots \\ 0 & 0 & \ldots & a'_{mn} \end{pmatrix},$$

then $|A| = k \cdot |A'| = k \cdot a'_{11} \cdots a'_{nn}$ for $k \in K$, $k \neq 0$. Now $\operatorname{rank}(A) = n$ if and only if A' has triangular form and all the elements a_{ii}, $i = 1, \ldots, n$, are different from 0. This is satisfied if and only if we have $|A| \neq 0$. □

Definition 6.3.13. Let A be a quadratic matrix of type (n,n) over $\mathcal{K}$. Then A is called regular if $\operatorname{rank}(A) = n$ or, equivalently, if $|A| \neq 0$.

In general, the matrix multiplication is not invertible, but here we have:

Theorem 6.3.14. *A matrix A is regular if and only if there exists a matrix X with $A \cdot X = X \cdot A = E$ where E is the identity matrix (all elements on the main diagonal are equal to the identity element $e \in K$ and all the other elements are $0 \in K$).*

Proof. For the proof, we assume that

$$A = \begin{pmatrix} a_{11} & a_{12} & \ldots & a_{1n} \\ \vdots & \vdots & & \vdots \\ a_{n1} & a_{n2} & \ldots & a_{nn} \end{pmatrix}$$

and

$$X = \begin{pmatrix} x_{11} & x_{12} & \dots & x_{1n} \\ \vdots & \vdots & & \vdots \\ x_{n1} & x_{n2} & \dots & x_{nn} \end{pmatrix}.$$

Then the equation $AX = E$ corresponds to

$$AX = \begin{pmatrix} \sum a_{1h}x_{h1} & \sum a_{1h}x_{h2} & \dots & \sum a_{1h}x_{hn} \\ \vdots & \vdots & & \vdots \\ \sum a_{nh}x_{h1} & \sum a_{nh}x_{h2} & \dots & a_{nh}x_{hn} \end{pmatrix} = \begin{pmatrix} 1 & 0 & \dots & 0 \\ \vdots & \vdots & & \vdots \\ 0 & 0 & \dots & 1 \end{pmatrix}.$$

This equation is solvable if and only if all equation systems

$$\begin{aligned}
a_{11}x_{11} + a_{12}x_{21} + \dots + a_{1n}x_{n1} &= 1, \\
a_{21}x_{11} + a_{22}x_{21} + \dots + a_{2n}x_{n1} &= 0, \\
&\vdots \\
a_{n1}x_{11} + a_{n2}x_{21} + \dots + a_{n1}x_{n1} &= 0, \\
\\
a_{11}x_{12} + a_{12}x_{22} + \dots + a_{1n}x_{n2} &= 0, \\
a_{21}x_{12} + a_{22}x_{22} + \dots + a_{2n}x_{n2} &= 0, \\
&\vdots \\
a_{21}x_{12} + a_{n2}x_{22} + \dots + a_{nn}x_{n2} &= 1, \\
&\vdots \\
a_{11}x_{1n} + a_{12}x_{2n} + \dots + a_{1n}x_{nn} &= 0, \\
a_{21}x_{1n} + a_{22}x_{2n} + \dots + a_{2n}x_{nn} &= 0, \\
&\vdots \\
a_{21}x_{1n} + a_{n2}x_{2n} + \dots + a_{nn}x_{nn} &= 1
\end{aligned}$$

are solvable. Because $\operatorname{rank}(A) = n$, all these equation systems are solvable as we will see in the next section. Therefore, there exists a matrix X with $AX = E$. Similarly, we show that there exists a matrix Y with $YA = E$. It is not difficult to prove that $X = Y$.

If, conversely, X is a solution of $AX = E$, we consider the linear equation system $A^T x = 0$. By multiplication with the transpose of X, we get $X^T(A^T x) = X^T 0$ and then $(AX)^T x = 0$, i. e., $E^T x = 0$, $Ex = 0$, and thus $x = 0$. Therefore, the linear equation system $A^T x = 0$ has a uniquely determined solution which is 0. In the next section, we will see that this means $\operatorname{rank}(A^T) = n$. Because $\operatorname{rank}(A) = \operatorname{rank}(A^T)$, we have $\operatorname{rank}(A) = n$ and A is regular. □

We will leave it as an exercise to the reader to show that the matrix equation $AX = E$ has a uniquely determined solution X (for details see e. g. [15]).

Definition 6.3.15. The uniquely determined matrix X with $AX = XA = E$ is said to be the inverse of the regular matrix A, written as A^{-1}.

Example 6.3.16. The inverse of the identity matrix E is E, since $EE = E$.
As a second example, we calculate the inverse of a diagonal matrix

$$D = \begin{pmatrix} d_{11} & 0 & \dots & 0 \\ \vdots & & & \vdots \\ 0 & 0 & \dots & d_{nn} \end{pmatrix}$$

with $d_{ii} \neq 0$ for $i = 1, \dots, n$. Then

$$D^{-1} = \begin{pmatrix} d_{11}^{-1} & 0 & \dots & 0 \\ \vdots & & & \vdots \\ 0 & 0 & \dots & d_{nn}^{-1} \end{pmatrix}.$$

Theorem 6.3.17. *Let A and B be regular matrices from $K_{n,n}$. Then $A \cdot B$ is regular and $(AB)^{-1} = B^{-1}A^{-1}$. Moreover, the inverse matrix is regular and satisfies $(A^{-1})^{-1} = A$.*

Proof. Since A and B are regular, the inverses exist. Now, we have $(AB)(B^{-1}A^{-1}) = A(BB^{-1})A^{-1} = AEA^{-1} = AA^{-1} = E$. Since the inverse matrix of AB is uniquely determined, there follows $(AB)^{-1} = B^{-1}A^{-1}$. From $A^{-1}(A^{-1})^{-1} = E$ and $A^{-1}A = E$, we get that the inverse of A^{-1} exists and is equal to A. Thus A^{-1} is regular. □

Altogether, we have

Corollary 6.3.18. *Let $L_{n,n}$ be the set of all regular matrices of type (n, n) over the field K. Then $(L_{n,n}; \cdot)$ is a group, the linear group.*

To calculate the inverse of a regular matrix, we can use the following formula:

$$A^{-1} = \frac{1}{D} \begin{pmatrix} A_{11} & \dots & A_{1n} \\ \vdots & & \vdots \\ A_{i1} & \dots & A_{in} \\ \vdots & & \vdots \\ A_{n1} & \dots & A_{nn} \end{pmatrix}^T.$$

Indeed, we have

$$A\cdot\frac{1}{D}\begin{pmatrix} A_{11} & \dots & A_{n1}\\ \vdots & & \vdots\\ A_{1i} & \dots & A_{ni}\\ \vdots & & \vdots\\ A_{1n} & \dots & A_{nn}\end{pmatrix} = \frac{1}{D}\begin{pmatrix} a_{11} & \dots & a_{1n}\\ \vdots & & \vdots\\ a_{i1} & \dots & a_{in}\\ \vdots & & \vdots\\ a_{n1} & \dots & a_{nn}\end{pmatrix}\begin{pmatrix} A_{11} & \dots & A_{n1}\\ \vdots & & \vdots\\ A_{i1} & \dots & A_{in}\\ \vdots & & \vdots\\ A_{1n} & \dots & A_{nn}\end{pmatrix}$$

$$= \frac{1}{D}\begin{pmatrix} \sum_{i=1}^{n} a_{1i}A_{1i} & \dots & \sum_{i=1}^{n} a_{1i}A_{ni}\\ \vdots & \ddots & \vdots\\ \sum_{i=1}^{n} a_{ni}A_{1i} & \dots & \sum_{i=1}^{n} a_{ni}A_{ni}\end{pmatrix}$$

$$= \frac{1}{D}\begin{pmatrix} D & 0 & \dots & 0\\ 0 & D & \dots & 0\\ & & \ddots & \\ 0 & 0 & \dots & D\end{pmatrix} = E.$$

The equation $A^{-1}\cdot A = E$ can be proved in a similar way.

One more simple consequence is

$$(A^{-1})^T = (A^T)^{-1}.$$

6.4 Systems of linear equations

In this section we will show how systems of linear equations can be solved using the Gaussian algorithm. We will prove the main theorems for homogeneous and inhomogeneous linear equation systems. Finally, we come back to the rank of a matrix, prove that the row rank equals the column rank, and prove Frobenius' rank criterion.

A system of linear equations consists of m linear equations in n unknowns $x_1, x_2, \dots, x_n$ of the form

$$\begin{aligned} & a_{11}x_1 + a_{12}x_2 + \dots + a_{1n}x_n = b_1, && \sum_{k=1}^{n} a_{1k}x_k = b_1,\\ (S)\quad & a_{21}x_1 + a_{22}x_2 + \dots + a_{2n}x_n = b_2, && \sum_{k=1}^{n} a_{2k}x_k = b_2,\\ & \qquad\vdots && \qquad\vdots\\ & a_{m1}x_1 + a_{m2}x_2 + \dots + a_{mn}x_n = b_m, && \sum_{k=1}^{n} a_{mk}x_k = b_m. \end{aligned}$$

The *coefficients* $a_{ij}, 1 \le i \le m, 1 \le j \le n$, are elements of a field $\mathcal{K}$ and $b_1, \dots, b_m$ are also elements of K. If only zeros are on the right-hand side, then the equation system is said to be *homogeneous*, otherwise it is called *inhomogeneous*.

For short, the system (S) can be written as $\sum_{k=1}^{n} a_{ik}x_k = b_i$, $i = 1, \dots, m$.

An n-tuple $(a_1, \dots, a_n)$ of elements of K is called a *solution* of S, if after substituting these elements we get m true statements. Let L_S be the set of all solutions of S.

We want to find answers to the following questions:

- Under which conditions is S solvable?
- Is there only one or are there many solutions?
- How can we determine L_S?

The idea is to bring the equation system without changing L_S in a "simpler" form, which allows recognizing the set of solutions.

Definition 6.4.1. Two linear equation systems S and S' are said to be equivalent ($S \sim S'$) if S and S' have the same set of solutions, $L_S = L_{S'}$.

Relation $\sim$ is an equivalence relation on the set of all equation systems.

Theorem 6.4.2. *From S an equivalent system S' can arise by the following steps:*

I. *Swapping the position of two rows,*
II. *Multiplying a row by a nonzero scalar $c \in K$, $c \neq 0$,*
III. *Adding to one row a scalar multiple of another.*

Proof. Step I is clear. To prove II, we multiply the ith equation with $c \in K$, $c \neq 0$, and obtain $c \cdot \sum_{k=1}^{n} a_{ik}x_k = c \cdot b_i$. If $(x_1, \dots, x_n)$ is a solution of the equation from which we started, then it is also a solution of the multiplied equation, and conversely. For the proof of III, we add the d-multiple of the ith equation to the jth equation and obtain $\sum_{k=1}^{n} a_{jk}x_k + d \cdot \sum_{k=1}^{n} a_{ik}x_k = b_j + d \cdot b_i$. If $(x_1, \dots, x_n)$ is a solution of the starting equation, then it is a solution of the new equation, and conversely. □

We will start with an example.

Example 6.4.3.

1. Consider the system of linear equations

$$\begin{aligned} 2x_1 + 2x_2 - 6x_3 &= 6, \downarrow \cdot 1 \quad \downarrow \cdot 1 \\ x_1 - 3x_2 + 2x_3 &= 1, \downarrow \cdot(-2) \\ x_1 + 5x_2 - 8x_3 &= 5. \qquad\quad \downarrow \cdot(-2) \end{aligned}$$

Applying the previous statements, we get the equivalent systems S_1' and S_1'':

$$(S_1') \qquad \begin{aligned} 2x_1 + 2x_2 - 6x_3 &= 6, \\ 8x_2 - 10x_3 &= 4, \\ -8x_2 + 10x_3 &= -4, \end{aligned}$$

$$(S_1'') \qquad \begin{aligned} 2x_1 + 2x_2 - 6x_3 &= 6, \\ 8x_2 - 10x_3 &= 4, \\ 0 &= 0. \end{aligned}$$

Here S_1'' has the same set of solutions as S_1.

2. Now we consider the equation system

$$(S_2)\quad \begin{aligned} 3x_1 - 4x_2 + 5x_3 &= 0, \downarrow \cdot 2 \downarrow \cdot 1 \\ 2x_1 + x_2 - 2x_3 &= -2, \downarrow \cdot(-3) \\ x_1 \quad 5x_2 + 7x_3 &= 3. \qquad \downarrow \cdot(-3) \end{aligned}$$

Then we get S_2' and S_2'', which are

$$(S_2')\quad \begin{aligned} 3x_1 - 4x_2 + 5x_3 &= 0, \\ -11x_2 + 16x_3 &= 6, \downarrow \cdot(-19) \\ -19x_2 - 16x_3 &= -9, \downarrow \cdot 11 \end{aligned}$$

$$(S_2'')\quad \begin{aligned} 3x_1 - 4x_2 + 5x_3 &= 0, \\ -11x_2 + 11x_3 &= 6, \\ -320x_3 &= -153. \end{aligned}$$

Then we calculate x_1, x_2, x_3 and obtain a uniquely determined solution of S_2.

If the calculations give a contradiction, then S_2'' is not solvable. Since S_2'' is equivalent to S_2, the equation system S_2 is not solvable.

The method which we used in these examples is based on an algorithm which is called *Gaussian elimination*. This algorithm will be described now for the general case. Consider an arbitrary linear equation system S consisting of m equations with n unknowns and coefficients a_{ij} from the field $\mathcal{K}$, namely

$$\begin{aligned} a_{11}x_1 + a_{12}x_2 + \cdots + a_{1n}x_n &= b_1, \downarrow a_{21} \downarrow a_{31} \cdots \\ a_{21}x_1 + a_{22}x_2 + \cdots + a_{2n}x_n &= b_2, \downarrow -a_{11} \\ \cdots \qquad \cdots \qquad & \cdots \\ a_{m1}x_1 + a_{m2}x_2 + \cdots + a_{mn}x_n &= b_m. \cdots \end{aligned}$$

In the first step, we eliminate the unknown x_1 from the second to the mth equation. In the second step, we consider the linear equation system consisting of $m-1$ equations in $n-1$ unknowns $x_2, \ldots, x_n$ arising after the first step. The first equation will be unchanged. This step will be repeated with the aim to eliminate x_2 from the new $m-1$ equations. After $n-1$ steps, we get an equation system of the following form:

$$\begin{aligned} a_{11}'x_1 + a_{12}'x_2 + \cdots + a_{1r}'x_r + \cdots + a_{1n}'x_n &= b_1', \\ \ddots \qquad & \\ a_{rr}'x_r + \cdots + a_{rn}'x_n &= b_r', \\ 0 &= b_{r+1}', \\ &\vdots \\ 0 &= b_m', \end{aligned}$$

with $a'_{11} \neq 0, \dots, a'_{rr} \neq 0$.

This equation system is equivalent to the system from which we started. It is characterized by the following properties:

1. For all $i > k$, we have $a'_{ik} = 0$,
2. There is a number r such that $a'_{11} \neq 0, \dots, a'_{rr} \neq 0$ and $a'_{ik} = 0$ for all $i > r$. The number r is said to be the *rank* of the linear equation systems S.

If one of the elements $b'_{r+1}, \dots, b'_m$ is different from 0, then the equation system is not solvable, otherwise it is solvable. If after the rth equation there are only equations of the form $0 = 0$ and in the rth equation on the left-hand side only the variable x_n, then x_n will be isolated and the result will be substituted into the previous equation to calculate x_{n-1}. Continuing in this way, we finally get x_1. In this case, the equation system is uniquely solvable. If the equation system is solvable and if the variables $x_r, \dots x_n$ occur in the rth equation, then we introduce for $x_r, \dots, x_{n-1}$ the so-called free parameters. For these free parameters, any element from K can be substituted. Then the rth equation will be solved for x_n. On the right-hand side, there appear $n - r$ free parameters. This expression for x_n will be substituted into the $(r-1)$th equation and, continuing in this way, we finally obtain x_1. We continue with S_1 from Example 6.4.3(1).

Example 6.4.4.

$$\begin{aligned} 2x_1 + 2x_2 - 6x_3 &= 6, \\ 8x_2 - 10x_3 &= 4, \\ 0 &= 0. \end{aligned}$$

Now S'_1 will be transformed into *diagonal form*,

$$\begin{aligned} 8x_1 \qquad &= 20 + 14x_3, \\ 8x_2 &= 4 + 10x_3. \end{aligned}$$

Introducing a free parameter $x_3 = t$ gives $x_1 = \frac{5}{2} + \frac{7}{4}t$, $x_2 = \frac{1}{2} + \frac{5}{4}t$ and

$$L_{S_1} = \left\{ \begin{pmatrix} x_1 \\ x_2 \end{pmatrix} \middle| \begin{pmatrix} x_1 \\ x_2 \end{pmatrix} = \begin{pmatrix} \frac{5}{2} \\ \frac{1}{2} \end{pmatrix} + t \begin{pmatrix} \frac{7}{4} \\ \frac{5}{4} \end{pmatrix} \text{ and } t \in \mathbb{R} \right\}.$$

There exist infinitely many solutions depending on one free parameter.

Now we summarize our procedure:

1. Each linear equation system S can be transformed into trapezoid form by applying the Gaussian elimination. If S is homogeneous, then the trapezoid form of S is homogeneous.
2. An equation system S is solvable if and only if there are no contradictions.
3. A solvable system has a uniquely determined solution if and only if $r = n$.
4. If $r < n$, then there are infinitely many solutions which depend on $s = n - r$ free parameters.

5. A homogeneous system is always solvable and has the *trivial* solution $(0,\dots,0)$. It is nontrivially solvable if and only if $r < n$. A homogeneous linear equation system with less equations than unknowns, i. e., with $m < n$, is nontrivially solvable because of $r \le m < n$.

6.5 Matrix equations

Using the matrix

$$A = \begin{pmatrix} a_{11} & a_{12} & \dots & a_{1n} \\ \vdots & & & \vdots \\ a_{m1} & a_{m2} & \dots & a_{mn} \end{pmatrix},$$

and the matrices

$$x = \begin{pmatrix} x_1 \\ x_2 \\ \vdots \\ x_n \end{pmatrix}, \quad B = \begin{pmatrix} b_1 \\ b_2 \\ \vdots \\ b_m \end{pmatrix},$$

the linear equation system

$$(S) \quad \begin{aligned} a_{11}x_1 &+ a_{12}x_2 + \dots + a_{1n}x_n = b_1, \\ a_{21}x_1 &+ a_{22}x_2 + \dots + a_{2n}x_n = b_2, \\ &\vdots \\ a_{m1}x_1 &+ a_{m2}x_2 + \dots + a_{mn}x_n = b_m \end{aligned}$$

can be written as $A \cdot x = B$.

Theorem 6.5.1 (Main theorem for homogeneous equation systems).

(i) *The solutions of a homogeneous linear equation system*

$$H : Ax = 0, \quad 0 = \begin{pmatrix} 0 \\ \dots \\ 0 \end{pmatrix}, \quad A \in K_{m,n}, \quad \operatorname{rank}(A) = r,$$

with m equations and n unknowns form a subspace (L_H) of the vector space $(K^n; +, \cdot)$ of dimension $s = n - r$. The homogeneous linear equation system H is nontrivially solvable if and only if r is smaller than n.

(ii) *A basis $\{x_1,\dots,x_s\}$ of L_H can be calculated using the Gaussian algorithm. Then*

$$L_H = \{x_H \mid x_H = t_1x_1 + \dots + t_sx_s,\ t_1,\dots,t_s \in K\}.$$

Proof. We apply the subspace criterion. The set of solutions L_H of the homogeneous linear equation system is nonempty, since the trivial solution 0 belongs to L_H. Also $Ax_1 = 0$ and $Ax_2 = 0$ imply $Ax_1 + Ax_2 = A(x_1 + x_2) = 0$. Therefore, the sum of two solutions is a solution. From $Ax = 0$ and $\lambda \in K$, there follows $\lambda(Ax) = 0$ and then $A(\lambda x) = 0$. Thus, with x also λx is a solution from L_H, and so L_H is a subspace of $(K^n; +, \circ)$.

By constructing a basis, we prove the statement concerning the rank of L_H.

After transforming H into trapezoid form, we have

$$\begin{aligned} a'_{11}x_1 + a'_{12}x_2 + \cdots + a'_{1r}x_r + \ldots + a'_{1n}x_n &= b'_1, \\ \ddots \qquad & \\ a'_{rr}x_r + \cdots + a'_{rn}x_n &= b'_r. \end{aligned}$$

Transformation into diagonal form by using Gaussian elimination yields

$$\begin{aligned} a'_{11}x_1 \qquad &= c_{11}x_{r+1} + \cdots + c_{1n-r}x_n, \\ &\vdots \\ a'_{rr}x_r &= c_{r1}x_{r+1} + \cdots + c_{rn-r}x_n. \end{aligned}$$

Introducing $s = n - r$ free parameters gives

$$\begin{aligned} x_1 &= d_{11}t_1 + d_{12}t_2 + \cdots + d_{1s}t_s, \\ &\vdots \\ x_r &= d_{r1}t_1 + d_{r2}t_2 + \cdots + d_{rs}t_s, \\ x_{r+1} &= t_1 + 0t_2 + \cdots + 0t_s, \\ &\vdots \\ x_n &= 0t_1 + 0t_2 + \cdots + 1t_s. \end{aligned}$$

To prove that

$$B = \left\{ \begin{pmatrix} d_{11} \\ \vdots \\ d_{r1} \\ 1 \\ \vdots \\ 0 \end{pmatrix}, \ldots, \begin{pmatrix} d_{1s} \\ \vdots \\ d_{rs} \\ 0 \\ \vdots \\ 1 \end{pmatrix} \right\}$$

is a basis of L_H, we have to prove the linear independence.

Consider the equation

$$\lambda_1 \begin{pmatrix} d_{11} \\ \vdots \\ d_{r1} \\ 1 \\ \vdots \\ 0 \end{pmatrix} + \cdots + \lambda_s \begin{pmatrix} d_{1s} \\ \vdots \\ d_{rs} \\ 0 \\ \vdots \\ 1 \end{pmatrix} = \begin{pmatrix} 0 \\ \vdots \\ 0 \end{pmatrix}.$$

There follow $\lambda_s = 0, \lambda_{s-1} = 0, \ldots, \lambda_1 = 0$. This shows that we have a basis of L_H and the dimension of L_H is $s = n - r$. Moreover,

$$n = r \Leftrightarrow \operatorname{rank}(L_H) = 0 \Leftrightarrow L_H = \{0\},$$

and the homogeneous equation system is only trivially solvable if $n = r$. Since $r \leq n$, we have

$$H \text{ is nontrivially solvable if and only if } r < n. \qquad \square$$

For inhomogeneous linear equation systems, we have

Theorem 6.5.2 (Main theorem for inhomogeneous linear equation systems). *Let $S : Ax = b$ with $A \in K_{m,n}$, $\operatorname{rank}(A) = r$ be an inhomogeneous equation system. Then:*

(i) *S is solvable if and only if $\operatorname{rank}(A) = \operatorname{rank}(A, b)$. (Here (A, b) is the matrix arising from A by adjoining the column b.)*

(ii) *L_S can be found as follows: If x_0 is a particular solution of S, then*

$$L_S = \{x \mid x = x_0 + x_H \wedge x_H \in L_H\}.$$

(iii) *S has exactly one solution if and only if $\operatorname{rank}(A) = \operatorname{rank}(A, b) = n$.*

Proof. (i) Consider $M_1 := \{a_1, \ldots, a_n\}$ and $M_2 := \{a_1, \ldots, a_n, b\}$, where $a_1, \ldots, a_n, b$ are the columns of matrix A and the right-hand sides of the equation system, respectively. Then S is solvable if and only if b is a linear combination of the vectors $a_1, \ldots, a_n$, i. e., if $b \in \langle M_1 \rangle$, where $\langle M_1 \rangle$ is the span of M_1. Now we prove that $b \in \langle M_1 \rangle$ if and only if $\operatorname{rank}(M_1) = \operatorname{rank}(M_2)$. If b belongs to the span of M_1, then b is a linear combination of $a_1, \ldots, a_n$ and plays no role in determining $\operatorname{rank}(M_2)$. Thus $\operatorname{rank}(M_1) = \operatorname{rank}(M_2)$. If, conversely, $\operatorname{rank}(M_1) = \operatorname{rank}(M_2)$, let B be a basis of M_1 and $b \in B$. Then b belongs to the linear closure of M_1 because $B \subseteq \langle B \rangle = M_1 \subseteq \langle M_1 \rangle$. If $b \notin B$, then $B \cup \{b\}$ is linearly dependent and $b \in \langle B \rangle \subseteq \langle M_1 \rangle$.

(ii) Let $x = x_0 + x_H$, where x_0 is a particular solution of the inhomogeneous system $Ax = b$ and where x_H is a general solution of the homogeneous system $Ax = 0$. Then

$$Ax = A(x_0 + x_H) = Ax_0 + Ax_H = \underline{0} + b = b \Rightarrow x_0 + x_H \in L_S.$$

If, conversely, $x \in L_S$ and $x_0 \in L_S$ is a particular solution, then $Ax = b$ and $Ax_0 = b$, i. e., $A(x - x_0) = 0$ and $x - x_0 \in L_H$ is a solution of the corresponding homogeneous system, thus $x = x_0 + x_H$. Altogether, we have equality.

(iii) By (ii), the inhomogeneous system L_S has a uniquely determined solution and x is the solution if and only if $L_H = \{0\}$, if and only if $\operatorname{rank}(A) = \operatorname{rank}(A, b) = n$. □

If the linear equation system is quadratic, then we can use the determinant of the quadratic matrix of coefficients. Then the solution is given by *Cramér's rule.*

Theorem 6.5.3. *If $D = |A| \neq 0$, then $Ax = b$ has the uniquely determined solution $x_1 = \frac{D_1}{D}$, …, $x_n = \frac{D_n}{D}$, where D_k is the determinant arising from D by replacing the kth column by the column b.*

Proof. We write the linear equation system in the form

$$(S) \quad \sum_{j=1}^{n} a_{ij} x_j = b_i, \quad i = 1, \ldots, n.$$

To apply the expansion theorem, we multiply S with the minors A_{ik} from the right and then we form

$$\sum_{i=1}^{n}\left(\sum_{j=1}^{n} a_{ij} x_j\right) A_{ik} = \sum_{i=1}^{n} b_i A_{ik} = \begin{vmatrix} a_{11} & \ldots & b_1 & \ldots & a_{1n} \\ \vdots & & \vdots & & \vdots \\ a_{n1} & \ldots & b_n & \ldots & a_{nn} \end{vmatrix} = D_k.$$

By reordering the summands, we obtain $\sum_{j=1}^{n} x_j \sum_{i=1}^{n} a_{ij} A_{ik} = D_k$. Since $\sum_{i=1}^{n} a_{ij} A_{ik} \neq 0$ only for $j = k$, there follows $x_k \sum_{i=1}^{n} a_{ik} A_{ik} = D_k$ and then $x_k = \frac{D_k}{D}$, using $D = \sum_{i=1}^{n} a_{ik} A_{ik}$ by the expansion theorem. □

Example 6.5.4. Let $A = \begin{pmatrix} 1 & 1 & 1 \\ 1 & 2 & 1 \\ 2 & 1 & 1 \end{pmatrix}$, $b = \begin{pmatrix} 1 \\ 2 \\ 3 \end{pmatrix}$. To solve the linear equation system $Ax = b$, we calculate the determinant $D = 2 + 2 + 1 - 1 - 1 - 4 \neq 0$ by Sarrus' rule. Therefore, the equation system is uniquely solvable. Moreover, $D_1 = -2$, $D_2 = -1$, $D_3 = 2$, and then $x_1 = \frac{D_1}{D} = 2$, $x_2 = \frac{D_2}{D} = 1$, $x_3 = \frac{D_3}{D} = -2$.

Now we will come back to Section 6.2, Theorem 6.2.29, and will prove that the rank of a matrix is the rank of the system of row vectors of a matrix. This is an application of the latter theorems. Actually, we will prove a bit more, namely that the rank of a matrix is equal to the largest order of the nonzero minors of the matrix. The latter statement is known under the name "rank criterion of Frobenius" (F. G. Frobenius (1849–1917)). The idea of the following proof can be found in [33].

Theorem 6.5.5. *The rank of a matrix (column rank) equals the row rank and both are equal to the largest order of the nonzero minors of the matrix.*

Proof. We have to prove the following statements:

1. The row rank of a matrix equals the largest order of the nonzero minors of the matrix.
2. In a similar way, only exchanging "row" and "column," one shows that the column rank equals the largest order of the nonzero minors of the matrix.
3. Then there follows that the rank of the matrix equals the largest order of nonzero minors of the matrix.

Suppose that s is the largest order of nonzero minors of the matrix

$$\begin{pmatrix} a_{11} & \cdots & a_{1n} \\ \vdots & \ddots & \vdots \\ a_{m1} & \cdots & a_{mm} \end{pmatrix}.$$

Then there exists a nonzero minor of order s. Therefore, either there does not exist any minor of order $s+1$ or all minors of order $s+1$ are zero. Without loss of generality, we may assume that there exists a quadratic submatrix F of A which has type (s, s) and $|F| \neq 0$, where F has the form

$$F = \begin{pmatrix} a_{11} & \cdots & a_{1s} \\ a_{21} & \cdots & a_{2s} \\ \vdots & & \vdots \\ a_{s1} & \cdots & a_{ss} \end{pmatrix} \quad \text{with} \quad \begin{vmatrix} a_{11} & \cdots & a_{1s} \\ a_{21} & \cdots & a_{2s} \\ \vdots & & \vdots \\ a_{s1} & \cdots & a_{ss} \end{vmatrix} \neq 0.$$

By Theorem 6.3.12, $\operatorname{rank}(F) = s$ and then $s \leq r = \operatorname{rank}(A)$. Because of $|F| = |F^T| \neq 0$ (Proposition 6.3.4), the row-rank ρ of A satisfies also $s \leq \rho$.

We will prove that the row-vectors $a^{(1)} = (a_{11}, a_{12}, \ldots, a_{1n}), \ldots, a^{(s)} = (a_{s1}, a_{s2} \ldots, a_{sn})$ are linearly independent and that any row-vector $a^{(i)}$ is a linear combination of $a^{(1)}, \ldots, a^{(s)}$.

If the row-vectors $a^{(1)}, \ldots, a^{(s)}$ were not linearly independent, there would exist $k_1, \ldots, k_s \in K$, not all zero, such that

$$k_1 a^{(1)} + \cdots + k_s a^{(s)} = 0.$$

Hence, also the row-vectors in F would be linearly dependent, a contradiction.

Now consider any $i, s+1 \leq i \leq m$. We will prove that the row-vectors $a^{(i)}$ are linear combinations of $a^{(1)}, \ldots, a^{(s)}$. Let $l \in \{s+1, \ldots, n\}$. Consider the quadratic matrices $F_{s+1}^{(l)}$ of type $(s+1, s+1)$,

$$F_{s+1}^{(l)} = \begin{pmatrix} a_{11} & a_{12} & \dots & a_{1s} & a_{1l} \\ a_{21} & a_{22} & \dots & a_{2s} & a_{2l} \\ \vdots & \vdots & \ddots & \vdots & \vdots \\ a_{s1} & a_{s2} & \dots & a_{ss} & a_{sl} \\ a_{i1} & a_{i2} & \dots & a_{is} & a_{il} \end{pmatrix}.$$

Since s is the largest order of nonzero minors of A, we have $|F_{s+1}^{(l)}| = 0$. We consider the following system of linear equations:

$$\text{(Eq)} \qquad \begin{aligned} a_{11}x_1 + \dots + a_{s1}x_s &= a_{i1}, \\ &\vdots \\ a_{1s}x_1 + \dots + a_{ss}x_s &= a_{is}. \end{aligned}$$

The coefficient matrix of the system Eq is F and since $|F| \neq 0$, by Cramér's rule, Eq is uniquely solvable and the unique solution is $(x_1^0, x_2^0 \dots, x_s^0)$.

Subtracting from the last row of $F_{s+1}^{(l)}$ the x_1^0-multiple of the first row, then subtracting the x_2^0-multiple of the second row, etc., and finally subtracting from the last row the x_s^0-multiple of the sth row gives

$$F_{s+1}^{(l)} = \begin{pmatrix} a_{11} & a_{12} & \dots & a_{1s} & a_{1l} \\ a_{21} & a_{22} & \dots & a_{2s} & a_{2l} \\ \vdots & \vdots & \ddots & \vdots & \vdots \\ a_{s1} & a_{s2} & \dots & a_{ss} & a_{sl} \\ 0 & 0 & \dots & 0 & b \end{pmatrix},$$

where

$$b = a_{il} - (a_{1l}x_1^0 + a_{2l}x_2^0 + \dots + a_{sl}x_s^0).$$

We calculate $|F_{s+1}^{(l)}|$ by expansion using the last row (Proposition 6.3.10) and obtain

$$|F_{s+1}^{(l)}| = b \cdot |F| = [a_{il} - (a_{1l}x_1^0 + \dots + a_{sl}x_s^0)] \cdot |F| = 0,$$

because $|F_{s+1}^{(l)}| = 0$. Since $|F| \neq 0$, we have

$$a_{il} = a_{1l}x_1^0 + \dots + a_{sl}x_s^0, \quad l = r+1, \dots, n.$$

Then, after replacing x_i by x_i^0, $i = 1, \dots, s$, the system Eq of linear equations can be completed to

$$\begin{aligned} x_1^0 a_{11} + \dots + x_s^0 a_{s1} &= a_{i1}, \\ &\vdots \\ x_1^0 a_{1n} + \dots + x_s^0 a_{sn} &= a_{in}, \end{aligned}$$

or

$$\alpha^{(i)} = x_1^0 \alpha^{(1)} + \cdots + x_s^0 \alpha^{(s)}.$$

This means that $\{\alpha^{(1]}, \dots \alpha^{(s)}\}$ forms a basis, i. e., a maximal linearly independent system of vectors and therefore $\rho \leq s$, and altogether $\rho = s$.

2. Repeating this proof and exchanging "row-rank ρ" by "$\text{rank}(A) = r$," we obtain

$$r = \text{rank}(A) = s.$$

3. Then we also get $\rho = r$. □

6.6 Historical remarks

The method of Gaussian elimination is very old and first appeared between 150 BC and AD 179 in the Chinese mathematics. In Chapter 8 of the Chinese book entitled "Chiu Chang Suan Ching," ("Nine chapters of the mathematical art"), a systematic method of solving linear equation systems was presented. During the Han dynasty, mathematicians considered problems leading to the linear equation system

$$\begin{aligned} 3x + 2y + z &= 39, \\ 2x + 3y + z &= 34, \\ x + 2y + 3z &= 26, \end{aligned}$$

written as a matrix equation with the coefficients as entries of the matrices. The solution was given as a matrix transformation. During this period, also matrices with negative numbers as entries were used. The method was commented by Liu Hui (225–295). Since several generations of Chinese students had to learn this text word by word, from this time on the method belonged to the traditions of Chinese mathematics. In Europe, the method of solving linear equation systems appeared in the late Renaissance. In 1670, Isaac Newton (1643–1727) complained that no algebra book which he knew contained methods for solving simultaneous equations which Newton then supplied and which was published in "Arihmetica Universalis" long time after Newton had finished his academic carrier. Arthur Cayley (1821–1895) was interested in calculations with matrices and noticed the commutativity of matrix addition and the noncommutativity of matrix multiplication. For more details, see, e. g., [59].

7 Universal algebra

In Chapter 2 we introduced several concrete algebraic structures: natural numbers, integers, rational numbers, and matrices. The sets $\mathbb{N}, \mathbb{Z}$, and $\mathbb{Q}$ form semigroups with respect to addition and multiplication. In Chapter 6 we proved that the matrices $\mathbb{Q}_{m,n}$ of type (m, n) form the semigroup $(\mathbb{Q}_{m,n}; +)$. The quadratic matrices $\mathbb{Q}_{n,n}$ form a semigroup with respect to matrix multiplication. Also $(\mathbb{N}; +, \cdot)$ is a semiring, while $(\mathbb{Z}; +, \cdot)$ and $(\mathbb{Q}; +, \cdot)$ are rings. Quadratic matrices over the rational numbers also form a ring $(\mathbb{Q}_{n,n}; +, \cdot)$. The integers form a group $(\mathbb{Z}; +)$ with respect to addition, and the rational numbers different from 0 form a multiplicative group $(\mathbb{Q} \setminus \{0\}; \cdot)$. The nonnegative rational numbers $\mathbb{Q}^+$ form the semifield $(\mathbb{Q}^+; +, \cdot)$, while $(\mathbb{Q}; +, \cdot)$ is a field. All these algebraic structures form universal algebras, the main topic of this chapter. We will introduce the concept of a universal algebra and study its properties. We define subalgebras, homomorphic images, prove the homomorphic image theorem as an application of the so-called diagram lemma (Theorem 7.4.17). Direct and subdirect products of universal algebras are constructions which lead to new algebras, constructed from given ones. Vector spaces are important examples of many-sorted (heterogeneous) algebras which extend the concept of a universal algebra. We generalize the concepts of subalgebras, homomorphic images, direct and subdirect products to this class of algebraic structures. Structures with operations which are not everywhere defined are called partial algebras. Since such algebras are useful in the study of production systems, we will define some basic concepts for partial algebras. Directed graphs, i. e., sets together with a binary relation defined on these sets, are special relational structures which we will generalize to algebraic systems that are sets together with indexed sets of operations, as well as indexed sets of relations. It turns out that the main topics of this book, i. e., numbers, graphs, and matrices, form algebraic structures which belong to the different classes studied in this chapter.

https://doi.org/10.1515/9783112234433-007

7.1 The concept of an algebra

An *algebra* is a nonempty set, the *base set*, *universe*, or *carrier set* of the algebra together with a set of *finitary* operations defined on this set. An n-ary operation, $n \geq 1$, is a function $f^A : A^n \to A$ from the nth Cartesian power of A into A. Let $O^n(A)$ be the set of all n-ary operations defined on A and let $O(A) := \bigcup_{n=1}^{\infty} O^n(A)$ be the set of all finitary operations defined on A.

Remark 7.1.1.

1. Any n-ary operation f^A on A can be regarded as an $(n+1)$-ary relation defined on A, called graph of f^A. The graph of f^A is defined as

$$\{(a_1, \ldots, a_n, a_{n+1}) \in A^{n+1} \mid f(a_1, \ldots, a_n) = a_{n+1}\}.$$

2. If $B \subseteq A$ is a nonempty subset of A, then the restriction $f^A|B$ of f^A onto B is defined by

$$(f^A|B)(a_1, \ldots, a_n) := f^A(a_1, \ldots, a_n) \quad \text{for all } a_1, \ldots, a_n \in B.$$

3. The definition of an n-ary operation may be extended to the special case that $n = 0$, i. e., for a nullary operation. At first, we define $A^0 := \{\emptyset\}$. A nullary operation is then a function $f^A : \{\emptyset\} \to A$. Thus, a nullary operation is uniquely determined by the element $f^A(\emptyset) \in A$. For every element $a \in A$, there is exactly one nullary operation with ${f^A}_a(\emptyset) = a$. If $A = \emptyset$ then there are no nullary operations on A.

The universe A can be finite or infinite. Since all sets of economical data are finite, operations on finite sets and finite algebras should play an important role in economics.

In Section 1.2 we introduced disjunction, conjunction, implication, and equivalence as binary operations on the two-element set $\{T, F\}$ of truth values:

$$\wedge, \vee, \Rightarrow, \Leftrightarrow: \{T, F\}^2 \to \{T, F\}.$$

If we identify the truth-value T with 1 and F with 0, then $\wedge, \vee, \Rightarrow, \Leftrightarrow \in O^2(\{0,1\})$. The functions from $O^n(\{0,1\})$ are said to be n-ary Boolean functions. The negation $\neg$ is a unary Boolean function which maps 0 to 1 and 1 to 0. Boolean functions can be given by tables, for instance,

	¬
0	1
1	0

∧	0	1
0	0	0
1	0	1

∨	0	1
0	0	1
1	1	1

⇒	0	1
0	1	1
1	0	1

⇔	0	1
0	1	0
1	0	1

There are two ways to view an algebra, in the nonindexed form as a pair $\mathcal{A} = (A; F^A)$ consisting of a set A and a set of operations defined on A or as an indexed algebra.

Definition 7.1.2. Let A be a nonempty set. Let I be some nonempty index set, and let $(f_i^A)_{i\in I}$ be a function which assigns to every element of I an n_i-ary operation f_i^A defined on A. Then the pair $\mathcal{A} = (A; (f_i^A)_{i\in I})$ is called an *(indexed) algebra* (indexed by the set I). The set A is called the *base, carrier set*, or *universe* of $\mathcal{A}$, and $(f_i^A)_{i\in I}$ is called the *sequence of fundamental operations* of $\mathcal{A}$. For each $i \in I$, the natural number n_i is called the *arity* of f_i^A. The sequence $\tau := (n_i)_{i\in I}$ of all the arities is called the *type* of the algebra $\mathcal{A}$. We use the name $\mathrm{Alg}(\tau)$ for the class of all algebras of a given type τ.

In our definition we do not allow the base set A of an algebra to be the empty set. It is possible to define an empty algebra with the empty set as universe if no nullary operations belong to the type. But mostly we will exclude this case. In the indexed case, the fundamental operations form a sequence, not a set, which means that repetitions are possible. We will give some examples for our definition.

7.2 Examples

An algebra $\mathcal{G} = (G; \cdot)$ of type (2), i. e., with one binary operation, is called a *groupoid*. The binary operation here is denoted by $\cdot$, and we will write $x \cdot y$, or xy for short. If the binary operation is commutative, the groupoid is said to be *abelian*. If the binary operation is associative, then the groupoid is called a *semigroup*. A semigroup with an additional nullary operation e is called a *monoid* $\mathcal{M} = (M; \cdot, e)$ if e is an identity element with respect to the multiplication $\cdot$, i. e., if for all $x \in M$ the equations $x \cdot e = e \cdot x = x$ are satisfied. Monoids are algebras of type (2, 0).

A group $\mathcal{G} = (G; \cdot)$ is a semigroup, which satisfies the axiom

$$\forall a, b \in G\, \exists x, y \in G\, (a \cdot x = b \text{ and } y \cdot a = b) \quad \text{(invertibility)}.$$

A group can also be regarded as an algebra $\mathcal{G} = (G; \cdot, ^{-1}, e)$ of type (2, 1, 0), where the associative law and the axioms

$$\forall x \in G\, (x \cdot x^{-1} = x^{-1} \cdot x = e)$$

and

$$\forall x \in G (x \cdot e = e \cdot x = x)$$

are satisfied.

An algebra $\mathcal{R} = (R; +, \cdot)$ of type (2, 2) is called a *semiring* (see Chapter 3, Definition 3.1.12) if both binary operations are associative, the addition is commutative, the *distributive laws*

$$\forall x, y, z \in R\, (x \cdot (y + z) = x \cdot y + x \cdot z)$$

and

$$\forall x,y,z \in R\,((x+y)\cdot z = x\cdot z + y\cdot z)$$

are satisfied, and if there exists a zero element 0 with $a + 0 = a$ and $0 \cdot a = a \cdot 0 = 0$ for all $a \in R$. We notice that there are different ways to define a semiring, sometimes commutativity of addition and the existence of a zero element are not requested.

Rings are semirings with invertible addition and a commutative ring $\mathcal{K} = (K; +, \cdot)$ is called a *field* if $(K \setminus \{0\}; \cdot)$ is a group, where 0 is the zero element, i. e., the neutral element with respect to addition.

Example 7.2.1. An algebra $\mathcal{V} = (V; \wedge, \vee)$ of type $(2,2)$ is called a *lattice* if the following equations are satisfied by its binary operations, which are usually called join and meet operations:

(V1) $\forall x,y \in V\,(x \vee y = y \vee x)$,
(V1′) $\forall x,y \in V\,(x \wedge y = y \wedge x)$,
(V2) $\forall x,y,z \in V\,(x \vee (y \vee z) = (x \vee y) \vee z)$,
(V2′) $\forall x,y,z \in V\,(x \wedge (y \wedge z) = (x \wedge y) \wedge z)$,
(V3) $\forall x \in V\,(x \vee x = x)$,
(V3′) $\forall x \in V\,(x \wedge x = x)$ (idempotency),
(V4) $\forall x,y \in V\,(x \vee (x \wedge y) = x)$,
(V4′) $\forall x,y \in V\,(x \wedge (x \vee y) = x)$ (absorption laws).

If, in addition, the lattice satisfies the following *distributive laws*:

(V5) $\forall x,y,z \in V\,(x \wedge (y \vee z) = (x \wedge y) \vee (x \wedge z))$,
(V5′) $\forall x,y,z \in V\,(x \vee (y \wedge z) = (x \vee y) \wedge (x \vee z))$,

then the lattice is said to be *distributive*. We remark that (*V*3) and (*V*3′) follow from (*V*4) and (*V*4′), respectively.

Lattices are important both as examples of a kind of algebra, and also in the study of all other kinds of algebras, too, since any algebra turns out to have some lattices associated with it. We shall study such associated lattices in the coming sections. For now, we show how lattices are connected to partially ordered sets. A binary relation $\leq$ on a set A is called a partial order on A if it is reflexive, antisymmetric, and transitive. There is a close connection between lattices and partially ordered sets, in the sense that each determines the other, as the following theorem shows.

Theorem 7.2.2. *Let $(V; \leq)$ be a partially ordered set in which for all $x,y \in V$ both the infimum $\bigwedge\{x,y\}$ and supremum $\bigvee\{x,y\}$ exist. Then the binary infimum and supremum operations make $(V; \wedge, \vee)$ a lattice. Conversely, every lattice defines a partially ordered set in which for all x,y the infimum $\bigwedge\{x,y\}$ and supremum $\bigvee\{x,y\}$ exist.*

Proof. Let $(V; \leq)$ be a partially ordered set, in which for any two elements $x,y \in V$ the infimum $\bigwedge\{x,y\}$ and supremum $\bigvee\{x,y\}$ exist. If we put $x \wedge y = \bigwedge\{x,y\}$ and $x \vee y = \bigvee\{x,y\}$, then the required identities (V1)–(V4) and (V1′)–(V4′) are easy to verify.

If, conversely, $(V; \wedge, \vee)$ is a lattice, then we define

$$x \le y :\Leftrightarrow x \wedge y = x,$$

and show that this gives a partial order relation on L. Reflexivity follows from (V3′), $x \wedge x = x$; for antisymmetry, we see that

$$(x \le y) \wedge (y \le x) \;\Leftrightarrow\; (x \wedge y = x) \wedge (y \wedge x = y),$$

and so from commutativity we get

$$x = x \wedge y = y \wedge x = y;$$

and transitivity follows from $(x \wedge y = x) \wedge (y \wedge z = y) \Rightarrow x = x \wedge y = x \wedge (y \wedge z) = (x \wedge y) \wedge z = x \wedge z \Leftrightarrow x \le z$. It is easy to check that $\bigwedge\{x, y\} = x \wedge y$ and $\bigvee\{x, y\} = x \vee y$ are satisfied. □

A lattice (or partially ordered set) V in which for all sets $B \subseteq L$ the infimum $\bigwedge B$ and supremum $\bigvee B$ exist is called a *complete lattice*. Obviously, any finite lattice is complete.

A *bounded lattice* $(V; \wedge, \vee, 0, 1)$ is an algebra of type $(2, 2, 0, 0)$ which is a lattice, with two additional nullary operations 0 and 1, which satisfy

(V6) $\forall x \in V\ (x \wedge 0 = 0)$ and

(V7) $\forall x \in V\ (x \vee 1 = 1)$.

These axioms tell us that in the partial order determined by the lattice, 0 acts as the least element and 1 as the greatest element.

An algebra $\mathcal{S} = (S; \cdot)$ of type (2) is called a *semilattice*, if the operation $\cdot$ is an associative, commutative, and idempotent (satisfies $x \cdot x = x$) binary operation on S. This means that a semilattice is a particular kind of semigroup.

An algebra $\mathcal{B} = (B; \wedge, \vee, \neg, 0, 1)$ of type $(2, 2, 1, 0, 0)$ is called a *Boolean algebra*, if $(B; \wedge, \vee, 0, 1)$ is a bounded distributive lattice with an additional unary operation called complementation, which also satisfies two identities:

(B1) $\forall x \in B\ (x \wedge \neg x = 0)$ and

(B2) $\forall x \in B\ (x \vee \neg x = 1)$ (complement laws).

As an example of a Boolean algebra, we recall the set of the binary Boolean functions $\wedge, \vee, \neg, 0, 1$ on the two-element set $\{0, 1\}$. This gives us the algebra $(\{0, 1\}; \wedge, \vee, \neg, 0, 1)$, usually denoted by 2_B. Another example of a Boolean algebra is the power set of a set A which gives the Boolean algebra $(\mathcal{P}(A); \cap, \cup, \emptyset, A)$ with intersection, union, complement, the empty set, and set A as operations.

7.3 Subalgebras

The set of all symmetry mappings of an equilateral triangle Φ described by permutations on the set $\{1, 2, 3\}$ is

$$D(\Phi) = \{(1), (123), (132), (12), (13), (23)\}.$$

The permutations (12), (13), and (23) correspond to reflections over the lines g_1, g_3, and g_2, respectively; and (1), (123), (132) correspond to rotations around the point O by 0, 120 and 240 degrees, respectively (Figure 7.1). We know already from Section 1.7 that $D(\Phi)$ forms a group with respect to the composition of these mappings, that is, with respect to the multiplication of the corresponding permutations.

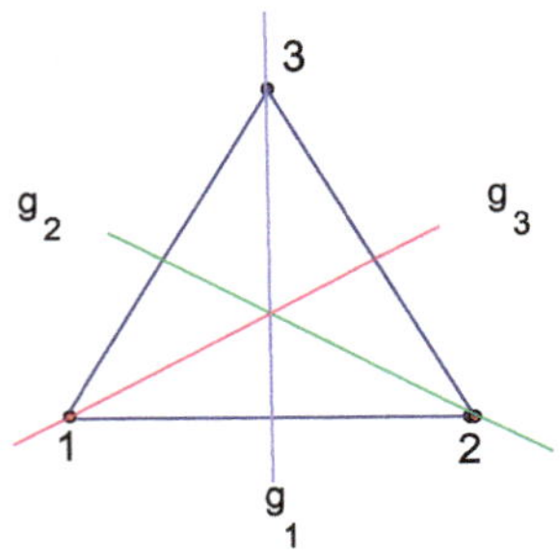

Figure 7.1: Symmetry images of an equilateral triangle.

Since the composition of two rotations around the point O is again a rotation around O, the operation preserves the subset of $D(\Phi)$ consisting of all rotations around O. Therefore, this subset is closed under the operation. This closure can also be seen from the multiplication table of the corresponding permutations:

$\circ$	(1)	(123)	(132)
(1)	(1)	(123)	(132)
(123)	(123)	(132)	(1)
(132)	(132)	(1)	(123)

As we can see, the composition of reflections over different lines gives a permutation which is not a reflection. Therefore, the set of reflections does not have this closure property. Our observation leads to the concept of a subalgebra.

Definition 7.3.1. Let $\mathcal{B} = (B; (f_i^B)_{i\in I})$ be an algebra of type τ. Then an algebra $\mathcal{A}$ is called a *subalgebra* of $\mathcal{B}$, written as $\mathcal{A} \subseteq \mathcal{B}$, if the following conditions are satisfied:

(i) $\mathcal{A} = (A; (f_i^A)_{i\in I})$ is an algebra of type τ;

(ii) $A \subseteq B$;

(iii) $\forall i \in I$, the graph of f_i^A is a subset of the graph of f_i^B.

Whether one algebra is a subalgebra of another can be checked by the following criterion:

Lemma 7.3.2 (Subalgebra criterion). *Let $\mathcal{B} = (B; (f_i^B)_{i\in I})$ be an algebra of type τ and let $A \subseteq B$ be a subset of B for which $f_i^A = f_i^B|A$ for all $i \in I$. Then $\mathcal{A} = (A, (f_i^A)_{i\in I})$ is a subalgebra of $\mathcal{B} = (B, (f_i^B)_{i\in I})$ iff A is closed with respect to all the operations f_i^B for $i \in I$, that is, iff $f_i^B(A^{n_i}) \subseteq A$ for all $i \in I$.*

Proof. Assume that $\mathcal{A} \subseteq \mathcal{B}$. Then $f_i^A = f_i^B|A$ is an operation in A for every $i \in I$, and so for all $(a_1, \ldots, a_{n_i}) \in A^{n_i}$ we have

$$f_i^B(a_1, \ldots, a_{n_i}) = f_i^A(a_1, \ldots, a_{n_i}) \in A.$$

Thus the application of any operation f_i^B of $\mathcal{B}$ to elements of A always gives elements of A.

If, conversely, A is closed with respect to f_i^B for all $i \in I$, then we have $f_i^B|A^{n_i} \subseteq A^{n_i} \times A$. This makes each $f_i^B|A^{n_i}$ an operation on A^{n_i} in A, and all the conditions of the definition are satisfied. □

The subalgebra relation has the following properties.

Corollary 7.3.3. *Let $\mathcal{A}$, $\mathcal{B}$, and $\mathcal{C}$ be algebras of type τ. Then:*
(i) $(\mathcal{A} \subseteq \mathcal{B}) \wedge (\mathcal{B} \subseteq \mathcal{C}) \Rightarrow \mathcal{A} \subseteq \mathcal{C}$;
(ii) $(A \subseteq B \subseteq C) \wedge (\mathcal{A} \subseteq \mathcal{C}) \wedge (\mathcal{B} \subseteq \mathcal{C}) \Rightarrow \mathcal{A} \subseteq \mathcal{B}$.

Proof. (i) For all $i \in I$, we have $f_i^A \subseteq f_i^B \subseteq f_i^C$ for the corresponding graphs, and thus $f_i^A \subseteq f_i^C$.

(ii) Applying the subalgebra criterion to an arbitrary n_i-tuple $(a_1, \ldots, a_{n_i}) \in A^{n_i}$, we have $f_i^B(a_1, \ldots, a_{n_i}) = f_i^C(a_1, \ldots, a_{n_i}) = f_i^A(a_1, \ldots, a_{n_i}) \in A$. Therefore, A is closed with respect to all operations of $\mathcal{B}$. □

If $\mathcal{B} = (B; (f_i^B)_{i\in I})$ is an algebra of type τ and if $\{\mathcal{A}_j \mid j \in J\}$ is a family of subalgebras of $\mathcal{B}$ with the nonempty intersection $A := \bigcap_{j\in J} A_j$ of its universes, then it is easy to see that $A = \bigcap_{j\in J} A_j$ is the universe of a subalgebra of $\mathcal{B}$ which is called the *intersection* of the family $\{\mathcal{A}_j \mid j \in J\}$. This allows us to consider the subalgebra

$$\langle X\rangle_{\mathcal{B}} := \bigcap\{\mathcal{A} \mid \mathcal{A} \subseteq \mathcal{B} \text{ and } X \subseteq A\}$$

of $\mathcal{B}$ generated by a subset $X \subseteq B$ of the universe. The set X is called a *generating system* of this algebra. The process of subalgebra generation satisfies three very important properties, called the *closure properties*. This makes the subalgebra generation process an example of a *closure operator*.

Example 7.3.4. As an example, we consider the group

$$\mathcal{Z}_6 = (\{[0]_6, [1]_6, [2]_6, [3]_6, [4]_6, [5]_6\}; +, -, [0]_6) \quad \text{(see Section 3.6).}$$

Here we have:

$$\begin{aligned}
\langle\{[0]_6\}\rangle_{\mathcal{Z}_6} &= (\{[0]_6\}; +, -, [0]_6), \quad \langle\{[1]_6\}\rangle_{\mathcal{Z}_6} = \langle\{[5]_6\}\rangle_{\mathcal{Z}_6} = \mathcal{Z}_6,\\
\langle\{[3]_6\}\rangle_{\mathcal{Z}_6} &= (\{[0]_6, [3]_6\}; +, -, [0]_6),\\
\langle\{[2]_6\}\rangle_{\mathcal{Z}_6} = \langle\{[4]_6\}\rangle_{\mathcal{Z}_6} &= (\{[0]_6, [2]_6, [4]_6\}; +, -, [0]_6),\\
\langle\{[2]_6, [3]_6\}\rangle_{\mathcal{Z}_6} = \langle\{[4]_6, [3]_6\}\rangle_{\mathcal{Z}_6} &= \mathcal{Z}_6.
\end{aligned}$$

Theorem 7.3.5. *Let $\mathcal{B}$ be an algebra. For all subsets X and Y of B, the following closure properties hold:*
(i) $X \subseteq \langle X\rangle_{\mathcal{B}}$ *(extensivity);*
(ii) $X \subseteq Y \Rightarrow \langle X\rangle_{\mathcal{B}} \subseteq \langle Y\rangle_{\mathcal{B}}$ *(monotonicity);*
(iii) $\langle X\rangle_{\mathcal{B}} = \langle\langle X\rangle_{\mathcal{B}}\rangle_{\mathcal{B}}$ *(idempotency).*

Proof. These properties follow immediately from the properties of the operator $\langle\cdot\rangle$; we leave the details to the reader. □

An important problem is, for a given algebra $\mathcal{B}$ of type τ and a subset X of B, to determine all elements of the carrier set of the subalgebra of $\mathcal{B}$ which is generated by X. To do this, we set

$$E(X) := X \cup \{f_i^B(a_1, \ldots, a_{n_i}) \mid i \in I,\ a_1, \ldots, a_{n_i} \in X\}.$$

Then we inductively define $E^0(X) := X$, and $E^{k+1}(X) := E(E^k(X))$, for all $k \in \mathbb{N}$. With this notation, we can describe the subalgebra.

Theorem 7.3.6. *For any algebra $\mathcal{B}$ of type τ and for any nonempty subset $X \subseteq B$, we have $\langle X\rangle_{\mathcal{B}} = \bigcup_{k=0}^{\infty} E^k(X)$.*

For the proof of this useful theorem, see [21].

If the intersection of two subalgebras of $\mathcal{B}$ is nonempty, then it is again a subalgebra of $\mathcal{B}$, and we can use this fact to define a binary operation on the set $\mathrm{Sub}(\mathcal{B})$ of all subalgebras of $\mathcal{B}$. A binary operation $\wedge$ on $\mathrm{Sub}(\mathcal{B})$ is defined by

$$\wedge : (\mathcal{A}_1, \mathcal{A}_2) \mapsto \mathcal{A}_1 \wedge \mathcal{A}_2 := \mathcal{A}_1 \cap \mathcal{A}_2.$$

Unfortunately, this operation is not defined on all of $\mathrm{Sub}(\mathcal{B})$ because it does not deal with the case when $A_1 \cap A_2 = \emptyset$. We can solve this problem by allowing the empty set to be considered as an algebra.

It is easy to see that the union $A_1 \cup A_2$ of the universes of two subalgebras of $\mathcal{B}$ is, in general, not a subalgebra of $\mathcal{B}$. But we can map the pair $(\mathcal{A}_1, \mathcal{A}_2)$ to the subalgebra generated by the union,

$$\vee : (\mathcal{A}_1, \mathcal{A}_2) \mapsto \mathcal{A}_1 \vee \mathcal{A}_2 := \langle A_1 \cup A_2 \rangle_{\mathcal{B}}.$$

Altogether, we obtain

Theorem 7.3.7. *For every algebra $\mathcal{B}$, the algebra* $(\mathrm{Sub}(\mathcal{B}); \wedge, \vee)$ *is a lattice, called the subalgebra lattice of $\mathcal{B}$.*

Proof. Here $\wedge$ and $\vee$ are binary operations on $\mathrm{Sub}(\mathcal{B})$. It is easy to check (V1)–(V4) and (V1′)–(V4′). □

7.4 Homomorphisms

Consider a function which maps a square matrix A to its determinant $|A|$. In Chapter 6, Section 6.3, we proved that

$$|A \cdot B| = |A| \cdot |B|,$$

where the multiplication on the left-hand side is matrix multiplication, and on the right-hand side we multiply two real numbers. On the left-hand side, we multiply first and map the result to the set $\mathbb{R}$ of real numbers, and on the right-hand side we first map each matrix to its determinant and then multiply the resulting real numbers. The equality shows that the result is independent of the order of multiplying and mapping.

Considering the mapping sgn which maps each permutation to its signature, i. e., either 1 or –1. We proved that

$$\mathrm{sgn}(s_1 \circ s_2) = \mathrm{sgn}(s_1) \cdot \mathrm{sgn}(s_2).$$

On the left-hand side, we multiply both permutations and map the product to $\{1, -1\}$. On the right-hand side, we assign to each s_1, s_2 their signatures and multiply them corresponding to the table

·	1	–1
1	1	–1
–1	–1	1

Again we can exchange the mapping with the operation and obtain the same result.

This compatibility property can be seen also in very "practical" situations. An industrial automaton is designed to construct from single parts the body of a car. If the parts are combined in a different order, represented by a mapping h, then the automaton will continue to work in the correct way, provided the mapping h has the compatibility property observed before (Figure 7.2).

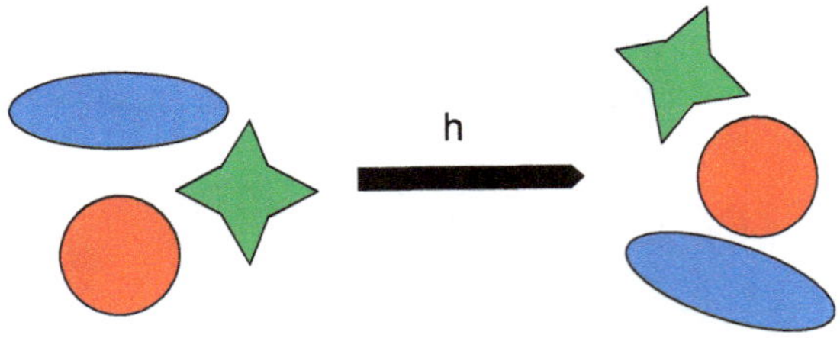

Figure 7.2: Artificial intelligence.

We now generalize our observations.

Definition 7.4.1. Let $\mathcal{A} = (A; (f_i^A)_{i \in I})$ and $\mathcal{B} = (B; (f_i^B)_{i \in I})$ be algebras of the same type τ. Then a function $h : A \to B$ is called a *homomorphism* $h : \mathcal{A} \to \mathcal{B}$ of $\mathcal{A}$ into $\mathcal{B}$ if for all $i \in I$ we have

$$h(f_i^A(a_1, \dots, a_{n_i})) = f_i^B(h(a_1), \dots, h(a_{n_i})),$$

for all $a_1, \dots, a_{n_i} \in A$. In the special case when $n_i = 0$, this equation means that $h(f_i^A(\emptyset)) = f_i^B(\emptyset)$. That is, the element designated by the nullary operation f_i^A in A must be mapped to the corresponding element f_i^B in B.

If the function h is bijective, that is, both one-to-one (injective) and onto (surjective), then the homomorphism $h : \mathcal{A} \to \mathcal{B}$ is called an *isomorphism* from $\mathcal{A}$ onto $\mathcal{B}$. An injective homomorphism from $\mathcal{A}$ into $\mathcal{B}$ is also called an *embedding* of $\mathcal{A}$ into $\mathcal{B}$.

A homomorphism $h : \mathcal{A} \to \mathcal{A}$ of an algebra $\mathcal{A}$ into itself is called an *endomorphism* of $\mathcal{A}$, and an isomorphism $h : \mathcal{A} \to \mathcal{A}$ from A onto A is called an *automorphism* of $\mathcal{A}$.

Example 7.4.2.
1. For every algebra $\mathcal{A}$, the identity mapping $\mathrm{id}_A : \mathcal{A} \to \mathcal{A}$, defined by $\mathrm{id}_A(x) = x$ for all $x \in A$, is an automorphism of $\mathcal{A}$.
2. Let $\mathcal{A}$, $\mathcal{B}$, and $\mathcal{C}$ be algebras of the same type, and let $h_1 : \mathcal{A} \to \mathcal{B}$ and $h_2 : \mathcal{B} \to \mathcal{C}$ be homomorphisms. The composition function $h_2 \circ h_1 : \mathcal{A} \to \mathcal{C}$ is defined by $(h_2 \circ h_1)(x) = h_2(h_1(x))$ for all $x \in A$. This composition is also a homomorphism, and when both h_1 and h_2 are surjective, injective, or bijective, then the composition has the same property.

We consider the behavior of subalgebras under homomorphic mappings.

Theorem 7.4.3. *Let $h : \mathcal{A} \to \mathcal{B}$ be a homomorphism of the algebra $\mathcal{A}$ of type τ into the algebra $\mathcal{B}$ of type τ. Then we have:*
(i) *The image $\mathcal{B}_1 = h(\mathcal{A}_1)$ of a subalgebra $\mathcal{A}_1$ of $\mathcal{A}$ under the homomorphism h is a subalgebra of $\mathcal{B}$.*
(ii) *The preimage $h^{-1}(\mathcal{B}') = \mathcal{A}'$ of a subalgebra $\mathcal{B}'$ of $h(\mathcal{A}) \subseteq \mathcal{B}$ is a subalgebra of $\mathcal{A}$.*
(iii) *For any subset $X \subseteq A$, we have $\langle h(X) \rangle_{\mathcal{B}} = h(\langle X \rangle_{\mathcal{A}})$.*

Proof. (i) By definition,

$$h(A_1) = \{b \in B \mid \exists a \in A_1\ (h(a) = b)\} \subseteq B.$$

Let f_i^B be an n_i-ary operation on $\mathcal{B}$, for $i \in I$, and let $(b_1, \ldots, b_{n_i}) \in h(A_1)^{n_i}$. Then for each $1 \le j \le n_i$, we have $b_j = h(a_j)$ for some a_j in A. Then

$$f_i^B(b_1, \ldots, b_{n_i}) = f_i^B(h(a_1), \ldots, h(a_{n_i})) = h(f_i^A(a_1, \ldots, a_{n_i})),$$

and the latter is in $h(A_1)$ since $f_i^A(a_1, \ldots, a_{n_i}) \in A_1$. An application of the subalgebra criterion thus proves that the image $\mathcal{B}_1$ is a subalgebra of $\mathcal{B}$.

(ii) Let $(a_1, \ldots, a_{n_i}) \in (h^{-1}(B'))^{n_i}$ and let f_i^A be n_i-ary. We have

$$h^{-1}(B') = \{a \in A \mid \exists b \in B'\ (h(a) = b)\};$$

so for each a_j in $h^{-1}(B')$, there is an element $b_j \in B'$ such that $a_j = h^{-1}(b_j)$. This means that

$$f_i^A(a_1, \ldots, a_{n_i}) = f_i^A(h^{-1}(b_1), \ldots, h^{-1}(b_{n_i})).$$

Then

$$h(f_i^A(h^{-1}(b_1), \ldots, h^{-1}(b_{n_i}))) = f_i^B(h(h^{-1}(b_1)), \ldots, h(h^{-1}(b_{n_i}))) = f_i^B(b_1, \ldots, b_{n_i}),$$

and this is in B' because all $b_1, \ldots, b_{n_i}$ are in B' and $\mathcal{B}'$ is a subalgebra of $\mathcal{B}$. From this, we get

$$f_i^A(h^{-1}(b_1), \ldots, h^{-1}(b_{n_i})) = f_i^A(a_1, \ldots, a_{n_i}) \in h^{-1}(B').$$

(iii) Let E be the operator used in the previous section, so

$$E(X) := X \cup \{f_i^A(a_1, \ldots, a_{n_i}) \mid i \in I,\ a_1, \ldots, a_{n_i} \in X\}.$$

We show that $E(h(X)) = h(E(X))$ for all $X \subseteq A$. The set $E(h(X))$ consists of all elements $h(y)$ with $y \in X$, plus elements of the form

$$f_i^B(h(y_1), \ldots, h(y_{n_i})), \quad \text{for } i \in I,\ y_1, \ldots, y_{n_i} \in X.$$

The set $h(E(X))$ also consists of the elements $h(y)$ with $y \in X$ plus the elements $h(f_i^A(y_1, \ldots, y_{n_i}))$, which agree with $f_i^B(h(y_1), \ldots, h(y_{n_i}))$.

By induction on k, we can prove that $E^k(h(X)) = h(E^k(X))$ for all $k \in \mathbb{N}$. Then we have

$$\langle h(X)\rangle_{\mathcal{B}} = \bigcup_{k=0}^{\infty} E^k(h(X)) = \bigcup_{k=0}^{\infty} h(E^k(X)) = h\left(\bigcup_{k=0}^{\infty} E^k(X)\right) = h(\langle X\rangle_{\mathcal{A}}).$$

□

It is interesting that the sets of automorphisms and of endomorphisms of an algebra can be equipped with operations to obtain algebras. The set of all automorphisms of the algebra $\mathcal{A}$ will be denoted by $\operatorname{Aut}\mathcal{A}$, and the set of all its endomorphisms is denoted by $\operatorname{End}\mathcal{A}$. Since the composition $\circ$ of two automorphisms (endomorphisms) of $\mathcal{A}$ is again an automorphism (endomorphism) of $\mathcal{A}$ and $\circ$ is associative, $(\operatorname{Aut}\mathcal{A}; \circ, \mathrm{id}_A)$ and $(\operatorname{End}\mathcal{A}; \circ, \mathrm{id}_A)$ are monoids. Clearly, any automorphism of $\mathcal{A}$ is also an endomorphism and it is not difficult to prove that $(\operatorname{Aut}\mathcal{A}; \circ, \mathrm{id}_A)$ is a submonoid of $(\operatorname{End}\mathcal{A}; \circ, \mathrm{id}_A)$. If $h : \mathcal{A} \to \mathcal{A}$ is an automorphism of $\mathcal{A}$, then $h^{-1} : \mathcal{A} \to \mathcal{A}$ is again an automorphism of $\mathcal{A}$. This can be checked as follows:

$$\begin{aligned} h^{-1}(f_i^A(b_1,\dots,b_{n_i})) &= h^{-1}(f_i^A(h(a_1),\dots,h(a_{n_i}))) \\ &= h^{-1}(h(f_i^A(a_1,\dots,a_{n_i}))) = f_i^A(a_1,\dots,a_{n_i}) \\ &= f_i^A(h^{-1}(b_1),\dots,h^{-1}(b_{n_i})). \end{aligned}$$

Altogether we obtain:

Lemma 7.4.4. *The set of all automorphisms of an algebra $\mathcal{A}$ forms a group,* $\operatorname{Aut}\mathcal{A} = (\operatorname{Aut}\mathcal{A}; \circ, {}^{-1}, \mathrm{id}_A)$, *called the automorphism group of $\mathcal{A}$.*

Let $\mathcal{A} = (A; (f_i^A)_{i\in I})$ be an algebra of type τ and let $h : \mathcal{A} \to \mathcal{A}$ be an automorphism of $\mathcal{A}$. Then $a \in A$ is called a *fixed point* of h if $h(a) = a$. Every $a \in A$ is a fixed point of the identity automorphism id_A on $\mathcal{A}$.

Lemma 7.4.5. *The set of all fixed points of an automorphism h of $\mathcal{A}$ is a subalgebra of $\mathcal{A}$.*

The proof is left as an exercise to the reader.

In Chapter 1, Section 1.7, we noticed already that the kernel of a function is an equivalence relation. This can be applied to homomorphisms. Let f_i^A, for some $i \in I$, be an n_i-ary fundamental operation of algebra $\mathcal{A}$ and let $(a_1, b_1), \dots, (a_{n_i}, b_{n_i})$ be in $\ker h$. This means that $h(a_1) = h(b_1), \dots, h(a_{n_i}) = h(b_{n_i})$. Now applying the operation f_i^B gives the equation

$$f_i^B(h(a_1),\dots,h(a_{n_i})) = f_i^B(h(b_1),\dots,h(b_{n_i})).$$

Since $h : \mathcal{A} \to \mathcal{B}$ is a homomorphism of $\mathcal{A}$ into $\mathcal{B}$, this gives us

$$h(f_i^A(a_1,\dots,a_{n_i})) = h(f_i^A(b_1,\dots,b_{n_i})),$$

and thus

$$(f_i^A(a_1,\dots,a_{n_i}), f_i^A(b_1,\dots,b_{n_i})) \in \ker h.$$

Therefore, the compatibility property of a homomorphism makes the kernel of this homomorphism compatible with the operations of the algebra. Equivalence relations on the universe of an algebra having this property are said to be *congruence relations* of the algebra.

Definition 7.4.6. Let A be a set, $\theta \subseteq A \times A$ an equivalence relation on A, and f an n-ary operation from $O^n(A)$. Then f is said to be *compatible* with θ, or to *preserve* θ, if for all $a_1, \dots, a_n, b_1, \dots, b_n \in A$,

$$(a_1, b_1) \in \theta, \ \dots, \ (a_n, b_n) \in \theta$$

implies

$$(f(a_1, \dots, a_n), f(b_1, \dots, b_n)) \in \theta.$$

Remark 7.4.7. We notice that this definition can also be applied to arbitrary relations. For instance, the monotone increasing real functions $f : \mathbb{R} \to \mathbb{R}$ are exactly the unary operations on $\mathbb{R}$ which are compatible with the relation $\leq$ on $\mathbb{R}$, since $x_1 \leq x_2 \Rightarrow f(x_1) \leq f(x_2)$ for all x_1, x_2 for which f is defined.

Definition 7.4.8. Let $\mathcal{A} = (A; (f_i^A)_{i\in I})$ be an algebra of type τ. An equivalence relation θ on A is called a *congruence relation* on $\mathcal{A}$ if all the fundamental operations f_i^A are compatible with θ. We denote by $\mathrm{Con}(\mathcal{A})$ the set of all congruence relations of the algebra $\mathcal{A}$. For every algebra $\mathcal{A} = (A; (f_i^A)_{i\in I})$, the trivial equivalence relations

$$\Delta_A := \{(a, a) \mid a \in A\} \quad \text{and} \quad \nabla_A = A \times A$$

are congruence relations. An algebra which has no congruence relations except for Δ_A and ∇_A is called *simple*.

Example 7.4.9.
1. Let $(\mathbb{Z}; +, \cdot, -, 0)$ be the ring of all integers and let $m \in \mathbb{Z}$ be an integer with $m \geq 0$. Then the congruence modulo m defined on $\mathbb{Z}$ is a congruence relation on $(\mathbb{Z}; +, \cdot, -, 0)$. For $m = 0$ or $m = 1$, we get the trivial relations on $\mathbb{Z}$.
2. On every set A, the constant operations $f_c : A^n \to A$, defined by $f_c(x_1, \dots, x_n) = c$ for all $x_1, \dots, x_n \in A$, and the identity map $\mathrm{id}_A : A \to A$ are compatible with all equivalence relations defined on A. The projections $e_i^n : A^n \to A$, defined by $e_i^n(a_1, \dots, a_n) = a_i$, for all $a_1, \dots, a_n \in A$ and $1 \leq i \leq n$, also have the same property.

To determine all congruence relations on $\mathcal{A}$, we consider all possible equivalence relations and look for those equivalence relations which are compatible with the fundamental operations of the algebra. Clearly, $A \times A$ and the diagonal $\Delta(A) := \{(a, a) \mid a \in A\}$ are always congruence relations of $\mathcal{A}$.

We want now to define lattice operations on the set $\mathrm{Con}(\mathcal{A})$ of all congruence relations of an algebra $\mathcal{A}$, as we did on the set $\mathrm{Sub}(\mathcal{A})$ for subalgebras. Our first observation here is the fact that the intersection of congruences is again a congruence.

Theorem 7.4.10. *The intersection* $\theta_1 \cap \theta_2$ *of two congruence relations on an algebra* $\mathcal{A} = (A; (f_i^A)_{i \in I})$ *is again a congruence relation on* $\mathcal{A}$.

Proof. Indeed, let $(a_1, b_1), \ldots, (a_{n_i}, b_{n_i})$ be pairs belonging to the intersection $\theta_1 \cap \theta_2$ and let f_i^A, for $i \in I$, be an arbitrary fundamental operation of $\mathcal{A}$. Then

$$(f_i^A(a_1, \ldots, a_{n_i}), f_i^A(b_1, \ldots, b_{n_i})) \in \theta_l, \quad \text{for } l = 1, 2,$$

and thus

$$(f_i^A(a_1, \ldots, a_{n_i}), f_i^A(b_1, \ldots, b_{n_i})) \in \theta_1 \cap \theta_2.$$

Therefore $\theta_1 \cap \theta_2$ is a congruence relation on $\mathcal{A}$. □

Remark 7.4.11. The latter theorem is also satisfied for arbitrary families of congruence relations on $\mathcal{A}$. But, in general, the union of two congruence relations of an algebra $\mathcal{A}$ is not a congruence relation, since this does not hold even for equivalence relations.

Although the union of two congruence relations θ_1 and θ_2 need not be a congruence relation, we can use intersections of congruences to define a smallest congruence generated by the union. This motivates the following definition.

Definition 7.4.12. Let $\mathcal{A}$ be an algebra, and let θ be a binary relation on A. We define the congruence relation $\langle\theta\rangle_{\mathrm{Con}(\mathcal{A})}$ on $\mathcal{A}$ generated by θ to be the intersection of all congruence relations θ' on $\mathcal{A}$ which contain θ:

$$\langle\theta\rangle_{\mathrm{Con}(\mathcal{A})} := \bigcap\{\theta' \mid \theta' \in \mathrm{Con}(\mathcal{A}) \text{ and } \theta \subseteq \theta'\}.$$

It is easy to see that $\langle\theta\rangle_{\mathrm{Con}(\mathcal{A})}$ has the three important properties of a closure operator:

$$\begin{aligned} &\theta \subseteq \langle\theta\rangle_{\mathrm{Con}(\mathcal{A})} && \text{(extensivity)},\\ &\theta_1 \subseteq \theta_2 \Rightarrow \langle\theta_1\rangle_{\mathrm{Con}(\mathcal{A})} \subseteq \langle\theta_2\rangle_{\mathrm{Con}(\mathcal{A})} && \text{(monotonicity)},\\ &\langle\langle\theta\rangle_{\mathrm{Con}(\mathcal{A})}\rangle_{\mathrm{Con}(\mathcal{A})} = \langle\theta\rangle_{\mathrm{Con}(\mathcal{A})} && \text{(idempotency)}. \end{aligned}$$

Remark 7.4.13. We ask how to derive, from a binary relation θ defined on A, the congruence relation of the algebra $\mathcal{A}$ generated by θ. First, we have to enlarge the relation θ to its reflexive and symmetric closure $\Delta_A \cup \theta \cup \theta^{-1}$. Then we form the compatible closure $\langle\Delta_A \cup \theta \cup \theta^{-1}\rangle$. Finally, the set produced in this way has to be completed by all pairs needed for transitivity. This is done by using the *transitive hull operator*, which maps any relation α to

$$\alpha_T := \bigcap\{\beta \subseteq A^2 \mid \beta \supseteq \alpha \text{ and } \beta \text{ is transitive}\}.$$

Altogether, then we need to use

$$\langle\theta\rangle_{\text{Con}(\mathcal{A})} = \langle\langle\Delta_A \cup \theta' \cup (\theta')^{-1}\rangle\rangle_T.$$

This construction allows us to produce, for two congruences on $\mathcal{A}$, the least congruence on $\mathcal{A}$ which contains them both. This gives us the two binary operations on $\text{Con}(\mathcal{A})$ we wanted:

$$\begin{aligned} \wedge &: \text{Con}(\mathcal{A}) \times \text{Con}(\mathcal{A}) \to \text{Con}(\mathcal{A}) \quad \text{defined by } (\theta_1, \theta_2) \mapsto \theta_1 \cap \theta_2, \\ \vee &: \text{Con}(\mathcal{A}) \times \text{Con}(\mathcal{A}) \to \text{Con}(\mathcal{A}) \quad \text{defined by } (\theta_1, \theta_2) \mapsto \langle\theta_1 \cup \theta_2\rangle_{\text{Con}(\mathcal{A})}. \end{aligned}$$

Now we must verify that these operations do make our set into a lattice.

Theorem 7.4.14. *For every algebra $\mathcal{A}$, the structure* $(\text{Con}(\mathcal{A}); \wedge, \vee)$ *is a lattice, called the congruence lattice* $\text{Con}(\mathcal{A})$ *of $\mathcal{A}$.*

The proof is left as an exercise to the reader.

Since congruence relations are equivalence relations, each congruence on an algebra defines the quotient set A/θ.

On the quotient set, for each $i \in I$, we define an n_i-ary operation $f_i^{A/\theta}$ on the quotient set A/θ, by

$$f_i^{A/\theta} : (A/\theta)^{n_i} \to A/\theta,$$

defined by

$$([a_1]_\theta, \ldots, [a_{n_i}]_\theta) \mapsto f_i^{A/\theta}([a_1]_\theta, \ldots, [a_{n_i}]_\theta) := [f_i^A(a_1, \ldots, a_{n_i})]_\theta.$$

Here we must verify that our operations are well defined, that is, that they are independent of the representatives chosen. As the following proof shows, this is exactly what the compatibility property of a congruence relation means.

Theorem 7.4.15. *For every algebra $\mathcal{A}$ of type τ and every congruence relation $\theta \in$* $\text{Con}(\mathcal{A})$, *the previous definition produces an algebra $\mathcal{A}/\theta$ of type τ, which is called the quotient algebra, or factor algebra, of $\mathcal{A}$ by θ.*

Proof. Let $i \in I$. We have to show that if $[a_j]_\theta = [b_j]_\theta$ for all $j = 1, \ldots, n_i$, then $[f_i^A(a_1, \ldots, a_{n_i})]_\theta = [f_i^A(b_1, \ldots, b_{n_i})]_\theta$. But this is satisfied by the compatibility property of a congruence. If $[a_j]_\theta = [b_j]_\theta$, then $(a_j, b_j) \in \theta$ for all $j = 1, \ldots, n_i$, and since θ is a congruence on the algebra $\mathcal{A}$, it follows that

$$(f_i^A(a_1, \ldots, a_{n_i}), f_i^A(b_1, \ldots, b_{n_i})) \in \theta,$$

and thus

$$[f_i^A(a_1, \ldots, a_{n_i})]_\theta = [f_i^A(b_1, \ldots, b_{n_i})]_\theta. \qquad \square$$

Example 7.4.16. Let $m \in \mathbb{N}$, and let $\mathbb{Z}/(m)$ be the set of all congruence classes modulo m. We define two binary operations on these classes by

$$[a]_m + [b]_m = [a+b]_m \quad \text{and} \quad [a]_m \cdot [b]_m = [a \cdot b]_m.$$

This makes $(\mathbb{Z}/(m); +, \cdot)$ an algebra of type $(2,2)$. In fact, it is a ring, called the *ring of all congruence classes modulo m*. If m is a prime number, this ring is also a field.

The diagram lemma from Chapter 1 shows that for a surjective mapping $h : A \to B$ and an equivalence relation θ on the set A, there exists a uniquely determined bijection $f : A/\theta \to B$ with $f \circ \mathrm{nat}(\ker h) = h$. If h is a surjective homomorphism from algebra $\mathcal{A}$ of type τ onto $\mathcal{B}$ of the same type, then the homomorphism theorem says that f is an isomorphism.

Theorem 7.4.17 (Homomorphic image theorem). *Let $h : \mathcal{A} \to \mathcal{B}$ be a surjective homomorphism. Then there exists a unique isomorphism f from $\mathcal{A}/\ker h$ onto $\mathcal{B}$ with $f \circ \mathrm{nat}(\ker h) = h$.*

Proof. It is easy to see that the natural mapping $\mathrm{nat} : A \to A/\ker h$ is a surjective homomorphism, the uniquely determined mapping $f : A/\ker h \to B$ is an isomorphism, and the three homomorphisms satisfy

$$f \circ \mathrm{nat}(\ker h) = h. \qquad \square$$

7.5 Direct and subdirect products

The formation of homomorphic images and subalgebras are methods to construct new algebras from given ones. The constructed algebras have the same type as the originals but are "smaller." The next constructions allow building "bigger" algebras

Definition 7.5.1. Let $(\mathcal{A}_j)_{j\in J}$ be a family of algebras of type τ. The *direct product* $\prod_{j\in J} \mathcal{A}_j$ of the $\mathcal{A}_j$ has the carrier set

$$P := \prod_{j\in J} A_j := \{(x_j)_{j\in J} \mid \forall j \in J\ (x_j \in A_j)\}$$

and the fundamental operations

$$(f_i^P(\underline{a}_1, \dots, \underline{a}_{n_i}))(j) = f_i^{A_j}(\underline{a}_1(j), \dots, \underline{a}_{n_i}(j)),$$

for $\underline{a}_1, \dots, \underline{a}_{n_i}$ in P; this means

$$f_i^P((a_{1j})_{j\in J}, \dots, (a_{n_ij})_{j\in J}) = (f_i^{A_j}(a_{1j}, \dots, a_{n_ij})_{j\in J}).$$

If $\mathcal{A}_j = \mathcal{A}$ for all $j \in J$, then we write $\mathcal{A}^J$ instead of $\prod_{j\in J} \mathcal{A}_j$. If $J = \emptyset$, then $\mathcal{A}^\emptyset$ is defined to be the one-element (trivial) algebra of type τ. If $J = \{1, \ldots, n\}$, then the direct product is written as $\mathcal{A}_1 \times \cdots \times \mathcal{A}_n$.

The *projections* of the direct product $\prod_{j\in J} \mathcal{A}_j$ are the mappings

$$p_k : \prod_{j\in J} A_j \to A_k \quad \text{defined by } (a_j)_{j\in J} \mapsto a_k.$$

It is easy to check that the projections of the direct product are, in fact, surjective homomorphisms.

Example 7.5.2. As an example, we consider the direct product of the two permutation groups $\mathcal{S}_2$ and $\mathcal{A}_3$. Here

$$\mathcal{S}_2 = \big(\{\tau_0, \tau_1\}; \circ, ^{-1}, (1)\big) \quad \text{and} \quad \mathcal{A}_3 = \big(\{\tau_0, \alpha_1, \alpha_2\}; \circ, ^{-1}, (1)\big),$$

with

$$\tau_0 := (1), \quad \tau_1 := (12), \quad \alpha_1 := (123), \quad \alpha_2 := (132).$$

For the Cartesian product set, we let

$$\gamma_{00} := \big((1),(1)\big), \quad \gamma_{01} := \big((1),(123)\big), \quad \gamma_{02} := \big((1),(132)\big),$$
$$\gamma_{10} := \big((12),(1)\big), \quad \gamma_{11} := \big((12),(123)\big), \quad \gamma_{12} := \big((12),(132)\big).$$

Then we have

$$S_2 \times A_3 = \{\gamma_{00}, \gamma_{01}, \gamma_{02}, \gamma_{10}, \gamma_{11}, \gamma_{12}\}.$$

The binary operation $\circ$ of the direct product is defined by the Cayley table

$\circ$	γ_{00}	γ_{01}	γ_{02}	γ_{10}	γ_{11}	γ_{12}
γ_{00}	γ_{00}	γ_{01}	γ_{02}	γ_{10}	γ_{11}	γ_{12}
γ_{01}	γ_{01}	γ_{02}	γ_{00}	γ_{11}	γ_{12}	γ_{10}
γ_{02}	γ_{02}	γ_{00}	γ_{01}	γ_{12}	γ_{10}	γ_{11}
γ_{10}	γ_{10}	γ_{11}	γ_{12}	γ_{00}	γ_{01}	γ_{02}
γ_{11}	γ_{11}	γ_{12}	γ_{10}	γ_{01}	γ_{02}	γ_{00}
γ_{12}	γ_{12}	γ_{10}	γ_{11}	γ_{02}	γ_{00}	γ_{01}

A special case is the direct product of two factors. Then we have two projection mappings, p_1 and p_2, each of which has a kernel which is a congruence relation on the product. These kernels have some special properties.

Two binary relations θ_1, θ_2 on A are called *permutable* if $\theta_1 \circ \theta_2 = \theta_2 \circ \theta_1$.

Lemma 7.5.3. *Let $\mathcal{A}_1$, $\mathcal{A}_2$ be two algebras of type τ and let $\mathcal{A}_1 \times \mathcal{A}_2$ be their direct product. Then:*

(i) $\ker p_1 \wedge \ker p_2 = \Delta_{A_1\times A_2}$;
(ii) $\ker p_1 \circ \ker p_2 = \ker p_2 \circ \ker p_1$;
(iii) $\ker p_1 \vee \ker p_2 = (A_1 \times A_2)^2$.

For the proof of this lemma, see [21].

Therefore any direct product of two factors defines two congruences with the three special properties of the previous lemma. Conversely, the next theorem shows that the direct product of two algebras can be characterized up to isomorphism by these three properties.

Theorem 7.5.4. *Let $\mathcal{A}$ be an algebra, and let $\theta_1, \theta_2 \in \mathrm{Con}(\mathcal{A})$ be a pair of congruence relations with the following properties:*
(i) $\theta_1 \wedge \theta_2 = \Delta_A$;
(ii) $\theta_1 \vee \theta_2 = A^2$;
(iii) $\theta_1 \circ \theta_2 = \theta_2 \circ \theta_1$.

Then $\mathcal{A}$ is isomorphic to the direct product $\mathcal{A}/\theta_1 \times \mathcal{A}/\theta_2$, by an isomorphism

$$\varphi : \mathcal{A} \to \mathcal{A}/\theta_1 \times \mathcal{A}/\theta_2$$

given by

$$\varphi(a) = ([a]_{\theta_1}, [a]_{\theta_2}), \quad a \in A.$$

Proof. Since φ is defined by two natural homomorphisms, it is also a homomorphism. We will show that it is a bijection. First, φ is injective. Indeed, if $\varphi(a) = \varphi(b)$, then $[a]_{\theta_1} = [b]_{\theta_1}$ and $[a]_{\theta_2} = [b]_{\theta_2}$, so it follows that $(a, b) \in \theta_1 \wedge \theta_2$ and $a = b$ by (i). To see that the map φ is also surjective, let (a, b) be any pair in A^2. Conditions (ii) and (iii) mean that there exists an element $c \in A$ with $(a, c) \in \theta_1$ and $(c, b) \in \theta_2$, and therefore

$$([a]_{\theta_1}, [b]_{\theta_2}) = ([c]_{\theta_1}, [c]_{\theta_2}) = \varphi(c).$$

□

Definition 7.5.5. An algebra $\mathcal{A}$ is called *directly irreducible* if whenever $\mathcal{A} \cong \mathcal{B}_1 \times \mathcal{B}_2$, either $|B_1| = 1$ or $|B_2| = 1$.

Corollary 7.5.6. *An algebra $\mathcal{A}$ is directly irreducible iff $(\Delta_A, A \times A)$ is the only pair of congruence relations on $\mathcal{A}$ which satisfies the conditions* (i)–(iii) *of Theorem* 7.5.4.

Proof. Let $\mathcal{A}$ be directly irreducible and assume that $\theta_1, \theta_2 \in \mathrm{Con}(\mathcal{A})$ satisfy the three conditions. Then $\mathcal{A} \cong \mathcal{A}/\theta_1 \times \mathcal{A}/\theta_2$, and the irreducibility means that one of the factors has cardinality one. Without loss of generality, suppose that $|A/\theta_1| = 1$. Then we must have $\theta_1 = A \times A$, and θ_2 must equal Δ_A by condition (i).

Assume now that, conversely, $(\Delta_A, A \times A)$ is the only pair with the properties (i)–(iii), and let $\mathcal{A} \cong \mathcal{A}_1 \times \mathcal{A}_2$. Then $(\Delta_{A_1 \times A_2}, (A_1 \times A_2)^2)$ is also the only pair of congruence relations on $\mathcal{A}_1 \times \mathcal{A}_2$ to satisfy conditions (i)–(iii). The kernels of the projection mappings p_1 and p_2 satisfy the three conditions. Therefore either $\ker p_1$ or $\ker p_2$ must equal $\Delta_{A_1 \times A_2}$, and thus either A_1 or A_2 must have cardinality one. □

Subdirect products are subalgebras of the direct product with particular properties.

Definition 7.5.7. Let $(\mathcal{A}_j)_{j\in J}$ be a family of algebras of type τ. A subalgebra $\mathcal{B} \subseteq \prod_{j\in J} \mathcal{A}_j$ of the direct product of the algebras $\mathcal{A}_j$ is called a *subdirect product* of the algebras $\mathcal{A}_j$ if for every projection mapping $p_k : \prod_{j\in J} \mathcal{A}_j \to \mathcal{A}_k$ we have

$$p_k(\mathcal{B}) = \mathcal{A}_k.$$

Example 7.5.8.
1. Every direct product is also a subdirect product.
2. For every algebra $\mathcal{A}$, the diagonal $\Delta_A = \{(a, a) \mid a \in A\}$ is the carrier set of a subalgebra $\boldsymbol{\Delta}_A$ of $\mathcal{A}\times\mathcal{A}$. Moreover, $p_1(\boldsymbol{\Delta}_A) = p_2(\boldsymbol{\Delta}_A) = \mathcal{A}$, so $\boldsymbol{\Delta}_A$ is a subdirect product of $\mathcal{A} \times \mathcal{A}$.

Theorem 7.5.9. *Let $\mathcal{B}$ be a subdirect product of the family $(\mathcal{A}_j)_{j\in J}$ of algebras of type τ. Then the projection mappings $p_k : \prod_{j\in J} \mathcal{A}_j \to \mathcal{A}_k$ satisfy the equation $\bigcap_{j\in J} \ker(p_j|B) = \Delta_B$.*

Proof. Let $(a, b) \in \bigcap_{j\in J} \ker(p_j|B)$. This implies that $p_k(a) = p_k(b)$ for all $k \in J$, therefore every component of a agrees with the corresponding component of b. Thus $a = b$ and then $(a, b) \in \Delta_B$. Conversely, it is clear that $\Delta_B \subseteq \ker(p_k|B)$ for all p_k. □

Our aim is to characterize subdirect products similar to the characterization of direct products.

Theorem 7.5.10. *Let $\mathcal{A}$ be an algebra. Let $\{\theta_j \mid j \in J\}$ be a family of congruence relations on $\mathcal{A}$, which satisfy the equation $\bigcap_{j\in J} \theta_j = \Delta_A$. Then $\mathcal{A}$ is isomorphic to a subdirect product of the algebras $\mathcal{A}/\theta_j$, for $j \in J$. The mapping $\varphi(a) := ([a]_{\theta_j} \mid j \in J)$ defines an embedding $\varphi : \mathcal{A} \to \prod_{j\in J}(\mathcal{A}/\theta_j)$, whose image $\varphi(\mathcal{A})$ is a subdirect product of the algebras $\mathcal{A}/\theta_j$.*

Proof. The map φ is a homomorphism determined by the natural homomorphism. Also φ is injective, since $\varphi(a) = \varphi(b)$ implies $[a]_{\theta_j} = [b]_{\theta_j}$ and thus $(a, b) \in \theta_j$ for all $j \in J$. Therefore $(a, b) \in \bigcap_{j\in J} \theta_j = \Delta_A$, and so $a = b$. This proves the isomorphism of $\mathcal{A}$ and $\varphi(\mathcal{A})$. Moreover, if $p_k : \prod_{j\in J}(\mathcal{A}/\theta_j) \to \mathcal{A}/\theta_k$ is the kth projection mapping, then, by the definition of φ, we have $p_k(\varphi(\mathcal{A})) = \mathcal{A}/\theta_k$ for all $k \in J$. Therefore $\varphi(\mathcal{A})$ is a subdirect product of the algebras $\mathcal{A}/\theta_j$. □

We remark that the converse of this theorem is also true. If $\mathcal{A}$ is isomorphic to a subdirect product of a family $(\mathcal{A}_j)_{j\in J}$ of algebras, then there exists a family of congruence relations on $\mathcal{A}$ whose intersection is the relation Δ_A.

We want to consider algebras which cannot be expressed as a subdirect product of other smaller algebras, except in trivial ways.

Definition 7.5.11. An algebra $\mathcal{A}$ of type τ is called *subdirectly irreducible* if the intersection of any family $\{\theta_j \mid j \in J\}$ of congruences on $\mathcal{A}$, none of which is equal to Δ_A, is different from Δ_A. (In this case, by Theorem 7.5.10, $\mathcal{A}$ cannot be represented as a subdirect product.)

Remark 7.5.12. An algebra $\mathcal{A}$ is subdirectly irreducible if and only if Δ_A has exactly one upper neighbor in the lattice $\mathrm{Con}(\mathcal{A})$ of all congruence relations on $\mathcal{A}$. Then the congruence lattice has the form shown in the diagram below. Here θ_0 is defined to be

$$\theta_0 := \bigcap(\mathrm{Con}(\mathcal{A}) \setminus \{\Delta_A\}).$$

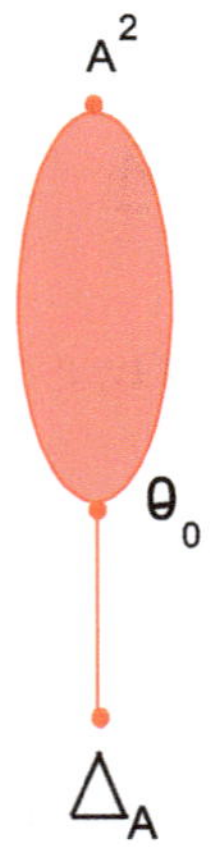

Figure 7.3: Congruence lattice of a subdirectly irreducible algebra.

Example 7.5.13.
1. Every simple algebra is subdirectly irreducible. Therefore, any two-element algebra is subdirectly irreducible.
2. A three-element algebra which has no more than one congruence different from A^2 and Δ_A is subdirectly irreducible.

The concept of a subdirect product is the right concept for a product, since G. Birkhoff proved the following result (for the proof, see [7]). We present this important result without proof.

Theorem 7.5.14. *Every algebra is isomorphic to a subdirect product of subdirectly irreducible algebras.*

Classes of algebras of the same type which are closed under taking of homomorphic images, subalgebras, and direct products are called *varieties*.

Definition 7.5.15. Let $K \subseteq \mathrm{Alg}(\tau)$ be a class of algebras of type τ. Then K is called a variety of algebras of type τ if K contains all homomorphic images, all subalgebras, and all direct products from algebras of K.

If K is the class of all semigroups, then it is not difficult to see that any subalgebra from an algebra in K, any homomorphic image from algebras in K, and any direct product from algebras from K satisfy the associative law, i. e., belong to K. So, the class of all semigroups forms a variety. In the next chapters we will describe an alternative approach to varieties.

The study of varieties of algebras of a fixed type is one of the central goals of universal algebra. But it turns out that for the applications which we want to describe here finiteness plays an important role. The formation of direct products produces arbitrarily large algebras, even from finite algebras. Therefore we consider only finite direct products and call a class of finite algebras of a fixed type which is closed under formation of subalgebras, quotient algebras, and finite direct products a *pseudovariety*.

7.6 Terms, trees, and term operations

Semigroups are algebras $(G;\cdot)$ of type $\tau = (2)$ where the associative law

$$x \cdot (y \cdot z) \approx (x \cdot y) \cdot z$$

is satisfied. Satisfaction of this law means that for all elements $x, y, z \in G$, the equation $x \cdot (y \cdot z) = (x \cdot y) \cdot z$ holds. To write this equation, we need variables x, y, and z. Variables are not elements of G; they are only symbols for which elements from G may be substituted. To write identities, we need a language, which must include variables, as well as symbols, to represent the operations. Let $n \geq 1$ be a natural number and let $X_n = \{x_1, \ldots, x_n\}$ be an n-element set. The set X_n is called an *alphabet* and its elements are called *variables*. Moreover, we need a set $\{f_i \mid i \in I\}$ of operation symbols, indexed by the set I. The sets X_n and $\{f_i \mid i \in I\}$ have to be disjoint. To every operation symbol f_i, we assign a natural number $n_i \geq 1$, called the arity of f_i. As in the definition of an algebra, the sequence $\tau = (n_i)_{i \in I}$ of all the arities is called the *type* of the language. With this notation for operation symbols and variables, we can define the terms of the language of type τ.

Definition 7.6.1. Let $n \geq 1$. The *n-ary terms* of type τ are defined in the following inductive way:

(i) Every variable $x_i \in X_n$ is an n-ary term.

(ii) If $t_1, \ldots, t_{n_i}$ are n-ary terms and f_i is an n_i-ary operation symbol, then $f_i(t_1, \ldots, t_{n_i})$ is an n-ary term.

(iii) The set $W_\tau(X_n) = W_\tau(x_1, \ldots, x_n)$ of all n-ary terms is the smallest set which contains $x_1, \ldots, x_n$ and is closed under finite application of (ii).

Remark 7.6.2. From the definition there follows that every n-ary term is also k-ary, for $k > n$.

Example 7.6.3. Let $\tau = (2)$, with one binary operation symbol f. Let $X_2 = \{x_1, x_2\}$. Then $f(f(x_1, x_2), x_2)$, $f(x_2, x_1)$, x_1, x_2 and $f(f(f(x_1, x_2), x_1), x_2)$ are binary terms. The expression $f(f(x_3, f(x_1, x_2)), x_4)$ is a quaternary (or 4-ary) term, but $f(f(x_1, x_2, x_3)$ is not a term since one bracket is missing.

The definition of a term is an inductive definition on the number of occurrences of operation symbols in the term. This number is called *complexity* of the term.

In computer science and other applications, it is very common to illustrate terms by tree diagrams, called *semantic trees*. The semantic tree of a term t was already considered in Section 5.3. We will repeat the definition in more details:

(i) If $t = x_i$, then the semantic tree of t consists only of one vertex which is labeled with x_i, and this vertex is called the root of the tree.

(ii) If $t = f_i(t_1, \dots, t_{n_i})$ then the semantic tree of t has as its root a vertex labeled with f_i, and has n_i edges which are incident with the vertex f_i; each of these edges is incident with the root of the corresponding term $t_1, \dots, t_{n_i}$ (ordered by $1 \leq 2 \leq \cdots \leq n_i$, starting from the left).

Consider, for example, the type $\tau = (2, 2)$ with two binary operation symbols f and g. Then the quaternary term

$$t = f(g(x_1, x_2), f(g(x_3, x_4), f(x_2, x_1)))$$

corresponds to the semantic tree shown in Figure 7.4.

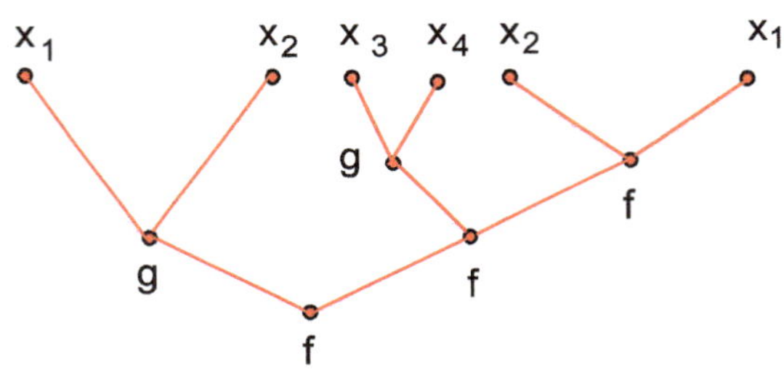

Figure 7.4: Semantic tree.

Let τ be a fixed type. Let X be the union of all sets X_n of variables, so $X = \{x_1, x_2, \dots\}$. We denote by $W_\tau(X)$ the set of all terms of type τ over the countably infinite alphabet X:

$$W_\tau(X) = \bigcup_{n=1}^{\infty} W_\tau(X_n).$$

Now we want to use this set $W_\tau(X)$ as the universe of some algebra, of the same type τ. For every $i \in I$, we define an n_i-ary operation $\bar{f}_i$ on $W_\tau(X)$, with

$$\bar{f}_i : W_\tau(X)^{n_i} \to W_\tau(X) \quad \text{defined by } (t_1, \dots, t_{n_i}) \mapsto f_i(t_1, \dots, t_{n_i}).$$

The second step of the term definition shows that $f_i(t_1, \dots, t_{n_i})$ belongs to $W_\tau(X)$, and so the operation $\bar{f}_i$ is well defined. In this way, $W_\tau(X)$ is the universe of an algebra of type $\tau = (n_i)_{i \in I}$, since for every operation symbol f_i we have a concrete operation $\bar{f}_i$ on $W_\tau(X)$.

Definition 7.6.4. The algebra $\mathcal{F}_\tau(X) := (W_\tau(X); (\bar{f}_i)_{i\in I})$ is called the *term algebra*, or the *absolutely free algebra*, of type τ over the set X.

As an easy consequence of the term definition we obtain

Lemma 7.6.5. *For any type τ, the term algebra $\mathcal{F}_\tau(X)$ is generated by the set X.*

Proof. Clearly, $X \subseteq W_\tau(X)$, and (ii) of the term definition gives $\langle X\rangle_{\mathcal{F}_\tau(X)} = \mathcal{F}_\tau(X)$. □

Instead of $\mathcal{F}_\tau(X)$, we can also consider the algebra $\mathcal{F}_\tau(X_n) := (W_\tau(X_n); (\bar{f}_i)_{i\in I})$. This algebra is called the *absolutely free algebra* or the *term algebra* of type τ over the set X_n of n generators. The algebra $\mathcal{F}_\tau(X_n)$ is generated by the set X_n.

The term algebra has the following property which justifies the name "absolutely free."

Theorem 7.6.6. *For every algebra $\mathcal{A} \in \mathrm{Alg}(\tau)$ and every mapping $f : X \to A$, there exists a uniquely determined homomorphism $\hat{f} : \mathcal{F}_\tau(X) \to \mathcal{A}$ which extends the mapping f and such that $\hat{f} \circ \varphi = f$, where $\varphi : X \to \mathcal{F}_\tau(X)$ is the embedding of X in $\mathcal{F}_\tau(X)$. This homomorphism makes the following diagram commute:*

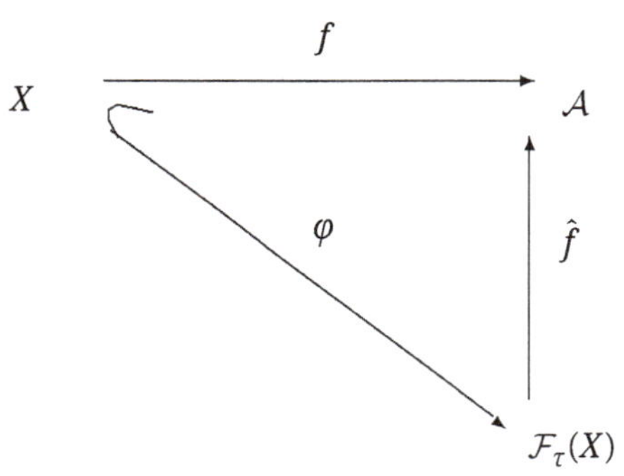

Figure 7.5: Absolute freeness.

Proof. We define $\hat{f}$ in the following way:

$$\hat{f}(t) := \begin{cases} f(t) & \text{if } t = x \in X \text{ is a variable,} \\ f_i^A(\hat{f}(t_1),\ldots,\hat{f}(t_{n_i})) & \text{if } t = f_i(t_1,\ldots,t_{n_i}). \end{cases}$$

Here f_i^A is the fundamental operation of the algebra $\mathcal{A}$ which corresponds to the operation symbol f_i. Clearly, $\hat{f}(\bar{f}_i(t_1,\ldots,t_{n_i})) = \hat{f}(f_i(t_1,\ldots,t_{n_i})) = f_i^A(\hat{f}(t_1),\ldots,\hat{f}(t_{n_i}))$ shows that $\hat{f}$ is a homomorphism, and it follows from the definition that $\hat{f}$ extends f. □

It is clear that the notation of the variables in the term can be changed. Instead of $x_1,\ldots,x_n$, we can also use $y_1,\ldots,y_n$ or something else. The reason is the following theorem which is easy to prove. The proof is left to the reader.

Theorem 7.6.7. *Let Y and Z be alphabets with the same cardinality. Then the term algebras $\mathcal{F}_\tau(Y)$ and $\mathcal{F}_\tau(Z)$ are isomorphic.*

Terms are formal expressions, but can be realized as concrete operations on algebras $\mathcal{A}$ of type τ.

Definition 7.6.8. Let $\mathcal{A}$ be an algebra of type τ and let t be an n-ary term of type τ over X. Then t induces an n-ary operation $t^{\mathcal{A}}$ on $\mathcal{A}$, called the *term operation induced by the term* t on the algebra $\mathcal{A}$, by the following steps:

(i) If $t = x_i \in X_n$, then $t^{\mathcal{A}} = x_j^{\mathcal{A}} = e_j^{n,A}$; here $e_j^{n,A}$ is the n-ary projection on A defined by $e_j^{n,A}(a_1, \dots, a_n) = a_j$ for all $a_1, \dots, a_n \in A$.

(ii) If $t = f_i(t_1, \dots, t_{n_i})$ is an n-ary term of type τ, and $t_1^{\mathcal{A}}, \dots, t_{n_i}^{\mathcal{A}}$ are the term operations which are induced by $t_1, \dots, t_{n_i}$, then $t^{\mathcal{A}} = f_i^{\mathcal{A}}(t_1^{\mathcal{A}}, \dots, t_{n_i}^{\mathcal{A}})$.

Term operations of an algebra have similar properties as fundamental operations as the following result shows.

Theorem 7.6.9. *Let $\mathcal{A}$ be an algebra of type τ and let $t^{\mathcal{A}}$ be the n-ary term operation on $\mathcal{A}$ induced by the n-ary term $t \in W_\tau(X)$.*

(i) *If $\mathcal{B}$ is a subalgebra of $\mathcal{A}$, then $t^{\mathcal{A}}(b_1, \dots, b_n) \in B$, for all $b_1, \dots, b_n \in B$.*

(ii) *If $\mathcal{B}$ is an algebra of type τ and $\varphi : \mathcal{A} \to \mathcal{B}$ is a homomorphism, then for all $a_1, \dots, a_n$ in A,*

$$\varphi(t^{\mathcal{A}}(a_1, \dots, a_n)) = t^{\mathcal{B}}(\varphi(a_1), \dots, \varphi(a_n)).$$

(iii) *If θ is a congruence relation on $\mathcal{A}$, then for all pairs $(a_1, b_1), \dots, (a_n, b_n)$ in θ, we have $(t^{\mathcal{A}}(a_1, \dots, a_n), t^{\mathcal{A}}(b_1, \dots, b_n)) \in \theta$.*

For the proof, we refer to [21].

7.7 Identities and models

We introduced the concept of a term to define equations and identities. An equation is a statement of the form $t_1 \approx t_2$, where t_1 and t_2 are terms. We will define what it means for an equation to be satisfied, or to be an identity in an algebra $\mathcal{A}$.

Definition 7.7.1. An *equation* of type τ is a pair of terms (s, t) from $W_\tau(X)$; such pairs are also written as $s \approx t$. An equation $s \approx t$ is said to be an *identity* in the algebra $\mathcal{A}$ of type τ if $s^{\mathcal{A}} = t^{\mathcal{A}}$, i. e., if the term operations induced by s and t on the algebra $\mathcal{A}$ are equal. In this case, we also say that the equation $s \approx t$ is *satisfied* or *modeled* by the algebra $\mathcal{A}$, and we write $\mathcal{A} \models s \approx t$.

Remark 7.7.2. Let $\mathcal{A}$ be an algebra and $s \approx t$ an equation of type τ. Let $X_n = \{x_1, \ldots, x_n\}$ be the set of variables occurring in the equation $s \approx t$. We know already that every map $f : X_n \to A$ has a unique extension $\hat{f} : \mathcal{F}_\tau(X_n) \to \mathcal{A}$ to the free term algebra on X_n. Then the equation $s^{\mathcal{A}} = t^{\mathcal{A}}$ means that for every mapping $f : X_n \to A$, we have $\hat{f}(s) = \hat{f}(t)$ or $(s,t) \in \ker \hat{f}$. That is, the pair (s,t) must belong to the intersection of the kernels of all these mappings $\hat{f}$. Thus an identity $s \approx t$ holds in an algebra $\mathcal{A}$ (or in a class K of algebras) iff (s,t) is in the intersection of the kernels of $\hat{f}$, for every map $f : X \to A$ (for every algebra $\mathcal{A}$ in K).

We now consider the class $\mathrm{Alg}(\tau)$ of all algebras of type τ, and the class $W_\tau(X) \times W_\tau(X)$ of all equations of type τ. Satisfaction of an equation by an algebra gives us a fundamental relation between these two sets. Formally, we have the relation $\models$ of all pairs $(\mathcal{A}, s \approx t)$ for which $\mathcal{A} \models s \approx t$.

For any subset $\Sigma \subseteq W_\tau(X) \times W_\tau(X)$ and any subclass $K \subseteq \mathrm{Alg}(\tau)$, we define

$$\begin{aligned} \mathrm{Mod}\,\Sigma &:= \{\mathcal{A} \in \mathrm{Alg}(\tau) \mid \forall s \approx t \in \Sigma,\ (\mathcal{A} \models s \approx t)\} \quad \text{and} \\ \mathrm{Id}\,K &:= \{s \approx t \in W_\tau(X)^2 \mid \forall \mathcal{A} \in K,\ (\mathcal{A} \models s \approx t)\}. \end{aligned}$$

The pair (Mod, Id) has the following properties::

Theorem 7.7.3. *Let τ be a fixed type.*

(i) *For all subsets Σ and Σ' of $W_\tau(X) \times W_\tau(X)$, and for all subclasses K and K' of* $\mathrm{Alg}(\tau)$, *we have*

$$\Sigma \subseteq \Sigma' \Rightarrow \mathrm{Mod}\,\Sigma \supseteq \mathrm{Mod}\,\Sigma' \quad \textit{and} \quad K \subseteq K' \Rightarrow \mathrm{Id}\,K \supseteq \mathrm{Id}\,K';$$

(ii) *For all subsets Σ of $W_\tau(X) \times W_\tau(X)$ and all subclasses K of* $\mathrm{Alg}(\tau)$, *we have* $\Sigma \subseteq \mathrm{Id}\,\mathrm{Mod}\,\Sigma$ *and* $K \subseteq \mathrm{Mod}\,\mathrm{Id}\,K$;

(iii) *The maps* Id Mod *and* Mod Id *are closure operators on $W_\tau(X) \times W_\tau(X)$ and* $\mathrm{Alg}(\tau)$, *respectively.*

(iv) *The sets closed under* Mod Id *are exactly the sets of the form* $\mathrm{Mod}\,\Sigma$, *for some* $\Sigma \subseteq W_\tau(X) \times W_\tau(X)$, *and the sets closed under* Id Mod *are exactly the sets of the form* $\mathrm{Id}\,K$, *for some* $K \subseteq \mathrm{Alg}(\tau)$.

The properties (i), (ii) define the concept of a Galois connection as a pair (ι, μ) of mappings between the power sets $\iota : \mathcal{P}(A) \to \mathcal{P}(B)$, $\mu : \mathcal{P}(B) \to \mathcal{P}(A)$ of two sets A, B which satisfy properties corresponding to (i) and (ii) (see, e. g., [23]).

Definition 7.7.4. A class $K \subseteq \mathrm{Alg}(\tau)$ is called an *equational class*, or is said to be *equationally definable*, if there is a set Σ of equations such that $K = \mathrm{Mod}\,\Sigma$. A set $\Sigma \subseteq W_\tau(X) \times W_\tau(X)$ is called an *equational theory* if there is a class $K \subseteq \mathrm{Alg}(\tau)$ such that $\Sigma = \mathrm{Id}\,K$.

From Theorem 7.7.3(iv), we see that the equational classes are exactly the fixed points with respect to the closure operator Mod Id, and, dually, the equational theories are exactly the fixed points with respect to the closure operator Id Mod. We mention that the collections of such fixed points form complete lattices.

Theorem 7.7.5. *The collection of all equational classes of type τ forms a complete lattice $\mathcal{L}(\tau)$, and the collection of all equational theories of type τ forms a complete lattice $\mathcal{E}(\tau)$. These lattices are dually isomorphic: there exists a bijection $\varphi : \mathcal{L}(\tau) \to \mathcal{E}(\tau)$, satisfying $\varphi(K_1 \vee K_2) = \varphi(K_1) \wedge \varphi(K_2)$ and $\varphi(K_1 \wedge K_2) = \varphi(K_1) \vee \varphi(K_2)$.*

Proof. That $\mathcal{L}(\tau)$ and $\mathcal{E}(\tau)$ are complete lattices follows from the properties of Galois connections and closure operators (see, e. g., [23]). For arbitrary subclasses K of $\mathcal{L}(\tau)$, the infimum of K is the set-theoretical intersection, $\bigwedge K = \bigcap K$, and the supremum is the variety which is generated by the set-theoretical union, so $\bigvee K = \bigcap\{K' \in \mathcal{L}(\tau) \mid K' \supseteq \bigcup K\}$. Moreover, we get $\bigvee K = \operatorname{Mod}\operatorname{Id}(\bigcup K)$. The meet and join for $\mathcal{E}(\tau)$ are obtained in a similar way.

We define a mapping $\varphi : \mathcal{L}(\tau) \to \mathcal{E}(\tau)$ by $K \mapsto \operatorname{Id} K$ for every equational class K of $\mathcal{L}(\tau)$. Then $\operatorname{Id} K$ is an equational theory of type τ and therefore φ is well defined. From Definition 7.7.4 and Theorem 7.7.3(iv), we know that φ is surjective. Furthermore, if $\operatorname{Id} K_1 = \operatorname{Id} K_2$ then $K_1 = \operatorname{Mod}\operatorname{Id} K_1 = \operatorname{Mod}\operatorname{Id} K_2 = K_2$, using Theorem 7.7.3(iii) and the fact that equational classes are exactly the fixed points with respect to the closure operator Mod Id. Therefore, φ is a bijection on $\mathcal{L}(\tau)$.

Now we show that φ has the property claimed on meets. By definition, we have $\varphi(K_1 \wedge K_2) = \operatorname{Id}(K_1 \cap K_2)$. Since Id reverses inclusions, the set $\operatorname{Id}(K_1 \cap K_2)$ contains both $\operatorname{Id} K_1$ and $\operatorname{Id} K_2$. Since it is an equational theory, it also contains their join, $\operatorname{Id} K_1 \vee \operatorname{Id} K_2$, which equals $\varphi(K_1) \vee \varphi(K_2)$. This gives the inclusion $\varphi(K_1 \wedge K_2) \supseteq \varphi(K_1) \vee \varphi(K_2)$.

For the opposite inclusion, we consider that for each $i = 1, 2$, we have $\operatorname{Id} K_i \subseteq \operatorname{Id} K_1 \vee \operatorname{Id} K_2$. Applying Mod and using $K_i = \operatorname{Mod}\operatorname{Id} K_i$, we have $K_i \supseteq \operatorname{Mod}(\operatorname{Id} K_1 \vee \operatorname{Id} K_2)$. Thus we have

$$\operatorname{Mod}(\operatorname{Id} K_1 \vee \operatorname{Id} K_2) \subseteq K_1 \cap K_2.$$

Applying Id to this gives

$$\operatorname{Id}\operatorname{Mod}(\operatorname{Id} K_1 \vee \operatorname{Id} K_2) \supseteq \operatorname{Id}(K_1 \cap K_2).$$

But $\operatorname{Id} K_1 \vee \operatorname{Id} K_2$ is an equational theory, and therefore it is closed under Id Mod. This gives

$$\operatorname{Id}(K_1 \cap K_2) \subseteq \operatorname{Id} K_1 \vee \operatorname{Id} K_2$$

and then $\varphi(K_1 \wedge K_2) \subseteq \varphi(K_1) \vee \varphi(K_2)$, hence the equality we needed.

Now we verify the claim for joins. Since φ reverses inclusions, the inclusion $K_i \subseteq K_1 \vee K_2$, for $i = 1, 2$, implies $\varphi(K_1 \vee K_2) \subseteq \varphi(K_i)$, for $i = 1, 2$. Then $\varphi(K_1 \vee K_2) \subseteq \varphi(K_1) \wedge \varphi(K_2)$. Conversely, $\operatorname{Id} K_1 \wedge \operatorname{Id} K_2 = \operatorname{Id} K_1 \cap \operatorname{Id} K_2 \subseteq \operatorname{Id} K_i$, for $i = 1, 2$ and applying Mod gives $\operatorname{Mod}(\operatorname{Id} K_1 \wedge \operatorname{Id} K_2) \supseteq \operatorname{Mod}\operatorname{Id} K_i$, for $i = 1, 2$. Since K_1 and K_2 are closed under Mod Id, we have $\operatorname{Mod}(\operatorname{Id} K_1 \wedge \operatorname{Id} K_2) \supseteq K_1 \vee K_2$. Applying Id gives $\operatorname{Id}\operatorname{Mod}(\operatorname{Id} K_1 \wedge \operatorname{Id} K_2) \subseteq \operatorname{Id}(K_1 \vee K_2)$. But $\operatorname{Id} K_1 \wedge \operatorname{Id} K_2$ is an equational theory and closed under Id Mod; so we have $\operatorname{Id} K_1 \wedge \operatorname{Id} K_2 \subseteq \operatorname{Id}(K_1 \vee K_2)$. This leads to $\varphi(K_1) \wedge \varphi(K_2) \subseteq \varphi(K_1 \vee K_2)$ and completes our proof of the equality $\varphi(K_1) \cap \varphi(K_2) = \varphi(K_1 \vee K_2)$. □

Clearly, this is true for the lattices of closed sets of any Galois connection since we used only properties of the Galois connection.

If K is an arbitrary class of algebras, then its closure under the closure operator Mod Id is the equational class Mod Id K. This class is called the *equational class generated by* K, and is usually denoted by $E(K) := \operatorname{Mod}\operatorname{Id} K$.

7.8 Fully invariant congruence relations

In this section we characterize equational theories, using the concept of a fully invariant congruence relation.

Definition 7.8.1. A congruence relation θ on an algebra $\mathcal{A}$ of type τ is said to be *fully invariant* if whenever $(x, y) \in \theta$, we also have $(\varphi(x), \varphi(y)) \in \theta$, for every endomorphism φ of $\mathcal{A}$; that is, if θ is compatible with all endomorphisms φ of $\mathcal{A}$.

Theorem 7.8.2. *Let $\Sigma \subseteq W_\tau(X) \times W_\tau(X)$ be a set of equations of type τ. Then Σ is an equational theory if and only if it is a fully invariant congruence relation on the term algebra $\mathcal{F}_\tau(X)$.*

Proof. Let Σ be an equational theory of type τ. Then there exists a class $K \subseteq \operatorname{Alg}(\tau)$ of algebras of type τ such that $\Sigma = \operatorname{Id} K$. Clearly, $\operatorname{Id} K$ is an equivalence relation on the set $W_\tau(X)$ since for any term $t \in W_\tau(X)$, $t \approx t$ belongs to $\operatorname{Id} K$, if $s \approx t \in \operatorname{Id} K$, then $t \approx s \in \operatorname{Id} K$, while if $s \approx t$, $t \approx u \in \operatorname{Id} K$, then $s \approx u \in \operatorname{Id} K$. The compatibility of identities with the operations of the term algebra is also easy to check. Therefore, $\operatorname{Id} K$ is a congruence relation of the term algebra. Any endomorphism $\varphi : \mathcal{F}_\tau(X) \to \mathcal{F}_\tau(X)$ maps variables to terms and this restriction of the given endomorphism onto X can be uniquely extended to the given endomorphism. But if we substitute for the variables occurring in an identity terms, the resulting equation is an identity in K. Therefore $\operatorname{Id} K$ is fully invariant.

Conversely, let θ be a fully invariant congruence relation on $\mathcal{F}_\tau(X)$. We show that θ is the equational theory $\operatorname{Id} K$, for K the class consisting of the quotient algebra $\mathcal{F}_\tau(X)/\theta$.

Let now $s \approx t \in \operatorname{Id} \mathcal{F}_\tau(X)/\theta$. Consider the mapping $f : X \to \mathcal{F}_\tau(X)/\theta$ defined by $f(x) = [x]_\theta$ for all $x \in X$. Since $\mathcal{F}_\tau(X)/\theta$ is an algebra of type τ, this mapping f has a unique homomorphic extension $\hat{f} : \mathcal{F}_\tau(X) \to \mathcal{F}_\tau(X)/\theta$. But the natural homomorphism $\operatorname{nat}\theta : \mathcal{F}_\tau(X) \to \mathcal{F}_\tau(X)/\theta$ has the same property, so the uniqueness of $\hat{f}$ gives $\hat{f} = \operatorname{nat}\theta$. Now our assumption that $s \approx t$ is an identity in $\mathcal{F}_\tau(X)/\theta$ means that (s, t) is in $\ker \hat{f}$. This gives $\operatorname{nat}\theta(s) = \operatorname{nat}\theta(t)$, so $(s, t) \in \theta$. Thus we have $\operatorname{Id}(\mathcal{F}_\tau(X)/\theta) \subseteq \theta$.

For the opposite inclusion, it suffices to show that for any map $f : X \to \mathcal{F}_\tau(X)/\theta$, we get $\theta \subseteq \ker \hat{f}$, where $\hat{f}$ is the unique extension of f. We define a mapping $g : X \to \mathcal{F}_\tau(X)$ by $x \mapsto s$ in such a way that we assign to x a representative $s \in \hat{f}(x) = [x]_\theta$. Then we get the commutative diagram shown in Figure 7.6, where φ is the natural embedding. Combining $\hat{g} \circ \varphi = g$ and $\mathrm{nat}\,\theta \circ g = \hat{f} \circ \varphi$ gives $\mathrm{nat}\,\theta \circ \hat{g} \circ \varphi = \hat{f} \circ \varphi$ and, since φ is an embedding, we have $\mathrm{nat}\,\theta \circ \hat{g} = \hat{f}$. Now for any $(s, t) \in \theta$, the full invariance of θ means that $(\hat{g}(s), \hat{g}(t)) \in \theta$. Therefore $\hat{f}(s) = [\hat{g}(s)]_\theta = [\hat{g}(t)]_\theta = \hat{f}(t)$, and $(s, t) \in \ker \hat{f}$. □

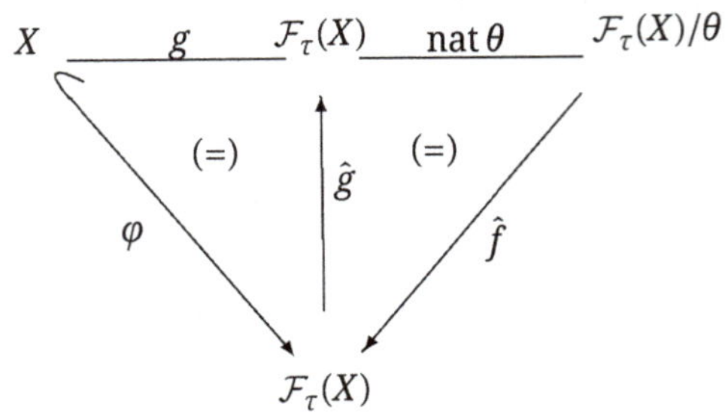

Figure 7.6: Equational theories and fully invariant congruence relations.

7.9 Relatively free algebras

Let Y be any nonempty set of variables, and let $\mathcal{F}_\tau(Y)$ be the absolutely free algebra of type τ generated by Y. Let $K \subseteq \mathrm{Alg}(\tau)$ be a class of algebras. Since $\mathrm{Id}\,K$ is a congruence relation on $\mathcal{F}_\tau(Y)$, we can form the quotient algebra.

Definition 7.9.1. Let Y be a nonempty set of variables, and K be a class of algebras of type τ. The algebra $\mathcal{F}_K(Y) := \mathcal{F}_\tau(Y)/\mathrm{Id}\,K$ is called the *K-free algebra over Y* or the *free algebra relative to K* generated by $\overline{Y} = Y/\mathrm{Id}\,K$.

Remark 7.9.2.

(i) $\mathcal{F}_K(Y)$ exists iff $\mathcal{F}_\tau(Y)$ exists, iff $Y \neq \emptyset$. Therefore, there is a K-free algebra of type τ over any nonempty set Y.

(ii) If $|Y| = |Z| \neq 0$, then $\mathcal{F}_K(Y) \cong \mathcal{F}_K(Z)$ under an isomorphism mapping $\overline{Y}$ to $\overline{Z}$.

(iii) If Y is the set $X_n = \{x_1, \dots, x_n\}$, we will write $\mathcal{F}_K(n)$ instead of $\mathcal{F}_K(X_n)$. In this case, our algebra is called the *K-free algebra on n generators*.

The algebra $\mathcal{F}_K(Y)$ satisfies a relative “freeness” property corresponding to the absolutely free property of the term algebra.

Theorem 7.9.3. *Let Y be any nonempty set of variables. For every algebra $\mathcal{A} \in K \subseteq \mathrm{Alg}(\tau)$ and every mapping $f : \overline{Y} \to A$, there exists a unique homomorphism $\hat{f} : \mathcal{F}_K(Y) \to \mathcal{A}$ which extends f.*

Proof. Let $\mathcal{A}$ be in K, with a map $f : \overline{Y} \to A$. Let φ be the inclusion mapping from $\overline{Y}$ to the absolutely free algebra $\mathcal{F}_\tau(Y)$. By the freeness of this algebra, there is a unique homomorphism $\overline{f} : \mathcal{F}_\tau(Y) \to \mathcal{A}$, which extends f, so that $\overline{f} \circ \varphi = f$. Now consider the homomorphism $\operatorname{nat} \operatorname{Id} K$ from $\mathcal{F}_\tau(Y)$ to $\mathcal{F}_K(Y)$. Its kernel is precisely $\operatorname{Id} K$, and this is contained in $\ker \overline{f}$. Since we have two homomorphisms defined on $\mathcal{F}_\tau(Y)$ with the kernel of one contained in the kernel of the other, we can use the general homomorphism theorem, a generalization of the homomorphism theorem (see [21]) to conclude that there is a homomorphism $\hat{f}$ from $\mathcal{F}_K(Y)$ to $\mathcal{A}$, with $\hat{f} \circ \operatorname{nat} \operatorname{Id} K = \overline{f}$ (see Figure 7.7). Moreover, $\hat{f} \circ \operatorname{nat} \operatorname{Id} K|Y = \hat{f} \circ \operatorname{nat} \operatorname{Id} K \circ \varphi = \hat{f} \circ \varphi = f$, and $\hat{f}$ extends f. By the homomorphism theorem, $\hat{f}$ is uniquely determined by K and f. □

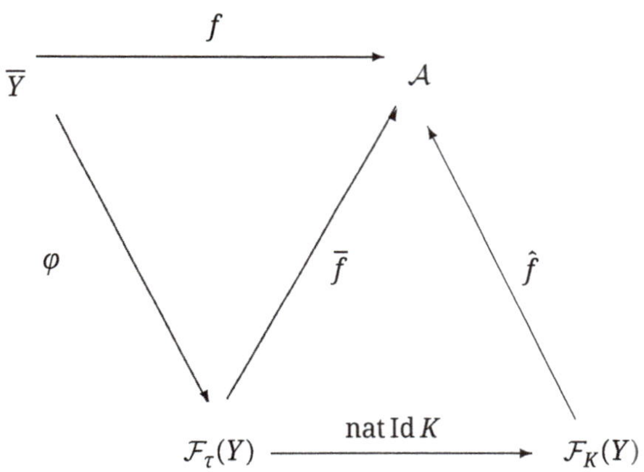

Figure 7.7: Relative freeness.

7.10 Varieties

Varieties of algebras of type τ are subclasses of $\operatorname{Alg}(\tau)$ which are closed under homomorphic images, subalgebras, and direct products. Another way to define classes of algebras of type τ is as models of sets of equations. Now we will show that these approaches are equivalent.

Definition 7.10.1. We define the following operators on $\operatorname{Alg}(\tau)$. For any class $K \subseteq \operatorname{Alg}(\tau)$,

- $\mathbf{S}(K)$ is the class of all subalgebras of algebras from K,
- $\mathbf{H}(K)$ is the class of all homomorphic images of algebras from K,
- $\mathbf{P}(K)$ is the class of all direct products of families of algebras from K,
- $\mathbf{I}(K)$ is the class of all algebras which are isomorphic to algebras from K,
- $\mathbf{P}_S(K)$ is the class of all subdirect products of families of algebras from K.

We verify that some of our operators are closure operators. Here **IP** means applying both operators, **P** and **I**.

Lemma 7.10.2. *The operators* **H**, **S**, *and* **IP** *are closure operators on the set* Alg(τ).

Proof. We have to show that the operators are extensive, monotone, and idempotent. From the definition follows that for any subclasses K and L of Alg(τ), $K \subseteq L$ implies $\mathbf{H}(K) \subseteq \mathbf{H}(L)$, and since any algebra is a homomorphic image of itself under the identity homomorphism, we have $K \subseteq \mathbf{H}(K)$. By the extensivity and monotonicity, we have $\mathbf{H}(K) \subseteq \mathbf{H}(\mathbf{H}(K))$. Conversely, let $\mathcal{A}$ be in $\mathbf{H}(\mathbf{H}(K))$. Then there exists an algebra $\mathcal{B} \in \mathbf{H}(K)$ and a surjective homomorphism $\varphi : \mathcal{B} \to \mathcal{A}$. For $\mathcal{B} \in \mathbf{H}(K)$, there exists an algebra $\mathcal{C} \in K$ and a surjective homomorphism $\psi : \mathcal{C} \to \mathcal{B}$. Then the composition $\varphi \circ \psi : \mathcal{C} \to \mathcal{A}$ is also a surjective homomorphism, and thus $\mathcal{A} \in \mathbf{H}(K)$. The proofs for the other operators are similar and left to the reader. □

Definition 7.10.3. A class $K \subseteq$ Alg(τ) is called a *variety* if K is closed under the operators **H**, **S**, and **P**, that is, if $\mathbf{H}(K) \subseteq K$, $\mathbf{S}(K) \subseteq K$ and $\mathbf{P}(K) \subseteq K$ (corresponding to 7.5.15).

It is not difficult to see that some combinations of two of the operators have the following properties [21].

Lemma 7.10.4.
(i) $\mathbf{SH}(K) \subseteq \mathbf{HS}(K)$,
(ii) $\mathbf{PS}(K) \subseteq \mathbf{SP}(K)$,
(iii) $\mathbf{PH}(K) \subseteq \mathbf{HP}(K)$.

Using Lemma 7.10.4, we can obtain a characterization of the smallest variety to contain a class K of algebras.

Theorem 7.10.5. *For any class K of algebras of type τ, the class* $\mathbf{HSP}(K)$ *is the least variety (with respect to set inclusion) which contains K.*

Proof. First, we show that $\mathbf{HSP}(K)$ is a variety and therefore closed under application of **H**, **S**, and **P**. We have $\mathbf{H}(\mathbf{HSP}(K)) = \mathbf{HSP}(K)$ by the idempotence of **H**, and $\mathbf{S}(\mathbf{HSP}(K)) \subseteq \mathbf{H}(\mathbf{SSP}(K)) = \mathbf{HSP}(K)$ by (i) and the idempotence of **S**. For **P**, we have $\mathbf{P}(\mathbf{HSP}(K)) \subseteq (\mathbf{HPSP}(K)) \subseteq \mathbf{HSPP}(K) \subseteq \mathbf{HSIPIP}(K) = \mathbf{HSIP}(K) \subseteq \mathbf{HSHP}(K) \subseteq \mathbf{HHSP}(K) = \mathbf{HSP}(K)$, using the closure properties and properties from Lemma 7.10.4. Thus $\mathbf{HSP}(K)$ is a variety. Now let K' be any variety which contains K. Then $\mathbf{HSP}(K) \subseteq \mathbf{HSP}(K') \subseteq K'$ since, as a variety, K' is closed under all three operators. □

When K is a variety, then the closure of K under each of **H**, **S**, and **P** gives us $\mathbf{HSP}(K) \subseteq K$ as well, and hence $\mathbf{HSP}(K) = K$ for K a variety. Conversely, whenever $\mathbf{HSP}(K) = K$, we must have K a variety, as shown before.

Corollary 7.10.6. *A class K of algebras of type τ is a variety if and only if* $\mathbf{HSP}(K) = K$.

Theorem 7.10.7. *Every algebra of a variety K is isomorphic to a subdirect product of subdirectly irreducible algebras from K.*

Proof. We know already that every algebra $\mathcal{A}$ in K is isomorphic to a subdirect product of some subdirectly irreducible algebras $\mathcal{A}_j$ (Theorem 7.5.14). Each $\mathcal{A}_j$ is, in fact, isomorphic to a quotient algebra of $\mathcal{A}$, and is thus a homomorphic image of $\mathcal{A}$. Since $\mathcal{A} \in K$ and K is a variety, we see that $\mathcal{A}_j \in \mathbf{H}(K) \subseteq K$, too. □

Now we have

Lemma 7.10.8. *Let K be an equationally definable class of algebras of type τ. Then K is a variety.*

Proof. Let K be an equationally definable class of algebras of type τ. This means that there exists a set Σ of equations of type τ, over the alphabet X, such that $K = \operatorname{Mod}\Sigma$. To see that $\operatorname{Mod}\Sigma$ is closed under $\mathbf{H}$, $\mathbf{S}$, and $\mathbf{P}$, let $s \approx t \in \Sigma$ be any identity in Σ. Then we have $s^{\mathcal{A}} = t^{\mathcal{A}}$ for any algebra $\mathcal{A} \in K$. If $\mathcal{B}$ is a subalgebra of an $\mathcal{A}$ in K, then $t^{\mathcal{B}} = t^{\mathcal{A}}|B$ and therefore $s^{\mathcal{B}} = t^{\mathcal{B}}$; thus $s \approx t$ holds in $\mathcal{B}$ and $\mathcal{B} \in K$. Therefore $\mathbf{S}(K) \subseteq K$. In a similar way, we can show that $\mathbf{H}(K) \subseteq K$. Lastly, suppose $\mathcal{A}_j \in K$ and that $\mathcal{A}_j$ satisfies $s \approx t$ for each $j \in J$. Then for $\underline{a_1}, \ldots, \underline{a_n} \in C = \prod_{j\in J} \mathcal{A}_j$ we have $s^{\mathcal{A}_j}(a_1(j), \ldots, a_n(j)) = t^{\mathcal{A}_j}(a_1(j), \ldots, a_n(j)) \Rightarrow (s^{\mathcal{C}}(\underline{a_1}, \ldots, \underline{a_n}))(j) = (t^{\mathcal{C}}(\underline{a_1}, \ldots, \underline{a_n}))(j)$ for all $j \in J$ and thus $s^{\mathcal{C}} = t^{\mathcal{C}}$. Therefore the product $\mathcal{C}$ of algebras in K is in K, which shows that $\mathbf{P}(K) \subseteq K$. □

One can show (see [21]) that for any class K of algebras of type τ, and for any set Y of variables, the free algebra $\mathcal{F}_K(Y)$ with respect to K belongs to $\mathbf{ISP}(K)$.

Theorem 7.10.9. *For every class $K \subseteq \operatorname{Alg}(\tau)$ and every nonempty set Y of variables, the relatively free algebra $\mathcal{F}_K(Y)$ is in $\mathbf{ISP}(K)$.*

Since varieties are closed under the operators $\mathbf{I}$, $\mathbf{S}$, and $\mathbf{P}$, it follows from this theorem that every variety K contains all the relatively free algebras $\mathcal{F}_K(Y)$, for each nonempty set Y. The identities satisfied in a class K of algebras of type τ are exactly the identities satisfied in the free algebra $\mathcal{F}_K(X)$, where X as usual is our countably infinite set of variables.

Lemma 7.10.10. *Let K be a variety of algebras of type τ, and let s and t be terms in $W_\tau(X)$. Then*

$$K \models s \approx t \Leftrightarrow \mathcal{F}_K(X) \models s \approx t.$$

Proof. Since K is a variety, we know by the remark just above that the relatively free algebra $\mathcal{F}_K(X)$ is in K. This means that any identity of K must in particular hold in $\mathcal{F}_K(X)$, giving us one direction of the claim. If, conversely, $\mathcal{F}_K(X)$ satisfies $s \approx t$, then $s^{\mathcal{F}_K(X)} = t^{\mathcal{F}_K(X)}$. Since $\mathcal{F}_K(X)$ is the quotient of $\mathcal{F}_\tau(X)$ by $\operatorname{Id} K$, this forces $[s]_{\operatorname{Id} K} = [t]_{\operatorname{Id} K}$; and from this, we have $(s, t) \in \ker \operatorname{nat} \operatorname{Id} K = \operatorname{Id} K$, and so K satisfies $s \approx t$. □

Now we can prove Birkhoff's theorem.

Theorem 7.10.11 (Main theorem of equational theory). *A class K of algebras of type τ is equationally definable if and only if it is a variety.*

Proof. We have already proved one direction of this theorem. We will now prove the converse, namely that any variety K is also an equational class. Specifically, we will show that K equals the equational class $K' := \operatorname{Mod}\operatorname{Id} K$. Clearly, this class is a variety, and we have $K \subseteq \operatorname{Mod}\operatorname{Id} K = K'$. Applying the operator Id to $K' = \operatorname{Mod}\operatorname{Id} K$ gives us $\operatorname{Id} K = \operatorname{Id}\operatorname{Mod}\operatorname{Id} K' = \operatorname{Id} K'$, since Id Mod is a closure operator. But then we have $\mathcal{F}_K(Y) = \mathcal{F}_\tau(Y)/\operatorname{Id} K = \mathcal{F}_\tau(Y)/\operatorname{Id} K' = \mathcal{F}_{K'}(Y)$, for every nonempty set Y of variables. Now let $\mathcal{A}$ be any algebra in K', and choose a set Y of variables such that $|Y| = |A|$. There will be a mapping $f : Y \to A$ which is surjective, and we know that this mapping has a unique extension to a homomorphism from $\mathcal{F}_{K'}(Y)$ to $\mathcal{A}$, also surjective. This makes $\mathcal{A}$ a homomorphic image of $\mathcal{F}_{K'}(Y)$, which in turn is equal to $\mathcal{F}_K(Y)$. But this relatively free algebra is in the variety K, and K is closed under homomorphic images, so we have $\mathcal{A}$ in K. This shows that $K' \subseteq K$, and finishes our proof that $K = K' = \operatorname{Mod}\operatorname{Id} K$. □

Without proofs, we want to mention some more facts:

(i) Let K be a variety of type τ and let Y be any nonempty variable set. The relatively free algebra $\mathcal{F}_K(Y)$ with respect to K over Y is isomorphic to a subdirect product of the algebras $\mathcal{F}_K(E)$, for $E \subseteq Y$ nonempty and finite.

(ii) For every variety K,

$$K = \mathbf{HSP}(\{\mathcal{F}_K(n) \mid n \in \mathbb{N}, n \geq 1\}) = \mathbf{HSP}(\{\mathcal{F}_K(X)\}).$$

7.11 Generalizations

In several mathematical systems, it is natural to have more than one set of objects and operations with inputs from one set and outputs from another. We want to give some examples of this situation and present some ideas of the theory of many-sorted algebras (which are also called heterogeneous or multi-based algebras).

Our first example is a K-vector space. We use two sets, the set V of vectors and the universe K of a field. As operations we have the addition of vectors $+ : V \times V \to V$ and the scalar multiplication $\cdot : K \times V \to V$ of an element from K with a vector. Then the structure $(V, K; +, \cdot)$ is an example of a heterogeneous algebra.

The algebraic theory of heterogeneous algebras can be developed in an analogous way to one-sorted algebras. The universes of subalgebras of a many-sorted algebra are sequences of subsets of the sets forming the sequence of the given algebra, which are closed under application of the heterogeneous operations. Homomorphisms are sequences of mappings which are compatible with the heterogeneous operations. The subalgebra criterion, the homomorphism theorem, and almost all other results for the one-based case are also valid in the multi-based case. Direct and subdirect products can be defined using the corresponding sets from the sequences of the universes of the many-sorted algebras whose products will be calculated.

Now we want to show a bit more of the formal aspects of the algebraic basic concepts on heterogeneous algebras and use the approach of Higgins [36]; see also [10] or [49].

I-sorted sets

Let I be a nonempty set and let A_i be sets indexed by elements $i \in I$. Then $A := (A_i)_{i\in I}$ is called an I-indexed family of sets and A_i is called the set of sort i. For I-sorted sets, we define the usual set-theoretic operations and relations in the following way:

I-sorted subsets

Let $(A_i)_{i\in I}$ and $(B_i)_{i\in I}$ be I-sorted sets. Then we define

$$A \subseteq B :\Leftrightarrow \forall i \in I\ (A_i \subseteq B_i).$$

I-sorted union, intersection, and Cartesian product

$$A \cap B := (A_i \cap B_i)_{i\in I}, \quad A \cup B := (A_i \cup B_i)_{i\in I}, \quad A \times B := (A_i \times B_i)_{i\in I}.$$

(Here the inclusions and operations on the right-hand side are the usual set-theoretical inclusions and operations.)

I-sorted relations, mappings, and quotient sets

Let θ_i be an n_i-ary relation on $A_i, i \in I$, i. e., $\theta_i \subseteq A_i^{n_i}$. Then $(\theta_i)_{i\in I}$ is called an I-sorted relation. If $\theta_i \subseteq A_i^2$ are equivalence relations, then $\theta = (\theta_i)_{i\in I}$ is said to be an I-sorted equivalence relation and $A/\theta := (A_i/\theta_i)_{i\in I}$ is called I-sorted factor algebra (quotient algebra). Furthermore, $f := (f_i)_{i\in I}$ is called an I-sorted mapping, written as $f : A \to B$ with $A := (A_i)_{i\in I}, B := (B_i)_{i\in I}$, if $f_i : A_i \to B_i$ are usual mappings, while $\ker f := (\ker f_i)_{i\in I}$ is the I-sorted kernel of the I-sorted mapping f.

With all these definitions in hand, it is not difficult and left as an exercise to the reader to formulate the I-sorted version of the diagram lemma.

I-sorted operations

Let I be a nonempty set and $n \in \mathbb{N}^+ = \mathbb{N} \setminus \{0\}$. We define $I^* := \bigcup_{n\in\mathbb{N}^+} I^n$ and $\Sigma \subseteq I^* \times I$. For each $\gamma \in \Sigma$, let K_γ be a set of indices. An n-ary operation on the I-sorted set $A = (A_i)_{i\in I}$ is a mapping

$$((f_\gamma)_k)^A : A_{k_1} \times \cdots \times A_{k_n} \to A_i$$

for $\gamma = (k_1, \ldots, k_n, i) \in \Sigma$, $k \in K_\gamma$.

Σ-algebras

A Σ-algebra (many-sorted algebra) is a pair $\mathcal{A} = (A; (((f_\gamma)_k)^{\mathcal{A}})_{k\in K_\gamma,\gamma\in\Sigma})$ where $A = (A_i)_{i\in I}$ is an I-sorted set and $(((f_\gamma)_k)^{\mathcal{A}})_{\gamma\in K_\gamma}$ is a sequence of operations on A indexed by a nonempty index set Σ. The set A is called the universe and $(((f_\gamma)_k)^{\mathcal{A}})_{k\in K_\gamma}$ is called the sequence of fundamental operations of $\mathcal{A}$. If $|\gamma|$ is the length of γ, then $|\gamma| - 1$ is the arity of $((f_\gamma)_k)^{\mathcal{A}}$. Let Alg(Σ) be the class of all Σ-algebras.

Σ-subalgebras

Let $\mathcal{A} = (A; (((f_\gamma)_k)^{\mathcal{A}})_{k\in K_\gamma,\gamma\in\Sigma})$ be a Σ-algebra. Then the Σ-algebra $\mathcal{B} = (B; (((f_\gamma)_k)^{\mathcal{B}})_{k\in K_\gamma,\gamma\in\Sigma})$ is said to be a Σ-subalgebra of $\mathcal{A}$, written as $\mathcal{B} \subseteq \mathcal{A}$ if

(i) $B \subseteq A$,

(ii) for each $k \in K_\gamma$, $\gamma \in \Sigma$,

$$((f_\gamma)_k)^{\mathcal{B}} = ((f_\gamma)_k)^{\mathcal{A}}|_{B_{k_1}\times\cdots\times B_{k_n}}$$

is the restriction of $((f_\gamma)_k)^{\mathcal{A}}$ onto the subset $B_{k_1} \times \cdots \times B_{k_n}$ of $A_{k_1} \times \cdots \times A_{k_n}$.

Σ-congruences

The I-sorted equivalence $\theta = (\theta_i)_{i\in I}$ is called a Σ-congruence of $\mathcal{A} = (A; (((f_\gamma)_k)^{\mathcal{A}})_{k\in K_\gamma,\gamma\in\Sigma})$ if

$$\begin{aligned}&(a_1, b_1) \in \theta_{k_1}, \dots, (a_n, b_n) \in \theta_{k_n} \Rightarrow \\ &\quad (((f_\gamma)_k)^{\mathcal{A}}(a_1, \dots, a_n), ((f_\gamma)_k)^{\mathcal{A}}(b_1, \dots, b_n)) \in \theta_i\end{aligned}$$

for all $a_q, b_q \in A_{k_q}$, $1 \le q \le n$, $n \in \mathbb{N} \setminus \{0\}$, $\gamma = (k_1, \dots, k_n, i)$. The quotient algebra is defined to be $\mathcal{A}/\theta = (A/\theta; (((f_\gamma)_k)^{\mathcal{A}}/\theta)_{k\in K_\gamma,\gamma\in\Sigma})$ with

$$((f_\gamma)_k)^{\mathcal{A}}/\theta([a_1]_{\theta_{k_1}}, \dots, [a_n]_{\theta_{k_n}}) = [((f_\gamma)_k)^{\mathcal{A}}(a_1, \dots, a_n)]_{\theta_i}.$$

Σ-homomorphisms

An I-sorted mapping $\varphi : A \to B$ is said to be a homomorphism from $\mathcal{A} = (A; (((f_\gamma)_k)^{\mathcal{A}})_{k\in K_\gamma,\gamma\in\Sigma})$ to $\mathcal{B} = (B; (((f_\gamma)_k)^{\mathcal{B}})_{k\in K_\gamma,\gamma\in\Sigma})$ if

$$\varphi((((f_\gamma)_k)^{\mathcal{A}}))(a_1, \dots, a_n) = (\varphi_{k_1}(a_1), \dots, \varphi_{k_n}(a_n))$$

for all $k \in K_\gamma$, $\gamma \in \Sigma$ and all $a_q \in A_{k_q}$, $1 \le q \le n$, $n \in \mathbb{N} \setminus \{0\}$, $\gamma = (k_1, \dots, k_n, i)$.

Σ-direct product

Let $\{\mathcal{A}_j \mid j \in J\}$ be a family of Σ-algebras. The direct product of this family is the Σ-algebra $\mathcal{P} := (P; (((f_\gamma)_k)^{\mathcal{P}})_{k\in K_\gamma, \gamma\in\Sigma})$ where $P := (P_i)_{i\in I}$, $P_i := \prod_{j\in J} A_{ij} = \{(a_{ij})_{j\in J} \mid a_{ij} \in A_{ij}\}$ is the Cartesian product of the A_{ij}, $i \in I, j \in J, k \in K_\gamma$, $\gamma \in \Sigma$, and $\gamma = (k_1, \ldots, k_n, i)$.

Now Σ-varieties can be defined as classes of Σ-algebras closed under Σ-subalgebras, Σ-homomorphisms, and Σ-direct products. To prove the multi-based variant of Birkhoff's theorem, we have to define Σ-terms, equations, identities, and equationally defined classes of Σ-algebras (see also [44]).

We will describe another important example of many-sorted algebras. In Section 7.1 we have already introduced the set $O^n(A)$ of all n-ary operations defined on set A. We recall the notation $O(A) = \bigcup_{n\geq 1} O^n(A)$ for the set of all finitary operations defined on A.

Definition 7.11.1. Let $O^n(A), n \geq 1$, be the set of all n-ary operations defined on set A. For each $n, m \geq 1$ we define an operation

$$S_m^{n,A} : O^n(A) \times \left(O^m(A)\right)^n \to O^m(A)$$

by

$$S_m^{n,A}(f^A, g_1^A, \ldots, g_n^A)(a_1, \ldots, a_m) := f^A(g_1^A(a_1, \ldots, a_m), \ldots, g_n^A(a_1, \ldots, a_m))$$

for all $a_1, \ldots, a_m \in A$.

The operations $S_m^{n,A}$ are called superposition operations on $O(A)$. Since $O(A)$ contains all finitary operations defined on A and the operations $S_m^{n,A}$ produce finitary operations, $O(A)$ is closed under the operations $S_m^{n,A}$. A clone of operations is a subset of $O(A)$ which is closed under all $S_m^{n,A}, m, n \in \mathbb{N} \setminus \{0\}$ and contains all projection operations $e_i^{n,A}, 1 \leq i \leq n$, defined by $e_i^{n,A}(a_1, \ldots, a_n) := a_i$ for all $a_1, \ldots, a_i, \ldots, a_n \in A$. Further, $O(A)$ is clearly a clone and is called the full clone of operations defined on A. The full clone can be regarded as many-sorted algebra

$$\text{clone}\, A := \left((O^n(A))_{n\geq 1}, (S_m^{n,A})_{m,n\in\mathbb{N}, m,n\geq 1}, (e_i^{n,A})_{n\in\mathbb{N}\setminus\{0\}, 1\leq i\leq n}\right).$$

We mention that the sorts are the sets $O^n(A)$, i. e., there are infinitely many sorts and also infinitely many operations $S_m^{n,A}$. The projections are nullary operations.

Definition 7.11.2. Let $C \subseteq O(A)$. The clone generated by C is the smallest subset of $O(A)$ which contains C and is closed under the operations $S_m^{n,A}$ and contains all projections $e_j^{n,A}, 1 \leq j \leq n$.

The set $W_\tau(X_n)$ of all n-ary terms of type τ was defined in Section 7.6. Now we define superposition operations S_m^n on sets $W_\tau(X_n)$,

$$S_m^n : W_\tau(X_n) \times \left(W_\tau(X_m)\right)^n \to W_\tau(X_m),$$

inductively by

(i) $S^n_m(x_i, t_1, \ldots, t_n) := t_i$ for n-ary variables and

(ii) $S^n_m(f_i(s_1, \ldots, s_{n_i}), t_1, \ldots, t_n) := f_i(S^n_m(s_1, t_1, \ldots, t_n), \ldots, S^n_m(s_{n_i}, t_1, \ldots, t_n))$ for n-ary composite terms.

Just as the full clone of all operations on a set A was viewed as a multi-based algebra, we can regard the sets of all n-ary terms as the universes of a multi-based algebra. This is the algebra

$$\operatorname{clone} \tau = ((W_\tau(X_n))_{n \in \mathbb{N}\setminus\{0\}}, (S^n_m)_{m,n \in \mathbb{N}\setminus\{0\}}, (x_i)_{1 \le i \le n, n \in \mathbb{N}\setminus\{0\}}),$$

called the term clone of type τ. (Actually, one speaks of a clone only if three identities (C1), (C2), (C3) are satisfied which generalize the defining identities of a monoid where (C1) is a generalization of the associative law; for the details, see [24].)

Remark 7.11.3. The set of term operations of an algebra $\mathcal{A} = (A; (f_i^{\mathcal{A}})_{i \in I})$ of type τ was defined as a set of all operations induced by terms of type τ on algebra $\mathcal{A}$. It can be proved that this set is a clone which is precisely the subclone of clone A generated by the fundamental operations of algebra $\mathcal{A}$.

There are easy examples of "correspondences" on A which are not everywhere defined, but if defined then they are unique mappings. The division $a : b, b \neq 0$, of natural numbers gives a result in $\mathbb{N}$ only if a is a multiple of b and is not defined for other cases. Let

$$f_i^A : A^n \multimap A$$

be an n-ary partial operation and let

$$P^n(A) := \{f^A \mid A^n \multimap A\}$$

be the set of all n-ary partial operations defined on A. By $\operatorname{dom} f^A$ we denote the domain of the partial operation f^A. The pair $\mathcal{A} = (A; (f_i^{\mathcal{A}})_{i \in I})$ with $f_i^{\mathcal{A}} \in P^{n_i}(A)$ is said to be a partial algebra of type τ. This is our second generalization of the concept of an algebra of type τ.

We recall some basic concepts for partial algebras. For them, there are several possibilities to define subalgebras, homomorphic images, and identities. For $f^A \in P^n(A)$, we define the *graph* of a partial operation f^A as

$$\operatorname{graph} f^A := \{(a_1, \ldots, a_n, a) \mid (a_1, \ldots, a_n) \in \operatorname{dom} f^A \text{ and } a = f^A(a_1, \ldots, a_n)\}.$$

Subalgebras

Let $\mathcal{A} = (A; (f_i^{\mathcal{A}})_{i \in I})$ and $\mathcal{B} = (B; (f_i^{\mathcal{B}})_{i \in I})$ be partial algebras of type τ. Then $\mathcal{B}$ is said to be a *weak subalgebra* of $\mathcal{A}$ if

$$\operatorname{graph} f_i^{\mathcal{B}} \subseteq \operatorname{graph} f_i^{\mathcal{A}}.$$

Also, $\mathcal{B}$ is said to be a *relative subalgebra* of $\mathcal{A}$ if

$$\operatorname{graph} f_i^{\mathcal{B}} = \operatorname{graph} f_i^{\mathcal{A}} \cap B^{n_i+1}.$$

Homomorphisms

Let $\mathcal{A} = (A; (f_i^{\mathcal{A}})_{i \in I})$ and $\mathcal{B} = (B; (f_i^{\mathcal{B}})_{i \in I})$ be partial algebras of type τ. A function $h : A \to B$ is called a *weak homomorphism* of $\mathcal{A}$ into $\mathcal{B}$ if the following is satisfied: if $(a_1, \ldots, a_{n_i}) \in \operatorname{dom} f_i^{\mathcal{A}}$ then $(h(a_1), \ldots, h(a_{n_i})) \in \operatorname{dom} f_i^{\mathcal{B}}$ and then

$$h(f_i^{\mathcal{A}}(a_1, \ldots, a_{n_i})) = f_i^{\mathcal{B}}(h(a_1), \ldots, h(a_{n_i})), \quad i \in I.$$

A weak homomorphism is called a *closed homomorphism* of $\mathcal{A}$ into $\mathcal{B}$ if for all $i \in I$ and for all $(a_1, \ldots, a_{n_i}) \in A^{n_i}$ there holds

$$\text{if } (h(a_1), \ldots, h(a_{n_i})) \in \operatorname{dom} f_i^{\mathcal{B}} \quad \text{then } (a_1, \ldots, a_{n_i}) \in \operatorname{dom} f_i^{\mathcal{A}}.$$

Direct products

Let $(\mathcal{A}_j)_{j \in J}$ be a family of partial algebras of type τ. Then the *direct product* $\prod_{j \in J} \mathcal{A}_j$ is a partial algebra with $\prod_{j \in J} A_j$ as universe and $f_i^{\prod_{j \in J} A_j}$ defined by

$$f_i^{\prod_{j \in J} \mathcal{A}_j}((a_{1j})_{j \in J}, \ldots, (a_{n_i j})_{j \in J}) = (f_i^{\mathcal{A}_j}(a_{1j}, \ldots, a_{n_i j}))_{j \in J}.$$

Here the left-hand side is defined if and only if for all $j \in J$ we have $(a_{1j}, \ldots, a_{n_i j}) \in \operatorname{dom} f_i^{\mathcal{A}_j}$.

Identities

The pair $s \approx t$ of terms of type τ is a *weak identity* of the partial algebra $\mathcal{A}$ of type τ if after evaluation of the terms s, t by term operations $s^{\mathcal{A}}$, $t^{\mathcal{A}}$ the following is satisfied: if one side of the equation is defined, then the other is defined and both sides are equal. This means that if the left-hand side of $s \approx t$ is defined, then the right-hand side is defined

or

if the right-hand side of $s \approx t$ is defined, then the left-hand side is defined, and after evaluation both sides are equal.

If we replace or by and, i. e., if the left-hand side of $s \approx t$ is defined, then the right-hand side is defined

and

if the right-hand side of $s \approx t$ is defined, then the left-hand side is defined, and after evaluation both sides are equal; $s \approx t$ is called a *strong identity* in the partial algebra $\mathcal{A}$.

From the definition, it becomes clear that each strong identity is a weak identity, but in general not conversely.

We consider the following examples of weak identities in a partial algebra. A term of type τ in which each variable occurs only once is called *linear*. For a formal definition, we use $\text{var}(t)$ for the set of variables occurring in t.

Definition 7.11.4. An n-ary linear term of type τ is defined in the following inductive way:

(i) $x_j \in X_n = \{x_1, \ldots, x_n\}$ is for any $j \in \{1, \ldots, n\}$ an n-ary linear term of type τ.

(ii) If $t_1, \ldots, t_{n_i}$ are n-ary linear terms of type τ, if f_i is an n_i-ary operation symbol of type τ and if $\text{var}(t_j) \cap \text{var}(t_k) = \emptyset$ for all $1 \leq j < k \leq n_i$, then $f_i(t_1, \ldots, t_{n_i})$ is an n-ary linear term of type τ.

(iii) The set $W_\tau^{\text{lin}}(X_n)$ of all n-ary linear terms of type τ is the least set which contains $x_1, \ldots, x_n$ and is closed under finite application of (ii).

For $m = n$, the many-sorted operation S_m^n defined on finitary sets of terms is an $(n+1)$-ary operation defined on set $W_\tau(X_n)$,

$$S_n^n : (W_\tau(X_n))^{n+1} \to W_\tau(X_n).$$

Together with the variables $x_1, \ldots, x_n$ regarded as nullary operations, one can form the algebra $(W_\tau(X_n); S_n^n, x_1, \ldots, x_n)$ of type $\tau = (n+1, 0, \ldots, 0)$ which is an example of a *unitary Menger algebra* of rank n since it satisfies the identities

(i) $S_n^n(t, S_n^n(t_1, s_1, \ldots, s_n), \ldots, S_n^n(t_n, s_1, \ldots, s_n)) \approx S_n^n(S_n^n(t, t_1, \ldots, t_n), s_1, \ldots, s_n)$;

(ii) $S_n^n(x_i, t_1, \ldots, t_n) \approx t_i$;

(iii) $S_n^n(t, x_1, \ldots, x_n) \approx t$.

(For more details, see [24]).

The set $W_\tau^{\text{lin}}(X_n)$ is in general not closed under S_n^n as the following example for $n = 2$ shows. Let $\tau = (2)$ and f be a binary operation symbol. Then $t = f(x_2, x_1)$, $s_1 = f(x_1, x_2)$, and $s_2 = x_1$ are linear, but $S_2^2(t, s_1, s_2) = S_2^2(f(x_2, x_1), f(x_1, x_2), x_1) = f(x_1, f(x_1, x_2))$ is not since it contains x_1 twice.

We define a partial operation

$$\bar{S}_n^n : (W_\tau^{\text{lin}}(X_n))^{n+1} \multimap W_\tau^{\text{lin}}(X_n)$$

by

$$\bar{S}_n^n(t, s_1, \ldots, s_n) := \begin{cases} S_n^n(t, s_1, \ldots, s_n) & \text{if } \text{var}(s_j) \cap \text{var}(s_k) = \emptyset,\ 1 \leq j < k \leq n, \\ \text{not defined} & \text{otherwise,} \end{cases}$$

and define a partial algebra

$$(W_\tau^{\mathrm{lin}}(X_n); \bar{S}_n^n, x_1, \dots, x_n)$$

of type $(n+1, 0, \dots, 0)$.

Now we will show that the partial algebra

$$(W_\tau^{\mathrm{lin}}(X_n); \bar{S}_n^n, x_1, \dots, x_n)$$

of type $(n+1, 0, \dots, 0)$ satisfies

(i') $\bar{S}_n^n(t, \bar{S}_n^n(t_1, s_1, \dots, s_n), \dots, \bar{S}_n^n(t_n, s_1, \dots, s_n)) \approx \bar{S}_n^n(\bar{S}_n^n(t, t_1, \dots, t_n), s_1, \dots, s_n)$;

(ii') $\bar{S}_n^n(x_i, t_1, \dots, t_n) \approx t_i$;

(iii') $\bar{S}_n^n(t, x_1, \dots, x_n) \approx t$

as weak identities and that the partial algebra $(W_\tau^{\mathrm{lin}}(X_n); \bar{S}_n^n)$ of type (n) satisfies (i') as a weak identity. Such algebras are called *unitary partial Menger algebras of rank n* (of type $(n+1, 0, \dots, 0)$) and *partial Menger algebras of rank n* of type $(n+1)$, respectively. If (i') is satisfied as a strong identity, then we speak of a *(unitary) strong partial Menger algebra of rank n.*

Now we prove

Theorem 7.11.5. *The partial algebra $(W_\tau^{\mathrm{lin}}(X_n); \bar{S}_n^n, x_1, \dots, x_n)$ of type $(n+1, 0, \dots, 0)$ is a unitary partial Menger algebra of rank n and the partial algebra $(W_\tau^{\mathrm{lin}}(X_n); \bar{S}_n^n)$ of type $(n+1)$ is a strong partial Menger algebra of rank n.*

Proof. Assume that the right-hand side of (i') is defined. Then $\bar{S}_m^n(t, t_1, \dots, t_n)$ is defined and $\mathrm{var}(t_j) \cap \mathrm{var}(t_k) = \mathrm{var}(s_j) \cap \mathrm{var}(s_k) = \emptyset$ for all $1 \le j < k \le n$. Then $\bar{S}_n^n(t_1, s_1, \dots, s_n), \dots, \bar{S}_n^n(t_n, s_1, \dots, s_n)$ are defined and, since $\mathrm{var}(t_j) \cap \mathrm{var}(t_k) = \emptyset$, we get $\mathrm{var}(\bar{S}_n^n(t_j, s_1, \dots, s_n)) \cap \mathrm{var}(\bar{S}_n^n(t_k, s_1, \dots, s_k)) = \emptyset$ for all $1 \le j < k \le n$ and thus the left-hand side of (i') is defined. By (i), both sides are equal and (i') is satisfied as a weak identity in both partial algebras. Clearly, (ii') and (iii') are satisfied as weak identities in the partial algebra $(W_\tau^{\mathrm{lin}}(X_n); \bar{S}_n^n, x_1, \dots, x_n)$ of type $(n+1, 0, \dots, 0)$. Indeed, if the left-hand sides are defined, then the right-hand sides are defined since they are always defined. By (ii) and (iii), the equations (ii') and (iii') are weak identities and thus the first algebra is a unitary partial Menger algebra of rank n. Further, (ii') and (iii') are not satisfied as strong identities in the first algebra, since the right-hand sides are defined even if the left-hand sides are not defined. To show that (i') is satisfies as a strong identity in both algebras, we assume that the left-hand side of (i') is defined. Then $\mathrm{var}(s_j) \cap \mathrm{var}(s_k) = \emptyset = \mathrm{var}(\bar{S}_n^n(t_j, s_1, \dots, s_n)) \cap \mathrm{var}(\bar{S}_n^n(t_k, s_1, \dots, s_k))$ for all $1 \le j < k \le n$. This is only possible if $\mathrm{var}(t_j) \cap \mathrm{var}(t_k) = \emptyset$ for all $1 \le j < k \le n$. Then the right-hand side is defined and because of (i) both sides are equal. This proves that (i') is satisfied as a strong identity in both algebras and that $(W_\tau^{\mathrm{lin}}(X_n); \bar{S}_n^n)$ of type $(n+1)$ is a strong partial Menger algebra of rank n. □

As generalizations of unitary partial Menger algebras of rank n, many-sorted partial algebras

$$\text{clone}_{\text{lin}}\tau = ((W_\tau(X_n))^{\text{lin}}_{n\in\mathbb{N}\setminus\{0\}}; (S^n_m)_{m,n\in\mathbb{N}\setminus\{0\}}, (x_i)_{1\le i\le n, n\in\mathbb{N}\setminus\{0\}})$$

can be considered.

It can be proved that these algebras satisfy (C1), (C2), (C3) as weak identities (see [24]) and are called *partial clones*.

In Chapter 5, Definition 5.1.15, we introduced directed graphs as pairs $G = (V, E)$ where E is the set of vertices and the set $E \subseteq V \times V$ of edges is a binary relation on V. As for algebras, one can define subgraphs, homomorphic images, and direct products. Therefore graphs are algebraic structures consisting of a set and a relation on this set. This can be generalized to algebraic systems of type (τ, τ') which are triples $\mathcal{A} = (A; (f_i^{\mathcal{A}})_{i\in I}, (\rho_j^{\mathcal{A}})_{j\in J})$ consisting of a set A, an indexed set of operations defined on A, where $f_i^{\mathcal{A}}$ is n_i-ary, and an indexed set of relations defined on A where $(\rho_j^{\mathcal{A}})_{j\in J}$ is n_j-ary. The theory of one-based algebraic systems can be developed similar to that of universal algebra (see, e. g., [50]). Other generalizations are partial algebraic systems, many-sorted algebraic systems, or partial many-sorted algebraic systems.

7.12 Historical remarks

The word "algebra" was mentioned first in the book "Hirab aljabr w'almuqabalah" of the Arab mathematician *Al-Hwarazmi*. He lived around 800 in Bagdad. This was the time of *Harun ar Raschid's* regency. Bagdad was a famous center of science and art. Mathematics and natural sciences were considered as important parts of human culture. The Arab mathematicians could follow the traces of Indian, Chinese, and Greek mathematicians. The representation of numbers by digits and the decimal system were already known. Al-Hwarazmi explained in his book the decimal system. This is why we speak up to our days of arabic digits. He mentioned also the word "algorithm." The Arab mathematicians could understand and solve quadratic equations and tried to solve algebraic equations of higher degrees. Around 1143, *Hermann of Carinthia* translated the mathematical books from Arabic into Latin. Only when the economical development demanded solutions of algebraic equations of higher degree, there was further progress. In the sixteenth century, the Italian mathematicians *N. Tartaglia* (1499–1557) and *G. Cardano* (1501–1576) solved algebraic equations of the third and fourth degree by radicals similar to the solutions of quadratic equations and discovered the complex numbers.

A milestone in the applications of algebraic methods in other fields of mathematics was the algebraization of geometry contained in *R. Descarte's* (1596–1650) work "Discours de la méthode." From then on, it was possible to describe geometrical objects by numbers and equations. The problem to solve algebraic equations of higher than fourth degree remained open. In the nineteenth century, the problem was attacked from a new direction. Based on former results of *Lagrange, Ruffing*, and *Gauss* in 1824, *N. H. Abel* (1802–1829) proved the impossibility to solve the general algebraic equation of degree five by radicals and only a short time later *E. Galois* (1811–1832) was able to give necessary and sufficient conditions to solve the general algebraic equation by radicals. This was the birth of group theory and Galois theory, two main areas of abstract algebra. As a result, algebra changed from being a theory of solutions of algebraic equations to the theory of algebraic structures. Galois' paper "Memoirs sur les conditions de résolubilités des équations par radicaux" is perhaps the most famous and most controversial paper in the history of algebra. Almost everybody dealing with the concept of a Galois connection cites the fundamental theorem of Galois theory as the classical prototype of such connections. However, that theorem, stated in its modern form as a dual isomorphism between the subfield lattice of a Galois extension and the subgroup lattice of the corresponding Galois group, was explicitly formulated quite some time after Galois. Galois was involved in the French revolution from 1830. Because of his political activities, he was arrested. Shortly after he became free and left prison, he fought in a duel and died, only twenty years old, of the wounds he suffered.

New classes of algebraic structures were considered by British algebraists in the nineteenth century. The discovery of the quaternions by *W. R. Hamilton* (1805–1865), the calculation with matrices by Cayley, and the definition of associative algebras by *R. S. Peirce* were important steps to general algebra. *G. Boole's* (1815–1865) books "The mathematical analysis of logic" and "An investigation of the laws of thought" linked algebra with propositional logic. Boolean algebras and the algebra of all Boolean functions are fundamentals of computer science.

In the first half of the twentieth century, important classes of algebraic structures such as groups, fields, rings, lattices, and semigroups were developed. In the 1920s and 1930s, many mathematicians became aware of a radical reorganization of algebra. This reorganization, specifically the transformation of algebra into a set-theoretical axiomatic science having as its primary object the study of algebraic operations performed on elements of an arbitrary set, had been prepared by the whole preceding development of algebra. It began at the end of the nineteenth century and continued with increasing momentum during the first decades of the twentieth. The publication of the two volumes of *B. L. van der Waerden's* (1903–1996) "Modern Algebra" [65] made the ideas, results, and methods of this new algebra accessible to all mathematicians, including nonalgebraists. It is common knowledge how decisive the impact of this modern algebra was on the development of other domains of mathematics. We mention here only topology and functional analysis. Van der Waerden himself decided, following a suggestion of H. Brandt, from the fourth edition on to change the title of his book into "Algebra." In his review of the book (Annual report of "Deutsche Mathematiker Vereinigung" 55) *H. Brandt* (1886–1954) wrote:

> Was den Titel anbetrifft, so würde ich es begrüßen, wenn in der vierten Auflage der schlichtere, aber kräftigere Titel "Algebra" gewählt würde. Ein Buch, das so viel an bester Mathematik bietet, wie sie war, ist und sein wird, sollte nicht durch den Titel den Verdacht erwecken, als ob es nur einer Modeströmung folgte, die gestern noch unbekannt war und vielleicht morgen vergessen sein wird.

During the following decades, the stormy development of algebra itself, as well as the discovery of several links with disciplines in the neighborhood, created generalizations of groups and rings and other "classical" algebraic structures. "Modern algebra" changed to "general algebra." During these decades, the older branches of general algebra – the theory of fields and of associative and associative–commutative rings, to which van der Waerden's book was mainly devoted – have undergone far-reaching changes. Even more crucial was the reorientation in the theory of groups, the oldest of all branches of general algebra. The theory of rings became more a theory of nonassociative rings, incorporating as a constituent part the theory of Lie rings and Lie algebras. Topological algebra sprang up and soon occupied a very prominent position, and a parallel development happened in the theory of ordered algebraic structures. Lattice theory appeared and developed rapidly. Within the framework of the classical parts of general algebra, independent trends arose: homological algebra leading to numerous results in topology and algebraic geometry, projective algebra, including the elements of projective geometry, and differential algebra. The theory of semigroups, that of quasigroups, and of n-groups ceased to be simply theories of "generalized groups" and found their own paths of development and their own areas of application. Eventually, the general theory of universal algebra came into being, as well as model theory, which is interwoven with mathematical logic. Category theory allows a wide overview on algebraic thinking. This corresponds with the need of algebraic-structural thinking in applied areas, especially in computer science, but also in economics.

8 An algebraic approach to production systems

8.1 Introduction

Production processes are traditionally described by input–output models. Production is a way that transforms inputs like hours of human labor, machine hours, raw material, energy, water, semifinished goods into outputs like final goods, intermediate goods, commodities, and others. We start with a simple example.

A factory of battery systems wants to produce 200 small and 100 big battery systems per day. One needs 2000 cells, 300 cell contact systems, 200 small casings, and 100 big casings, as well as 7000 working minutes per day. The corresponding production system can be visualized by an input–output graph (see Figure 8.1) as follows.

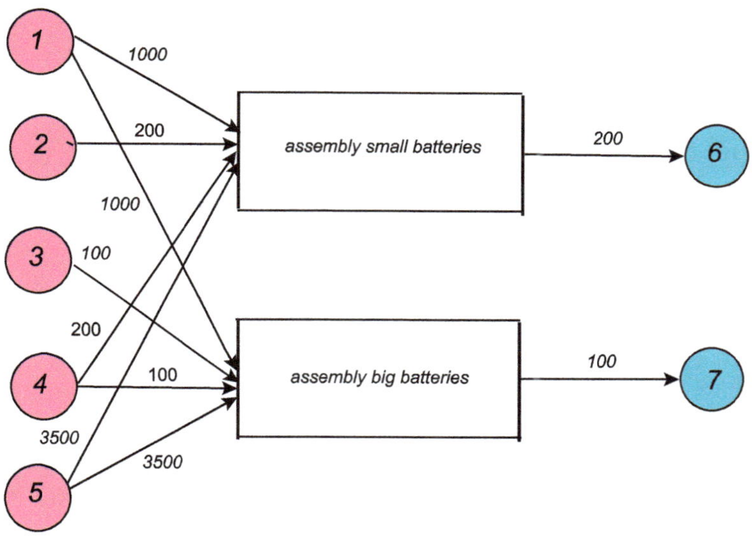

Figure 8.1: Input–output system.

1 – cells
2 – small casings
3 – big casings
4 – cell contact systems
5 – minutes of human labor
6 – small batteries
7 – big batteries

https://doi.org/10.1515/9783112234433-008

8.2 Input–output graphs and matrices

In Chapter 5 we introduced the basic concepts of graphs. We mentioned that an abstract graph is a pair $G = (V, E)$, where V is an arbitrary set and E is a set of two-element subsets of V, i. e.,

$$E \subseteq \{\{v, w\} \mid v, w \in E, v \neq w\}.$$

The elements of V are said to be vertices and the elements of E are called edges. A directed graph is a pair $G = (V, P)$, where V is a set and P is a set of ordered pairs of elements of V, i. e.,

$$P \subseteq \{(v, w) \mid v, w \in E\}.$$

For the application in the theory of production, we consider directed graphs. If V is a finite set then the graph is said to be finite. Vertices and edges of a graph can be valued by integers. A sequence $(v_0, \dots, v_n)$ of vertices of G is said to be a path if no vertex appears more than once and a path which starts and ends with the same vertex is said to be a cycle.

Let $G = (V, P)$ be a directed graph. A directed graph $G' = (V', P')$ is said to be a subgraph of G if $V' \subseteq V$ and $P' \subseteq P$. The subgraph G' of G is said to be induced by $V' \subseteq V$ if for all $i, j \in V'$,

$$(i, j) \in P \Rightarrow (i, j) \in P'.$$

If $G_i = (V_i, P_i)$, $i = 1, 2$, are two directed graphs, then $G_1 \cup G_2 = (V_1 \cup V_2, P_1 \cup P_2)$ is called union of G_1 and G_2.

The following concept of an input–output graph was used in [28].

Definition 8.2.1. An abstract input–output graph G_{I-S-O} is a finite directed edge-valued graph where the set of vertices is the union of three nonempty sets, $I \cup S \cup O$ with $I \cap S \neq \emptyset$, $I \cap O \neq \emptyset$, and $S \cap O \neq \emptyset$, and where the directed edges are subsets of $I \times S$ or subsets of $S \times O$, i. e., connect vertices from I with vertices from S, or vertices from S with vertices from O.

Clearly, any input–output graph G_{I-S-O} is a 3-partite graph in the sense of Chapter 5, Definition 5.4.1.

Definition 8.2.2. The input–output graph $G_{I'-S'-O'}$ is said to be subgraph of the input–output graph G_{I-S-O} if $I' \subseteq I$, $S' \subseteq S$, $O' \subseteq O$, and if all edges from $G_{I'-S'-O'}$ are edges from G_{I-S-O}. The subgraph $G_{I'-S'-O'}$ is said to be induced by (I', S', O') if for all $i \in I'$, j, $k \in S'$, $l \in O'$,

$$(i, j), (k, l) \text{ are edges of } G_{I-S-O} \Rightarrow (i, j), (k, l) \text{ are edges of } G_{I'-S'-O'}$$

holds.

We use the following terms:
- vertices from I – input vertices,
- vertices from O – output vertices,
- vertices from S – process vertices.

If we allow that I or O is empty, then we may consider the following induced subgraphs of G_{I-S-O}:
- G_{I-S} with $I \cup S$ as set of vertices and
$$E_{\text{in}} := \{(i,j) \mid i \in I, j \in S, \text{ and } (i,j) \text{ is an edge of } G_{I-S-O}\}$$
as set of edges, and
- G_{S-O} with $S \cup O$ as set of vertices and
$$E_{\text{out}} := \{(k,l) \mid k \in S, l \in O, \text{ and } (k,l) \text{ is an edge of } G_{I-S-O}\}$$
as set of edges.

Here $G_{I-S} = G_{\text{in}} := (V_{\text{in}}, E_{\text{in}})$ with $V_{\text{in}} = I \cup S$ is called the *input graph* of G_{I-S-O} and is the subgraph of G_{I-S-O} induced by $(I, S, \emptyset)$ and $G_{S-O} = G_{\text{out}} := (V_{\text{out}}, E_{\text{out}})$ with $V_{\text{out}} = S \cup O$ is called the *output graph* of G_{I-S-O} and is the subgraph of G_{I-S-O} induced by $(\emptyset, S, O)$.

Clearly, G_{in} and G_{out} are bipartite graphs. By Definition 8.2.1, the input–output graph G_{I-S-O} is edge-valued. Then the input graph G_{in} and the output graph G_{out} are also edge-valued graphs. Let $(i,j) \in E_{\text{in}}$ and $(k,l) \in E_{\text{out}}$. Let $\text{val}(i,j), \text{val}(k,l) \in \mathbb{Q}^+$ denote the values of (i,j), (k,l). This defines the valuation function
$$\text{val} : E_{\text{in}} \cup E_{\text{out}} \to \mathbb{Q}^+.$$

The sets
$$\text{Val}(G_{\text{in}}) := \text{val}(E_{\text{in}}), \quad \text{Val}(G_{\text{out}}) := \text{val}(E_{\text{out}}),$$
and
$$\text{Val}(G_{I-S-O}) := \text{val}(E_{\text{in}}) \cup \text{val}(E_{\text{out}}) = \text{Val}(G_{\text{in}}) \cup \text{Val}(G_{\text{out}})$$
are sets of positive rational numbers. Let $\mathcal{G}_{I-S-O}$ be the set of all abstract input–output graphs.

By Chapter 5, Definition 5.4.3, a bipartite graph which contains all possible edges is called complete.

Definition 8.2.3. An input–output graph G_{I-S-O} is said to be complete if the bipartite graphs G_{in} and G_{out} are complete.

Remark 8.2.4. We remark that a complete input–output graph is not complete as a 3-partite graph in the sense of Chapter 5, Definition 5.4.3, since not all vertices from different sets of the partition $I \cup S \cup O$ are adjacent, but only vertices from I and S, as well as from S and O, but not from I and O.

Figure 8.2 gives the complete input–output graph with m input vertices, p process vertices, and n output vertices.

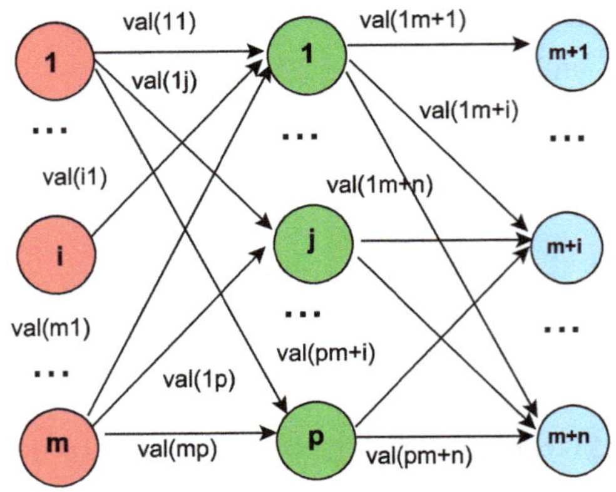

Figure 8.2: Complete input–output graph.

Using matrices, we will propose a different way to describe input–output systems. Since production systems are finite, we can restrict ourselves to finite matrices. Let R and T be finite and let A be an arbitrary nonempty set. A family

$$M_{R,T}(A) := (a_{ij})_{(i,j)\in R\times T}$$

of elements from A is said to be an $(R \times T)$-matrix on A. For $R = \{1, \dots, i, \dots, m\}$, $T = \{1, \dots, j, \dots, n\}$, we will also write $M_{m,n}(A)$. In this case, the pair (R, T) or (m, n) is said to be the type of the matrix $M_{m,n}(A)$. One can also write

$$M_{m,n}(A) = \begin{pmatrix} a_{11} & \cdots & a_{1n} \\ \vdots & \ddots & \vdots \\ a_{m1} & \cdots & a_{mn} \end{pmatrix} = (a_{ij}), \quad i = 1, \dots m,\ j = 1 \dots, n, a_{ij} \in A.$$

If A is a set of real numbers which contains with any $a, b \in A$ also the sum $a + b$, then on the set $M_{m,n}(A)$ of matrices an addition by

$$(a_{i,j}) + (b_{i,j}) := (a_{ij} + b_{i,j})_{(i,j)\in R\times T}$$

is defined (see also Chapter 6.1).

If for $a \in A$ also the opposite $-a$ belongs to A, then the opposite matrix

$$-(a_{i,j})_{(i,j)\in R\times T} := (-a_{i,j})_{(i,j)\in R\times T}$$

can be defined.

The first equation is satisfied if $A = \mathbb{Q}^+$ is the set of nonnegative rational numbers. This is the situation for applications in production systems. Both conditions are satisfied if A is the set of rational numbers.

Definition 8.2.5. If $M_{R,T}(A)$ is an $R \times T$-matrix and $R' \subseteq R$, $T' \subseteq T$, then $M_{R',T'}(A)$ is said to be a submatrix of $M_{R,T}(A)$.

Let $(i,j) \in I \times S$. We define a nonnegative rational number as follows:

$$a_{ij} := \begin{cases} \mathrm{val}(i,j) & \text{if } (i,j) \in E_{\mathrm{in}}, \\ 0 & \text{if } (i,j) \notin E_{\mathrm{in}}. \end{cases}$$

Let $(k,l) \in S \times O$. Then a nonnegative rational number is defined as follows:

$$b_{lk} := \begin{cases} \mathrm{val}(k,l) & \text{if } (k,l) \in E_{\mathrm{out}}, \\ 0 & \text{if } (k,l) \notin E_{\mathrm{in}}. \end{cases}$$

If G_{I-S-O} is an input–output graph, then we may consider the *input matrix* $M_{I,S}(\mathrm{Val}(G_{I-S})) := (a_{ij})_{i\in I, j\in S}$ of type (m,p) and the *output matrix* $M_{S,O}(\mathrm{Val}(G_{S-O})) := (b_{lk})_{k\in O, l\in S}$ of type $(n-m,p)$. (Here $I = \{1,\dots,m\}$, $S = \{1,\dots,p\}$, $O = \{m+1,\dots,n\}$.)

Definition 8.2.6.

$$M_{I,S,O}(\mathrm{Val}(G_{I-S-O})) := \begin{pmatrix} -M_{I,S}(\mathrm{Val}(G_{I-S})) \\ M_{S,O}(\mathrm{Val}(G_{S-O})) \end{pmatrix}$$

is said to be an input–output matrix of type $I\times S\times O$. (Clearly, $-M_{I,S}(\mathrm{Val}(G_{I-S}))$ is a matrix over the set of nonpositive rational numbers.)

Clearly, for $I = \{1,\dots,m\}$, $S = \{1,\dots,p\}$, and $O = \{m+1,\dots,m+n\}$, the matrix $M_{I,S,O}(\mathrm{Val}(G_{I-S-O}))$ has type $(m+n,p)$.

Let $\mathcal{TM}_{I,S,O}(\mathrm{Val}(G_{I-S-O}))$ denote the set of all input–output matrices of type $I\times S\times O$.

Theorem 8.2.7. *Let $\mathcal{G}_{I-S-O}$ be the set of all abstract input–output graphs. Then there is a bijection φ between the set $\mathcal{TM}_{I,S,O}(\mathrm{Val}(G_{I-S-O}))$ and the set $\mathcal{G}_{I-S-O}$.*

Proof. We show that there is a one-to-one correspondence between the set $\mathcal{G}_{I-S-O}$ of all input–output graphs and the set of all matrices of type $(m+n,p)$ of the form

$$M_{I,S,O}(\mathrm{Val}(G_{I-S-O})) := \begin{pmatrix} -M_{I,S}(\mathrm{Val}(G_{I-S})) \\ M_{S,O}(\mathrm{Val}(G_{S-O})) \end{pmatrix}.$$

If $(i,j) \in I \times S$ is a directed edge of G_{I-S-O}, then its label $a_{ij} \in \mathbb{Q}^+ \setminus \{0\}$ is an element of the matrix $M_{I,S}(\mathrm{Val}(G_{I-S}))$ and if $(i,j) \in I \times S$ is not an edge of the directed graph G_{I-S-O}, then $a_{ij} = 0$. If $(k,l) \in S \times O$ is a directed edge of G_{I-S-O}, then its label $b_{lk} \in \mathbb{Q}^+ \setminus \{0\}$ is an element of the matrix $M_{I,S}(\mathrm{Val}(G_{I-S}))$ and if $(i,j) \in S \times O$ is not an edge of the directed graph G_{I-S-O}, then $b_{lk} = 0$. Therefore, the abstract input–output graph with edge-labels from $\mathbb{Q}^+ \setminus \{0\}$ uniquely determines the pair $M_{I,S}(\mathrm{Val}(G_{I-S})), M_{S,O}(\mathrm{Val}(G_{S-O}))$ of matrices and therefore also the matrix

$$M_{I,S,O}(\mathrm{Val}(G_{I-S-O})) := \begin{pmatrix} -M_{I,S}(\mathrm{Val}(G_{I-S})) \\ M_{S,O}(\mathrm{Val}(G_{S-O})) \end{pmatrix}.$$

If, conversely, the matrix

$$M_{I,S,O}(\mathrm{Val}(G_{I-S-O})) := \begin{pmatrix} -M_{I,S}(\mathrm{Val}(G_{I-S})) \\ M_{S,O}(\mathrm{Val}(G_{S-O})) \end{pmatrix}$$

is given, then the graph $G_{I,S,O}$ is uniquely determined by $V = I \cup S \cup O$ as the set of its vertices and by

$$E = \{(i,j) \in I \times S \mid a_{ij} \neq 0\} \cup \{(k,l), k \in S, l \in O \mid b_{lk} \neq 0\}$$

as the set of its directed edges. Clearly, by this construction the graph G_{I-S-O} is an abstract input–output graph with edge-labels in $\mathbb{Q}^+ \setminus \{0\}$ and is uniquely determined by $M_{I,S,O}(\mathrm{Val}(G_{I-S-O}))$. □

It is not difficult to give the function

$$\varphi : \mathcal{TM}_{I,S,O}(\mathrm{Val}(G_{I-S-O})) \to \mathcal{G}_{I-S-O},$$

which was described in the second part of the proof, in an analytic form since the vertices of the image graph are given by the sets I, S, and O, and the edges are given by the indices of the matrix elements a_{ij}, b_{kl} which are different from 0. The valuations of the edges are the elements which have this index. Showing that φ is a bijection means simply showing that φ is injective and surjective. The details of the proof are left as an exercise to the reader. In the proof, we described at first the inverse mapping, the bijection φ^{-1}.

Remark 8.2.8. If we consider the two directed, edge-valued graphs $G_{\mathrm{in}} := (V_{\mathrm{in}}, E_{\mathrm{in}})$ with $V_{\mathrm{in}} := I \cup S$, $E_{\mathrm{in}} := \{(i,j) \in I \times S \mid a_{ij} \in \mathbb{Q}^+ \setminus \{0\}$ (where a_{ij} is the value of the edge (i,j))$\}$ and $G_{\mathrm{out}} := (V_{\mathrm{out}}, E_{\mathrm{out}})$ with $V_{\mathrm{out}} := S \cup O$, $E_{\mathrm{out}} := \{(k,l) \in S \times O \mid b_{lk} \in \mathbb{Q}^+ \setminus \{0\}$ (b_{lk} is the value of edge (k,l))$\}$, then clearly there is a bijection between G_{in} and the matrix $-M_{I,S}(\mathrm{Val}(G_{I-S}))$ and a bijection between G_{out} and $M_{S,O}(\mathrm{Val}(G_{S-O}))$ and $G_{I-S-O} = G_{\mathrm{in}} \cup G_{\mathrm{out}}$ corresponds to the sum

$$M_{I,S,O}(\mathrm{Val}(G_{I-S})) := \begin{pmatrix} -M_{I,S}(\mathrm{Val}(G_{I-S})) \\ M_{S,O}(\mathrm{Val}(G_{S-O})) \end{pmatrix}$$
$$= \begin{pmatrix} -M_{I,S}(\mathrm{Val}(G_{I-S})) \\ \mathbf{0} \end{pmatrix} + \begin{pmatrix} \mathbf{0} \\ M_{S,O}(\mathrm{Val}(G_{S-O})) \end{pmatrix}.$$

Here **0** are zero-matrices of the corresponding types.

We noticed already that G_{I-S}, G_{S-O} can be regarded as induced input–output subgraphs of the input–output graph G_{I-S-O} and the corresponding matrices $M_{I,S}(\mathrm{Val}(G_{I-S}))$ and $M_{S,O}(\mathrm{Val}(G_{S-O}))$ are submatrices of $M_{I,S,O}(\mathrm{Val}(G_{I-S-O}))$. This observation can be generalized.

Theorem 8.2.9. *Let G_{I-S-O} be an abstract input–output graph. Then to each induced input–output subgraph $G'_{I'-S'-O'}$ (induced by $(I' \subseteq I, S' \subseteq S, O' \subseteq O)$) there belongs a submatrix $M_{I',S',O'}(\mathrm{Val}(G'_{I'-S'-O'}))$ of $M_{I,S,O}(\mathrm{Val}(G_{I-S-O}))$, and to each submatrix M of $M_{I,S,O}(\mathrm{Val}(G_{I-S-O}))$ there is an induced subgraph $G'_{I'-S'-O'}$ of G_{I-S-O}.*

Proof. If $i \in I', j \in S'$, and $(i,j) \in E_{\text{in}}$ of G_{I-S-O}, then $(i,j) \in E'_{\text{in}}$ of $G_{I'-S'-O'}$ since $G'_{I'-S'-O'}$ is a subgraph of G_{I-S-O} induced by (I', S', O'). Then for $i \in I', j \in S'$,

$$a_{ij} = \begin{cases} a_{ij} \in \mathbb{Q}^+ \setminus \{0\} & \text{if } (i,j) \in E_{\text{in}}, \\ 0 & \text{otherwise.} \end{cases}$$

Therefore

$$M_{I',S'}(\mathrm{Val}(G'_{I'-S'})) = (a_{i,j})_{i \in I', j \in S'}$$

is a submatrix of $M_{I,S}(\mathrm{Val}(G_{I-S}))$. In a similar way, one proves that

$$M_{S',O'}(\mathrm{Val}(G'_{S'-O'})) = (b_{lk})_{k \in S', l \in O'}$$

is a submatrix of $M_{S,O}(\mathrm{Val}(G_{S-O}))$. Since $M_{I,S}(\mathrm{Val}(G_{I-S}))$ and $M_{S,O}(\mathrm{Val}(G_{S-O}))$ are submatrices of $M_{I-S-O}(\mathrm{Val}(G_{I-S-O}))$, the matrix

$$M_{I'-S'-O'}(\mathrm{Val}(G_{I'-S'-O'})) = \begin{pmatrix} -M_{I',S'}(\mathrm{Val}(G'_{I'-S'})) \\ M_{S',O'}(\mathrm{Val}(G'_{S'-O'})) \end{pmatrix}$$

is a submatrix of $M_{I-S-O}(\mathrm{Val}(G_{I-S-O}))$.

Let now M be a submatrix of $M_{I,S,O}(\mathrm{Val}(G_{I-S-O}))$. Then $I' = \{i \mid \exists a_{ij} \in M(a_{ij} < 0)\} \subseteq I$, $O' = \{k \mid \exists b_{lk} \in M(b_{lk} > 0)\} \subseteq O$, and $S' = \{j \mid \exists a_{ij} \in M(a_{ij} < 0)\} \cup \{l \mid \exists b_{kl} \in M(b_{kl} > 0)\} \subseteq S$ are subsets of I, O, and S, respectively.

We consider the following cases:

1. $O' = \emptyset$. Then $G'_{I'-S'}$ is an input graph which is a subgraph of G_{I-S} since $I' \subseteq I, S' \subseteq S$, $O' \subseteq O$, which is induced by (I', O') since for all $i \in I', j \in S'$ there hold $a_{ij} \in M$ and $a_{ij} < 0$, and thus $(i,j) \in E'_{\text{in}}$. Since G_{I-S} is a subgraph of G_{I-S-O} induced by

$(I, S, \emptyset) \subseteq (I, S, O)$, the special input–output graph $G_{I-S} = G_{I-S-\emptyset}$ is a subgraph of $G_{I,S,O}$ induced by $(I', S', \emptyset)$.

2. In a similar way, one shows that $G'_{S'-O'} = G'_{\emptyset-S'-O'}$ is a subgraph of $G_{I,S,O}$ induced by $(\emptyset, S', O')$.
3. Now we assume that $I' \neq \emptyset$, $O' \neq \emptyset$. If $a_{ij} \in M$ and $a_{ij} < 0$, then $i \in I'$, $j \in S'$, and $G'_{\text{in}} = G'_{I'-S'}$ is a subgraph of G_{I-S} induced by (I', S'). If $b_{lk} \in M$, $b_{lk} > 0$, then $k \in O'$, $l \in S'$, and $G'_{\text{out}} = G'_{S'-O'}$ is a subgraph of G_{S-O} induced by (S', O'). Altogether, $G'_{I'-S'-O'}$ is a subgraph of G_{I-S-O} induced by (I', S', O'). □

We notice that by Theorem 8.2.9 the submatrix $M_{I',S',O'}(\text{Val}(G'_{I'-S'-O'}))$ of $M_{I,S,O}(\text{Val}(G_{I-S-O}))$ belonging to $G'_{I'-S'-O'}$ is uniquely determined. Since for a given submatrix M of $M_{I,S,O}(\text{Val}(G_{I-S-O}))$ the sets I', S', O' defined in the second part of the proof are uniquely determined, we expect that the corresponding subgraph $G'_{I'-S'-O'}$ of G_{I-S-O} is uniquely determined. Indeed, we have

Corollary 8.2.10. *If for the submatrix M of the input–output matrix $M_{I,S,O}(\text{Val}(G_{I-S-O}))$ the subsets I', S', O' are defined as in the second part of the proof of Theorem* 8.2.9, *then*

$$M = M_{I',S',O'}(\text{Val}(G'_{I'-S'-O'})).$$

Therefore, we have

Corollary 8.2.11. *Any submatrix of an input–output matrix is an input–output matrix.*

Clearly, not every matrix is an input–output matrix. For instance,

$$\begin{pmatrix} -1 & 2 & 3 \\ 4 & -4 & 5 \\ 3 & 4 & -6 \end{pmatrix}$$

is not an input–output matrix. If we apply a permutation of the 9-element set of elements which exchanges 2 with −4, 3 with −6, and fixes all other elements, we obtain

$$\begin{pmatrix} -1 & -4 & -6 \\ 4 & 2 & 5 \\ 3 & 4 & 3 \end{pmatrix}$$

and an input–output matrix. The corresponding input–output graph is given in Figure 8.3.

The bijection φ from Theorem 8.2.7 allows expressing properties of input–output matrices by properties of the corresponding input–output graphs, and conversely. This is an example of a very common method to connect two different scientific disciplines. Interdisciplinary research combines knowledge and expertise from different fields to address a problem.

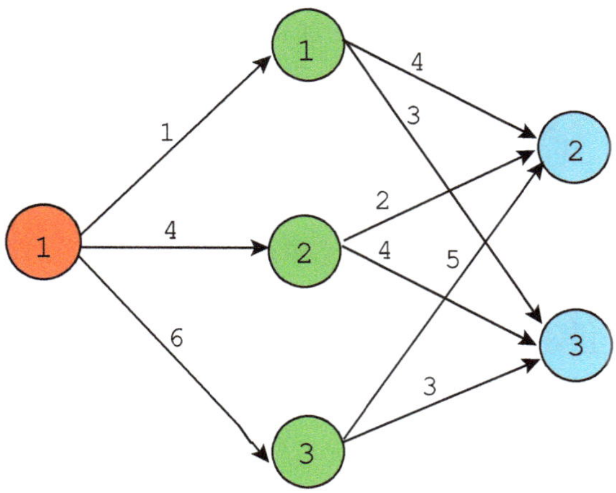

Figure 8.3: Input–output graph.

We will give an example.

Proposition 8.2.12. *The input–output graph G_{I-S-O} is complete if and only if all entries of the matrix $M_{I-S-O}(\mathrm{Val}(G_{I-S-O}))$ are different from* 0.

Proof. Without loss of generality, we assume that $I = \{1, \ldots, m\}$, $S = \{1, \ldots, p\}$, and $O = \{m+1, \ldots, m+n\}$. Since no entry of $M_{I-S}(\mathrm{Val}(G_{I-S}))$ is 0, the bijection φ maps each entry a_{ij} of this matrix to the directed edge (i, j), where $i \in \{1, \ldots, m\}$ denotes a vertex from I and $j \in \{1, \ldots, p\}$ denotes a vertex from S. Therefore, each vertex from I is adjacent with a vertex from S. Using the matrix $M_{S-O}(\mathrm{Val}(G_{S-O}))$, the second part of matrix $M_{I-S-O}(\mathrm{Val}(G_{G-S-O}))$, one proves that each vertex from S is adjacent with each vertex from O. Thus G_{I-S-O} is complete. If, conversely, G_{I-S-O} is complete, then the bijection φ^{-1} maps each edge (i, j) to an element $a_{ij} \neq 0$ and each edge (k, l) to an element $b_{kl} \neq 0$. But these are the entries of the matrix $M_{I-S-O}(\mathrm{Val}(G_{I-S-O}))$. □

In the next section, we consider input–output graphs and matrices with one-element sets of process vertices.

8.3 Activities and techniques

As we mentioned already in Chapter 1.6, the activity analysis of production was considered first by Koopmanns in 1951 [40] (see also [18] and [28]).

Let $\mathbb{Q}^n$, $n \geq 1$, be the set of all n-tuples of rational numbers. The elements of $\mathbb{Q}^n$ can be added by

$$\begin{pmatrix} a_1 \\ \vdots \\ a_n \end{pmatrix} + \begin{pmatrix} b_1 \\ \vdots \\ b_n \end{pmatrix} := \begin{pmatrix} a_1 + b_1 \\ \vdots \\ a_n + b_n \end{pmatrix},$$

where the addition on the right-hand side is the addition in $\mathbb{Q}$.

If $\alpha \in \mathbb{Q}$, then

$$\alpha \circ \begin{pmatrix} a_1 \\ \vdots \\ a_n \end{pmatrix} := \begin{pmatrix} \alpha a_1 \\ \vdots \\ \alpha a_n \end{pmatrix}$$

is the application of rational numbers as operators on n-tuples.

The subtraction of n-tuples is defined similar to addition:

$$\begin{pmatrix} a_1 \\ \vdots \\ a_n \end{pmatrix} - \begin{pmatrix} b_1 \\ \vdots \\ b_n \end{pmatrix} := \begin{pmatrix} a_1 - b_1 \\ \vdots \\ a_n - b_n \end{pmatrix},$$

where the subtraction on the right-hand side is the usual subtraction of n-tuples from $\mathbb{Q}^n$ (see also Chapter 6.1).

Definition 8.3.1 ([28]). The difference

$$\begin{pmatrix} z_1 \\ \vdots \\ z_n \end{pmatrix} := \begin{pmatrix} y_1 \\ \vdots \\ y_n \end{pmatrix} - \begin{pmatrix} x_1 \\ \vdots \\ x_n \end{pmatrix}$$

is said to be an activity.

In [28], the following notation is used. Let $z_1, \dots, z_l$ be input objects and let $z_{l+1} \dots, z_n$ be output objects of a production system. We set $z_1 := -x_1, \dots, z_l := -x_l$, $z_{l+1} := y_{l+1}, \dots,$ $z_n := y_n$ with $x_1, \dots, x_n, y_{l+1}, \dots, y_n \in \mathbb{Q}^+$.

Definition 8.3.2 ([28]). The activity

$$\begin{pmatrix} 0 \\ \vdots \\ 0 \\ y_{l+1} \\ \vdots \\ y_n \end{pmatrix} - \begin{pmatrix} x_1 \\ \vdots \\ x_l \\ 0 \\ \vdots \\ 0 \end{pmatrix} = \begin{pmatrix} -x_1 \\ \vdots \\ -x_l \\ y_{l+1} \\ \vdots \\ y_n \end{pmatrix}$$

is said to be an admissible activity of the production system with these kinds of input–output objects.

Example 8.3.3. By Definition 8.3.2, our starting example can be written as

$$\begin{pmatrix} z_1 \\ z_2 \\ z_3 \\ z_4 \\ z_5 \\ z_6 \\ z_7 \end{pmatrix} = \begin{pmatrix} -2000 \\ -200 \\ -100 \\ -300 \\ -7000 \\ 200 \\ 100 \end{pmatrix}.$$

Since there are five kinds of input objects and two kinds of output objects, we have $l = 5$ and $n = 7$.

Since the quantities in real production systems are nonnegative rational numbers, the admissible activities can be written as vectors of the n-dimensional rational vector space $\underline{\mathbb{Q}}^n = (\mathbb{Q}^n, +, \circ)$.

Remark 8.3.4. If the processes of a production system are not specified, i. e., appear as a "black box," each admissible activity

$$\begin{pmatrix} -x_1 \\ \vdots \\ -x_l \\ y_{l+1} \\ \vdots \\ y_n \end{pmatrix}, x_1, \ldots, x_l, y_{l+1}, \ldots, y_n \in \mathbb{Q}^+$$

corresponds to an input–output graph in the sense of Section 8.2, which has the form of Figure 8.4, with $I = \{-x_1, \ldots, -x_l\}$, $O = \{y_{l+1}, \ldots, y_n\}$, and $S = \{1\}$. Here we identify the edge-values with the elements of the matrices since no entry can be 0 using the fact that the corresponding input–output graphs are complete (Proposition 8.2.12). The corresponding input–output matrix is

$$M_{I,S,O}(\mathbb{Q}^+) = \begin{pmatrix} -M_{I,S}(\mathbb{Q}^+) \\ M_{S,O}(\mathbb{Q}^+) \end{pmatrix},$$

with

$$M_{I,S}(\mathbb{Q}^+) = \begin{pmatrix} x_1 \\ \vdots \\ x_l \end{pmatrix} \quad \text{and} \quad M_{S,O}(\mathbb{Q}^+) = \begin{pmatrix} y_{l+1} \\ \vdots \\ y_n \end{pmatrix}.$$

This shows that admissible activities are input–output matrices of type $(n, 1)$.

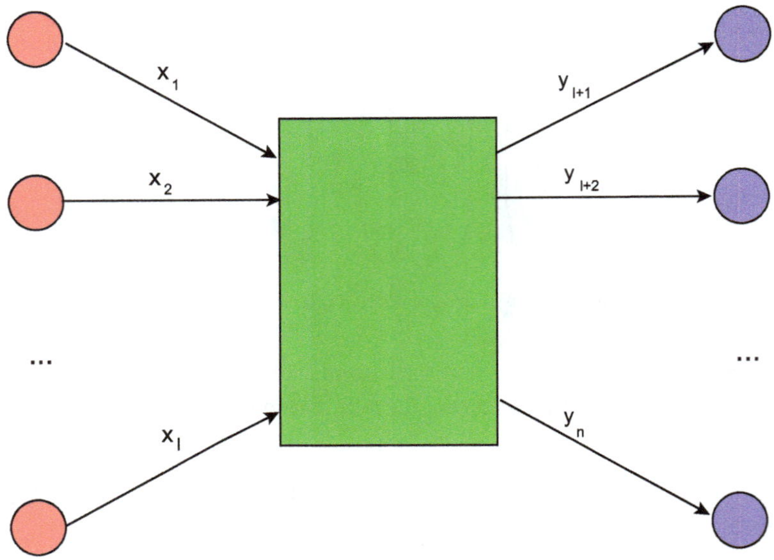

Figure 8.4: Admissible activities as input–output graphs.

Sometimes, a production system requires more admissions which cannot be specified in a general form and depend on the specific production. This leads to the following definition:

Definition 8.3.5.

$$T^{\mathrm{ad}} := \left\{ \underline{z} = \begin{pmatrix} -x_1 \\ \vdots \\ -x_l \\ y_{l+1} \\ \vdots \\ y_n \end{pmatrix}, x_1, \ldots, x_l, y_{l+1}, \ldots, y_n \in \mathbb{Q}^+ \;\middle|\; \begin{array}{l} \underline{z} \text{ is an admissible activity} \\ \text{of the production system} \end{array} \right\}$$

is said to be a technique (more precisely, the technique of the production systems under consideration). If there are no more requirements except that $\underline{z}$ is an input–output activity, then we define

$$T^{\text{in-out}} := \left\{ \underline{z} = \begin{pmatrix} -x_1 \\ \vdots \\ -x_l \\ y_{l+1} \\ \vdots \\ y_n \end{pmatrix}, x_1, \ldots, x_l, y_{l+1}, \ldots, y_n \in \mathbb{Q}^+ \right\}.$$

Clearly, $T^{\mathrm{ad}} \subseteq T^{\text{in-out}}$. Now we are interested in the algebraic properties of $T^{\text{in-out}}$. Especially, we ask under which operations $T^{\text{in-out}}$ is closed.

Proposition 8.3.6. *Technique $T^{\text{in-out}}$ is closed with respect to the addition of vectors from the n-dimensional vector space $\underline{\mathbb{Q}}^n$.*

Proof. Let

$$\begin{pmatrix} -x_{11} \\ \vdots \\ -x_{l1} \\ y_{l+11} \\ \vdots \\ y_{n1} \end{pmatrix}, x_{11},\dots,x_{l1}, y_{l+11},\dots,y_{n1} \in \mathbb{Q}^+, \begin{pmatrix} -x_{12} \\ \vdots \\ -x_{l2} \\ y_{l+12} \\ \vdots \\ y_{n2} \end{pmatrix}, x_{12},\dots,x_{l2},$$

$y_{l+12},\dots,y_{n2} \in \mathbb{Q}^+$ be two elements of $T^{\text{in-out}}$. Then the sum

$$\begin{pmatrix} -(x_{11}+x_{12}) \\ \vdots \\ -(x_{l1}+x_{l2}) \\ (y_{l+11}+y_{l+12}) \\ \vdots \\ (y_{n1}+y_{n2}) \end{pmatrix}$$

is in $T^{\text{in-out}}$ since $x_{11}+x_{12},\dots,x_{l1}+x_{l2}, y_{l+11}+y_{l+12},\dots,y_{n1}+y_{n2}$ are in $\mathbb{Q}^+$. □

The zero-element $\underline{0} = \begin{pmatrix} 0 \\ \vdots \\ 0 \end{pmatrix}$ belongs to $T^{\text{in-out}}$, but the opposite of

$$\begin{pmatrix} -x_1 \\ \vdots \\ -x_l \\ y_{l+1} \\ \vdots \\ y_n \end{pmatrix}, \quad \text{i. e.,} \quad \begin{pmatrix} x_1 \\ \vdots \\ x_l \\ -y_{l+1} \\ \vdots \\ -y_n \end{pmatrix}$$

does not belong to $T^{\text{in-out}}$. This means that, different from the situation in $\underline{\mathbb{Q}}^n$, the addition is not invertible and $T^{\text{in-out}}$ does not form a group with respect to addition. Since the addition of vectors from $\underline{Q}^n$ is associative and commutative, this is also valid in $T^{\text{in-out}}$. Selection of the zero-element can be regarded as a nullary operation with

$$\underline{z} + \underline{0} = \underline{0} + \underline{z} = \underline{z}.$$

Such algebraic structures are called commutative monoids.

Theorem 8.3.7. *Technique $\mathcal{T}^{\text{in-out}} := (T^{\text{in-out}}; +, \underline{0})$ is a commutative monoid.*

Since $T^{\mathrm{ad}} \subseteq T^{\mathrm{in-out}}$ and the inclusion can be proper, not every technique must be closed with respect to addition, and the following definition makes sense (see [37] or [28]).

Definition 8.3.8. A technique T^{ad} is said to be additive if the sum of two admissible activities is admissible.

The zero-vector $\underline{0}$ can be regarded as an admissible activity: if all inputs are 0, i. e., if there are no inputs, then there are no outputs, i. e., all outputs are 0. (Having no time delay between input and output is one of the properties of our model). Therefore, the additive techniques form also a commutative monoid.

Not any technique is closed under the application of nonnegative rational numbers on admissible activities. Therefore, it makes sense to state

Definition 8.3.9. A technique is said to be size-proportional if for all admissible activities and for all $\lambda \in \mathbb{Q}^+$,

$$\underline{v} \in T^{\mathrm{ad}} \Rightarrow \lambda \circ \underline{v} \in T^{\mathrm{ad}}.$$

If, in particular, $\lambda > 1$, then $\lambda \circ \underline{v}$ is called *scale-increasing*, and for $0 \leq \lambda < 1$ it is called *scale-decreasing*. A technique is said to be *linear* if it is both additive and size-proportional.

Any linear technique T^{ad} can be written as a two-based algebra with two base sets T^{ad} and $\mathbb{Q}^+$ and with two operations,

$$+ : T^{\mathrm{ad}} \times T^{\mathrm{ad}} \to T^{\mathrm{ad}}, \quad \circ : \mathbb{Q}^+ \times T^{\mathrm{ad}} \to T^{\mathrm{ad}},$$

i. e.,

$$(T^{\mathrm{ad}}, \mathbb{Q}^+; +, \circ),$$

as we can do with rational or real vector spaces. But the two-based algebra $(T^{\mathrm{ad}}, \mathbb{Q}^+; +, \circ)$ is not a vector space since addition is not invertible. To be completely precise, it is incorrect to call the activities $\underline{v} \in T^{\mathrm{ad}}$ vectors despite of the fact that this two-based algebraic structure has several properties of a vector space.

Theorem 8.3.10. *Let T^{ad} be a linear technique and let $\lambda, \mu \in \mathbb{Q}^+$. Then for all $\lambda, \mu \in \mathbb{Q}^+$ and all $\underline{u}, \underline{v} \in T^{\mathrm{ad}}$, the following equations are satisfied:*
(i) $(\lambda\mu) \circ \underline{v} \approx \lambda \circ (\mu \circ \underline{v})$;
(ii) $(\lambda + \mu) \circ \underline{v} \approx \lambda \circ \underline{v} + \mu \circ \underline{v}$;
(iii) $\lambda \circ (\underline{u} + \underline{v}) \approx \lambda \circ \underline{u} + \lambda \circ \underline{v}$;
(iv) $1 \circ \underline{u} \approx \underline{u}$;
(v) $\lambda \circ \underline{0} \approx 0 \circ \underline{u} \approx \underline{0}$.

Proof. (i) Since T^{ad} is linear, with

$$\underline{v} = \begin{pmatrix} -x_1 \\ \vdots \\ -x_l \\ y_{l+1} \\ \vdots \\ y_n \end{pmatrix} \in T^{\text{ad}}$$

and $\mu \in \mathbb{Q}^+$, we have

$$\mu \circ \underline{v} = \begin{pmatrix} -\mu x_1 \\ \vdots \\ -\mu x_l \\ \mu y_{l+1} \\ \vdots \\ \mu y_n \end{pmatrix} \in T^{\text{ad}}.$$

Moreover,

$$\lambda \circ (\mu \circ \underline{v}) = \begin{pmatrix} -\lambda\mu x_1 \\ \vdots \\ -\lambda\mu x_l \\ \lambda\mu y_{l+1} \\ \vdots \\ \lambda\mu y_n \end{pmatrix} = (\lambda\mu) \circ \underline{v}.$$

(ii) Since $\lambda + \mu \in \mathbb{Q}^+$, (ii) can be proved similarly to (i).
(iii) follows by an application of the distributive law in $\mathbb{Q}^+$.
(iv), (v) are also clear. □

Remark 8.3.11. In many applications, λ, μ are natural numbers. Because of

$$\lambda \circ \underline{v} = \underbrace{\underline{v} + \cdots + \underline{v}}_{n\text{-times}}$$

the operation $\circ$ can be replaced by iterated addition of elements in $\underline{v} \in T^{\text{ad}}$. Therefore, the commutative monoid $(T^{\text{ad}}; +, \underline{0})$ describes the algebraic properties of the activities from T^{ad} completely.

We noticed already that in general a linear technique does not form a vector space since $(T^{\text{ad}}; +, \underline{0})$ is not a group, but only a monoid. If the second set is a *semiring SR* with identity element, i. e., is closed under two binary operations, addition and multiplication, and if the equations of Theorem 8.3.10 are satisfied, the algebraic structure $(T^{\mathrm{ad}}; SR; +, \circ)$ is said to be a semimodule. The following definition goes back to Golan [34].

Definition 8.3.12 ([34]). Let $\mathcal{S}R$ be a semiring (with additive identity 0) and a multiplicative identity 1. A left-SR-semimodule, also called a semimodule over $\mathcal{S}R$, is a commutative monoid $(M; +, \underline{0})$ with additive identity (zero-element) $\underline{0}$ and multiplicative identity 1 such there is a function $S \times M \to M$ defined by $(\lambda, \underline{v}) \mapsto \lambda \circ \underline{v}$, called scalar multiplication, which satisfies equations (i)–(v) from Theorem 8.3.10.

The nonnegative rational numbers $\mathbb{Q}^+$ form a semiring with the rational numbers 0, 1 as identities with respect to addition and multiplication, respectively. Since the multiplication of elements in $\mathbb{Q}^+ \setminus \{0\}$ is invertible, the nonnegative rational numbers even form a semifield. Therefore, any linear technique is a semimodule over a semifield. Since $T^{\text{in-out}}$ is linear, contains $\underline{0} = \begin{pmatrix} 0 \\ \vdots \\ 0 \end{pmatrix}$, and satisfies (i)–(v) from Theorem 8.3.10, $(T^{\text{in-out}}, \mathbb{Q}^+; +, \circ)$ is a semimodule over a semifield (because of (iv) we need an identity element, but $1 \in \mathbb{Q}^+$ is it).

Clearly, all linear combinations $\sum_{k=1}^{n} \lambda_k \circ \underline{v_k}$ belong to $T^{\text{in-out}}$ and $T^{\text{in-out}}$ is the set of all these linear combinations. Since for real production systems $T^{\text{in-out}}$ is finite, every activity from $T^{\text{in-out}}$ can be generated by finitely many activities forming linear combinations.

Motivated by these remarks, we mention a few important concepts and properties of a S-semimodule $\mathcal{M}$. Let $A \subseteq M$. The set

$$\left\{ \sum_{k=1}^{n} \lambda_k \circ \underline{v_k} \mid \lambda_k \in S, \underline{v_k} \in M, n \in \mathbb{N} \setminus \{0\} \right\}$$

of all linear combinations of elements from A is said to be Span(A). If Span$(A) = M$, then A is said to be *generating set* of $\mathcal{M}$. A semimodule with a finite generating set is called *finitely generated.*

Definition 8.3.13. Let $\mathcal{S}R$ be a semiring and let $\mathcal{M}$ be an SR-semimodule. A nonempty subset $A \subseteq M$ is called linearly independent if $\underline{v} \in \text{Span}(A \setminus \{\underline{v}\})$ for any $\underline{v} \in A$. If A is not linearly independent, then it is called linearly dependent.

It is not difficult to see that A is linearly independent if and only if the following implication is satisfied.

Assume that $\underline{v_1}, \ldots, \underline{v_n} \in A$ are not all equal to the zero-vector $\underline{0}$ and $\lambda_1, \ldots, \lambda_n \in S$. Then

$$\sum_{k=1}^{n} \lambda_k \circ \underline{v_k} = \underline{0} \Rightarrow \lambda_1 = 0, \ldots, \lambda_n = 0.$$

Definition 8.3.14. Let $\mathcal{SR}$ be a semiring and let $\mathcal{M}$ be an $\mathcal{SR}$-semimodule. A linearly independent generating set for $\mathcal{M}$ is called a basis for $\mathcal{M}$. The set A is called free if each element from $\mathcal{M}$ can be represented in at most one way as a linear combination of elements from A.

If the $\mathcal{SR}$-semimodule is finitely generated, then, as for vector spaces, $\mathcal{M}$ has a basis if and only if each vector can be uniquely represented as a linear combination of the basis, i. e., any free generating system is a basis. But in contrast to vector spaces, different bases for an S-semimodule need not have the same cardinality.

If $A = \{\underline{v_1}, \dots, \underline{v_n}\}$ is a basis of a linear technique $T^{\mathrm{ad}}(\mathbb{Q}^+)$, the matrix $(\underline{v_1}, \dots, \underline{v_n})$ is said to be the *technique matrix* of the production system.

To recognize the economical meaning of the basic activities, we go back to our starting example.

Example 8.3.15. The activity "production of batteries" is given by

$$\begin{pmatrix} z_1 \\ z_2 \\ z_3 \\ z_4 \\ z_5 \\ z_6 \\ z_7 \end{pmatrix} = \begin{pmatrix} 0 \\ 0 \\ 0 \\ 0 \\ 0 \\ 200 \\ 100 \end{pmatrix} - \begin{pmatrix} 2000 \\ 200 \\ 100 \\ 1300 \\ 7000 \\ 0 \\ 0 \end{pmatrix} = \begin{pmatrix} -2000 \\ -200 \\ -100 \\ -300 \\ -7000 \\ 200 \\ 100 \end{pmatrix}.$$

Small and big battery systems can be produced independently. The basic activities to produce small and big battery systems are

$$\underline{v_1} = \begin{pmatrix} -5 \\ -1 \\ 0 \\ -1 \\ -20 \\ 1 \\ 0 \end{pmatrix} \quad \text{and} \quad \underline{v_2} = \begin{pmatrix} -10 \\ 0 \\ -1 \\ -1 \\ -30 \\ 0 \\ 1 \end{pmatrix},$$

respectively. The daily production can be written as a linear combination

$$\underline{z} = 200 \circ \begin{pmatrix} -5 \\ -1 \\ 0 \\ -1 \\ -20 \\ 1 \\ 0 \end{pmatrix} + 100 \circ \begin{pmatrix} -10 \\ 0 \\ -1 \\ -1 \\ -30 \\ 0 \\ 1 \end{pmatrix} = \begin{pmatrix} -2000 \\ -200 \\ -100 \\ -300 \\ -7000 \\ 200 \\ 100 \end{pmatrix}.$$

Because of

$$\lambda_1 \circ \begin{pmatrix} -5 \\ -1 \\ 0 \\ -1 \\ -20 \\ 1 \\ 0 \end{pmatrix} + \lambda_2 \circ \begin{pmatrix} -10 \\ 0 \\ -1 \\ -1 \\ -30 \\ 0 \\ 1 \end{pmatrix} = \begin{pmatrix} -5\lambda_1 - 10\lambda_2 \\ -\lambda_1 \\ -\lambda_2 \\ -\lambda_1 - \lambda_2 \\ -20\lambda_1 - 30\lambda_2 \\ \lambda_1 \\ \lambda_2 \end{pmatrix} = \begin{pmatrix} 0 \\ 0 \\ 0 \\ 0 \\ 0 \\ 0 \\ 0 \end{pmatrix}$$

$$\iff \lambda_1 = \lambda_2 = 0,$$

$\{\underline{v_1}, \underline{v_2}\}$ is linearly independent and, since it is a generating system, it is a basis of the production system, while

$$\begin{pmatrix} -5 & -10 \\ -1 & 0 \\ 0 & -1 \\ -1 & -1 \\ -20 & -30 \\ 1 & 0 \\ 0 & 1 \end{pmatrix}$$

is the technique matrix of the production system.

8.4 The area of restrictions

Bottlenecks on procurement markets, supply problems, or supplier obligations lead to restrictions on the input and output sides. For a mathematical description of those restrictions on a set $T^{\text{in-out}}$ of input–output activities, we define a binary relation componentwise by

$$\begin{pmatrix} -x_{11} \\ -x_{21} \\ \vdots \\ -x_{l1} \\ y_{l+11} \\ \vdots \\ y_{n1} \end{pmatrix} \leq \begin{pmatrix} -x_{12} \\ -x_{22} \\ \vdots \\ -x_{l2} \\ y_{l+12} \\ \vdots \\ y_{n2} \end{pmatrix} :\iff x_{11} \geq x_{12}, \dots, x_{l1} \geq x_{l2}, \dots,$$

$y_{l+11} \leq y_{l+12}, \dots, y_{n1} \leq y_{n2}$. (For simplicity, we use the same notation for the different binary relations on the left- and right-hand side.)

Theorem 8.4.1. *The binary relation $\leq$ on $T^{\text{in-out}}$ is an order relation which is monotone with respect to addition $+$ and scalar multiplication $\circ$ of the $\mathbb{Q}^+$-semimodule $(T^{\text{in-out}}, \mathbb{Q}^+; +, \circ)$.*

Proof. We have to prove that the order relation $\leq$ on $T^{\text{in-out}}$ is reflexive, antisymmetric, and transitive.

(Reflexivity) We have

$$\begin{pmatrix} -x_{11} \\ -x_{21} \\ \vdots \\ -x_{l1} \\ y_{l+11} \\ \vdots \\ y_{n1} \end{pmatrix} \leq \begin{pmatrix} -x_{11} \\ -x_{21} \\ \vdots \\ -x_{l1} \\ y_{l+11} \\ \vdots \\ y_{n1} \end{pmatrix}$$

because of $x_{11} \geq x_{11}, \ldots, x_{l1} \geq x_{l1}, y_{l+11} \leq y_{l+11}, \ldots, y_{n1} \leq y_{n1}$.

(Transitivity) Relation $\leq$ on $T^{\text{in-out}}$ is transitive since from

$$\begin{pmatrix} -x_{11} \\ -x_{21} \\ \vdots \\ -x_{l1} \\ y_{l+11} \\ \vdots \\ y_{n1} \end{pmatrix} \leq \begin{pmatrix} -x_{12} \\ -x_{22} \\ \vdots \\ -x_{l2} \\ y_{l+12} \\ \vdots \\ y_{n2} \end{pmatrix} \leq \begin{pmatrix} -x_{13} \\ -x_{23} \\ \vdots \\ -x_{l3} \\ y_{l+13} \\ \vdots \\ y_{n3} \end{pmatrix}$$

there follows $x_{11} \geq x_{12} \geq x_{13}, \ldots, x_{l1} \geq x_{l2} \geq x_{l3}, y_{l+11} \leq y_{l+12} \leq y_{l+13}, \ldots, y_{n1} \leq y_{n2} \leq y_{n3}$ and, because of the transitivity of $\leq$ $(\geq)$ on $\mathbb{Q}^+$, there follows $x_{11} \geq x_{13}, \ldots, x_{l1} \geq x_{l3}$, $y_{l+11} \leq y_{l+13}, \ldots, y_{n1} \leq y_{n3}$, and thus

$$\begin{pmatrix} -x_{11} \\ -x_{21} \\ \vdots \\ -x_{l1} \\ y_{l+11} \\ \vdots \\ y_{n1} \end{pmatrix} \leq \begin{pmatrix} -x_{13} \\ -x_{23} \\ \vdots \\ -x_{l1} \\ y_{l+13} \\ \vdots \\ y_{n3} \end{pmatrix}.$$

Moreover, $\leq$ on $T^{\text{in-out}}$ is antisymmetric, i. e.,

$$\begin{pmatrix} -x_{11} \\ -x_{21} \\ \vdots \\ -x_{l1} \\ y_{l+11} \\ \vdots \\ y_{n1} \end{pmatrix} \leq \begin{pmatrix} -x_{12} \\ -x_{22} \\ \vdots \\ -x_{l1} \\ y_{l+12} \\ \vdots \\ y_{n2} \end{pmatrix} \quad \text{and} \quad \begin{pmatrix} -x_{12} \\ -x_{22} \\ \vdots \\ -x_{l2} \\ y_{l+12} \\ \vdots \\ y_{n2} \end{pmatrix} \leq \begin{pmatrix} -x_{11} \\ -x_{21} \\ \vdots \\ -x_{l1} \\ y_{l+11} \\ \vdots \\ y_{n1} \end{pmatrix}$$

imply

$$\begin{pmatrix} -x_{11} \\ -x_{21} \\ \vdots \\ -x_{l1} \\ y_{l+11} \\ \vdots \\ y_{n1} \end{pmatrix} = \begin{pmatrix} -x_{12} \\ -x_{22} \\ \vdots \\ -x_{l1} \\ y_{l+12} \\ \vdots \\ y_{n2} \end{pmatrix},$$

as can be proved in a similar way.

The relation $\leq$ is monotone with respect to addition, i. e.,

$$\begin{pmatrix} -x_{11} \\ -x_{21} \\ \vdots \\ -x_{l1} \\ y_{l+11} \\ \vdots \\ y_{n1} \end{pmatrix} \leq \begin{pmatrix} -x_{12} \\ -x_{22} \\ \vdots \\ -x_{l2} \\ y_{l+12} \\ \vdots \\ y_{n2} \end{pmatrix} \quad \text{and} \quad \begin{pmatrix} -x_{13} \\ -x_{23} \\ \vdots \\ -x_{l3} \\ y_{l+13} \\ \vdots \\ y_{n3} \end{pmatrix} \leq \begin{pmatrix} -x_{14} \\ -x_{24} \\ \vdots \\ -x_{l4} \\ y_{l+14} \\ \vdots \\ y_{n4} \end{pmatrix}$$

imply

$$\begin{pmatrix} -x_{11} \\ -x_{21} \\ \vdots \\ -x_{l1} \\ y_{l+11} \\ \vdots \\ y_{n1} \end{pmatrix} + \begin{pmatrix} -x_{13} \\ -x_{23} \\ \vdots \\ -x_{l3} \\ y_{l+13} \\ \vdots \\ y_{n3} \end{pmatrix} \leq \begin{pmatrix} -x_{12} \\ -x_{22} \\ \vdots \\ -x_{l2} \\ y_{l+12} \\ \vdots \\ y_{n2} \end{pmatrix} + \begin{pmatrix} -x_{14} \\ -x_{24} \\ \vdots \\ -x_{l4} \\ y_{l+14} \\ \vdots \\ y_{n4} \end{pmatrix}.$$

This follows immediately from the corresponding property of ≤ (≥) on $\mathbb{Q}^+$. In a similar way, one can check that

$$\begin{pmatrix} -x_{11} \\ -x_{21} \\ \vdots \\ -x_{l1} \\ y_{l+11} \\ \vdots \\ y_{n1} \end{pmatrix} \leq \begin{pmatrix} -x_{12} \\ -x_{22} \\ \vdots \\ -x_{l2} \\ y_{l+12} \\ \vdots \\ y_{n2} \end{pmatrix}$$

and $\alpha \in \mathbb{Q}^+$ imply

$$\alpha \circ \begin{pmatrix} -x_{11} \\ -x_{21} \\ \vdots \\ -x_{l1} \\ y_{l+11} \\ \vdots \\ y_{n1} \end{pmatrix} \leq \alpha \circ \begin{pmatrix} -x_{12} \\ -x_{22} \\ \vdots \\ -x_{l2} \\ y_{l+12} \\ \vdots \\ y_{n2} \end{pmatrix}. \qquad \square$$

Adding the binary relation ≤ to the operations of the (left-)$\mathbb{Q}^+$-semimodule $(T^{\text{in-out}}, \mathbb{Q}^+; +, \circ)$, one obtains an ordered left-$\mathbb{Q}^+$-semimodule $(T^{\text{in-out}}, \mathbb{Q}^+; +, \circ, \leq)$. This algebraic structure is a two-sorted algebraic system in the sense of I. A. Mal'cev (see [50]).

Denote by $\underline{z_n}^{\min}$ and $\underline{z_n}^{\max}$ the least and greatest activity with respect to the order relation ≤ of the considered production system. The set

$$R := \{\underline{z} \in T^{\text{in-out}} \mid \exists \underline{z_n}^{\min}, \underline{z_n}^{\max} (\underline{z_n}^{\min} \leq \underline{z} \leq \underline{z_n}^{\max})\}$$

is said to be the *area of restrictions* and the intersection $T^R := T^{\text{in-out}} \cap R$ is said to be the *production space.*

For our example "production of batteries," we consider the following restrictions:

- *restrictions concerning input objects*: at most 2500 cells, at most 250 small casings, at most 150 big casings, at most 350 cells contact systems, and at most 8000 human labor minutes;
- *supplier obligations*: 100 small battery systems and 50 big battery systems. These conditions give

$$\underline{z_7}^{\min} = \begin{pmatrix} -2500 \\ -250 \\ -150 \\ -350 \\ -8000 \\ 100 \\ 50 \end{pmatrix} \quad \text{and} \quad \underline{z_7}^{\max} = \begin{pmatrix} \infty \\ \infty \\ \infty \\ \infty \\ \infty \\ \infty \\ \infty \end{pmatrix}.$$

Since there are no more restrictions, for any activity $\underline{z}$ we have

$$\underline{z} = \lambda_1 \begin{pmatrix} -5 \\ -1 \\ 0 \\ -1 \\ -20 \\ 1 \\ 0 \end{pmatrix} + \lambda_2 \begin{pmatrix} -10 \\ 0 \\ -1 \\ -1 \\ -30 \\ 0 \\ 1 \end{pmatrix} \geq \begin{pmatrix} -2500 \\ -250 \\ -150 \\ -350 \\ -8000 \\ 100 \\ 50 \end{pmatrix}.$$

With

$$\underline{z} = \begin{pmatrix} -x_1 \\ -x_2 \\ -x_3 \\ -x_4 \\ -x_5 \\ y_6 \\ y_7 \end{pmatrix},$$

we obtain the following system of linear inequalities:

$$\begin{aligned} x_1 &= 5\lambda_1 + 10\lambda_2 \leq 2500, \\ x_2 &= \lambda_1 \leq 250, \\ x_3 &= \lambda_2 \leq 150, \\ x_4 &= \lambda_1 + \lambda_2 \leq 350, \\ x_5 &= 20\lambda_1 + 30\lambda_2 \leq 8000, \\ y_6 &= \lambda_1 \geq 100, \\ y_7 &= \lambda_2 \geq 50. \end{aligned}$$

Then $100 \leq y_6 \leq 250$, $50 \leq y_7 \leq 150$,

$$\begin{aligned} y_6 + 2y_7 &\leq 500, \\ y_6 + y_7 &\leq 350, \\ 2y_6 + 3y_7 &\leq 800, \end{aligned}$$

and thus

$$y_7 \le -\frac{1}{2}y_6 + 250, \quad y_7 \le -y_6 + 350, \quad y_7 \le -\frac{2}{3}y_6 + \frac{800}{3}.$$

The production space is given by Figure 8.5.

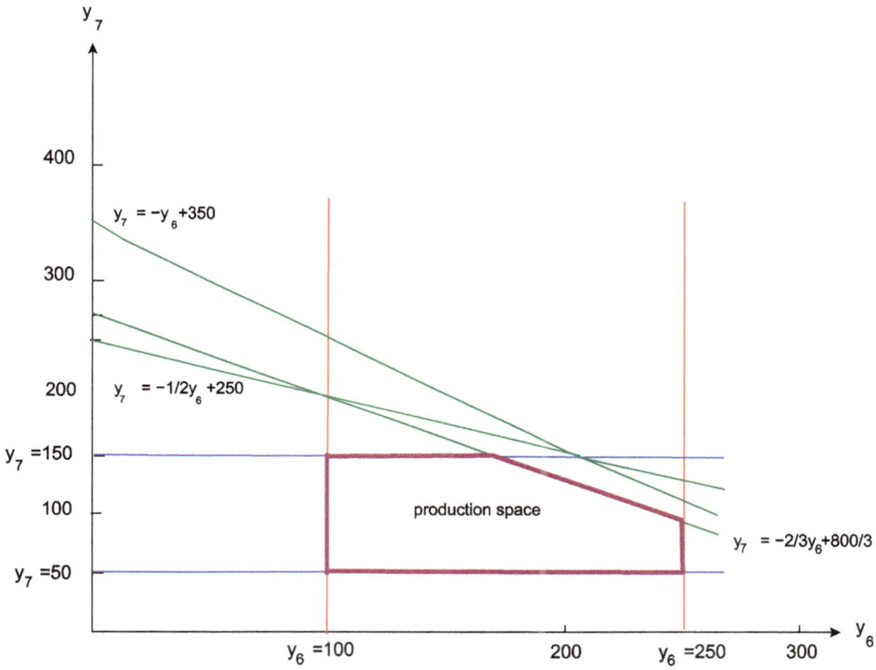

Figure 8.5: Production space: battery systems.

8.5 One-based, one-step techniques

As in Definition 8.3.2, we consider again the case that the input–output matrix has the form

$$M_{I,\{1\},O}(\mathbb{Q}^+) = \begin{pmatrix} -M_{I,\{1\}}(\mathbb{Q}^+) \\ M_{O,\{1\}}(\mathbb{Q}^+) \end{pmatrix}$$

with

$$M_{I,\{1\}}(\mathbb{Q}^+) = \begin{pmatrix} a_{11} \\ \vdots \\ a_{m1} \end{pmatrix}$$

as the input matrix and

$$M_{O,\{1\}}(\mathbb{Q}^+) = \begin{pmatrix} b_{m+11} \\ \vdots \\ b_{m+n1} \end{pmatrix}$$

as the output matrix. There is only one basic activity, and the one-step technique is given by

$$T = \left\{ \underline{z} \mid \underline{z} = \begin{pmatrix} -a_{11} \\ \vdots \\ -a_{m1} \\ b_{m+11} \\ \vdots \\ b_{m+n1} \end{pmatrix}, \lambda, a_{i1}, b_{m+k1} \in \mathbb{Q}^+, 1 \le i \le m, 1 \le k \le n \right\}.$$

Then T is said to be an *elementary technique*. Different from the general case of T^{ad} and $T^{\text{in-out}}$, elementary techniques consist of all multiples of one activity. For I, S, and O, we consider the following special cases:

	I	O	S
(i)	$\{1\}$	$\{m+1\}$	$\{1\}$
(ii)	$\{1\}$	$\{m+1,\dots,m+n\}$	$\{1\}$
(iii)	$\{1,\dots,m\}$	$\{m+1\}$	$\{1\}$
(iv)	$\{1,\dots,m\}$	$\{m+1,\dots,m+n\}$	$\{1\}$

To these four cases there correspond the following matrices and the input–output graphs given in Figure 8.6:

(1) $\begin{pmatrix} (-a_{11}) \\ (b_{m+11}) \end{pmatrix} = \begin{pmatrix} -a_{11} \\ b_{m+11} \end{pmatrix}$,

(2) $\begin{pmatrix} (-a_{11}) \\ \begin{pmatrix} b_{m+11} \\ \vdots \\ b_{m+n1} \end{pmatrix} \end{pmatrix} = \begin{pmatrix} -a_{11} \\ b_{m+11} \\ \vdots \\ b_{m+n1} \end{pmatrix}$,

(3) $\begin{pmatrix} \begin{pmatrix} -a_{11} \\ \vdots \\ -a_{m1} \end{pmatrix} \\ (b_{m+11}) \end{pmatrix} = \begin{pmatrix} -a_{11} \\ \vdots \\ -a_{m1} \\ b_{m+11} \end{pmatrix}$,

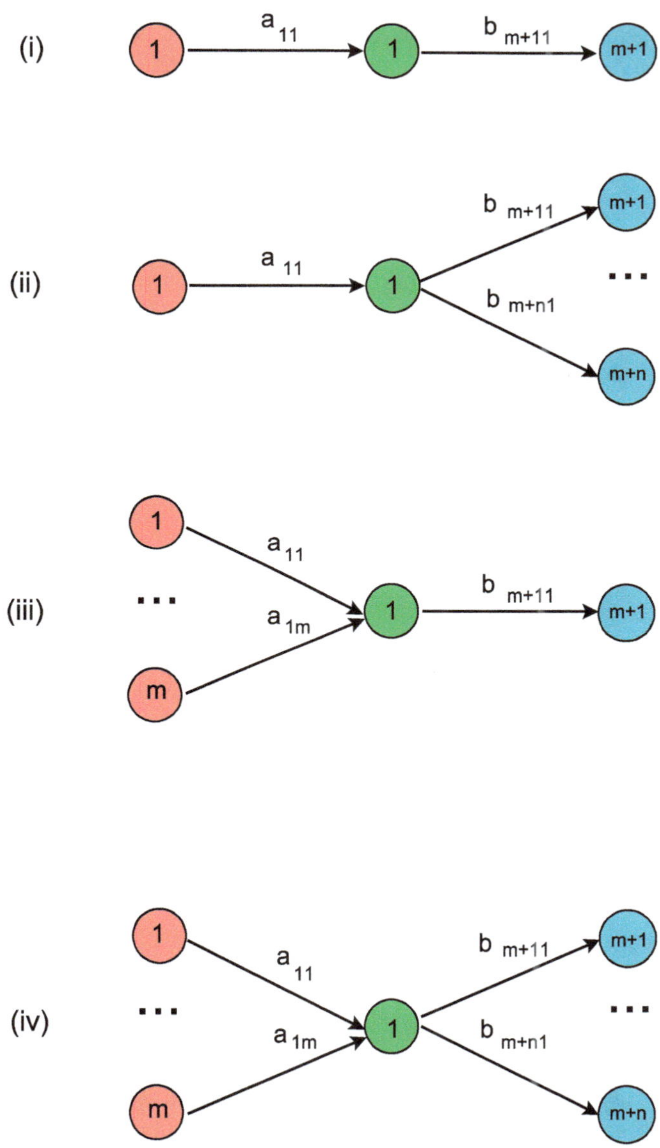

Figure 8.6: One-based techniques.

$$(4)\quad \begin{pmatrix} \begin{pmatrix} -a_{11} \\ \vdots \\ -a_{m1} \end{pmatrix} \\ \begin{pmatrix} b_{m+11} \\ \vdots \\ b_{m+n1} \end{pmatrix} \end{pmatrix} = \begin{pmatrix} -a_{11} \\ \vdots \\ -a_{1m} \\ b_{m+11} \\ \vdots \\ b_{m+n1} \end{pmatrix}.$$

(i) In the first case, one speaks of a *smooth production*. Planing a board is an example of a smooth production.
ii) The second case is a *divergent production*, for instance, when a workpiece is divided into different parts.
(iii) In the third case, we have a *convergent production*. Several individual parts are combined into one unit.
(iv) In the fourth case, the *regrouping production* means producing from several individual parts several other individual parts.

A very special case and at a first glance without any practical meaning is a production system without output, i. e., when $O = \emptyset$. In this case the matrix is

$$\begin{pmatrix} -a_{11} \\ \vdots \\ -a_{m1} \end{pmatrix}$$

and the graph is given by Figure 8.7.

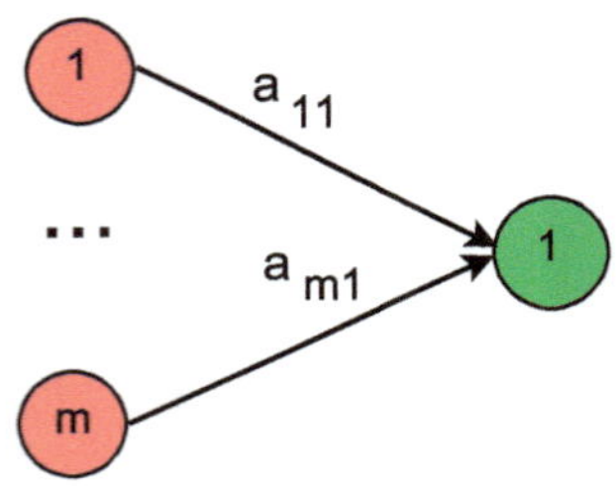

Figure 8.7: Technique without output.

This is a situation similar to a finite deterministic automaton with inputs, one inner state, but no outputs. The question here is whether the automaton accepts the input signals or not. This happens, for instance, if the syntax of a formal language is checked.

For the arbitrary case, i. e., if there is more than one basis activity and m, n, p are arbitrary positive natural numbers, the following cases, given by the input–output graphs, are of interest.

Using the algorithm leading to the one-to-one correspondence between input–output graphs (Figure 8.8) and matrices, we get:

(1) $n = p$ and the input-output matrices are given by

$$M_{I,S}(\mathbb{Q}^+) = \begin{pmatrix} a_{11} & \dots & a_{1p} \\ \vdots & \ddots & \vdots \\ a_{m1} & \dots & a_{mp} \end{pmatrix}$$

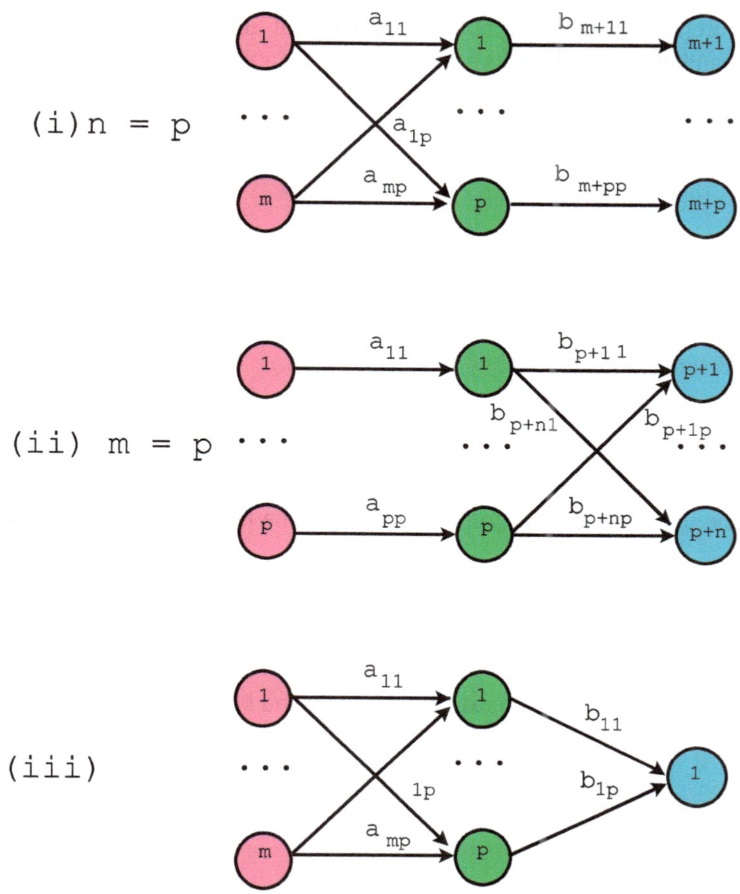

Figure 8.8: Case $n = p$ and $m = p$.

and $M_{O,S}(\mathbb{Q}^+)$ which has the following form: for an arbitrary permutation s on $\{1,\ldots,p\}$, the element in the $(m+i)$th row for $i,j \in \{1,\ldots,p\}$ is $b_{m+is(j)}$. All other elements of matrix $M_{O,S}(\mathbb{Q}^+)$ are 0. If s is the identity permutation, we get the matrix

$$M_{O,S}(\mathbb{Q}^+) = \begin{pmatrix} b_{m+11} & \cdots & 0 \\ \vdots & \ddots & \vdots \\ 0 & \cdots & b_{m+pp} \end{pmatrix}.$$

This kind of production is said to be *production determined by the output side.*

(2) $m = p$ and for the input and output matrices we have $M_{O,S}(\mathbb{Q}^+)$ as an arbitrary matrix of type (n,p) and $M_{I,S}(\mathbb{Q}^+)$ as a quadratic matrix of type (m,m) where for a

permutation s on $\{1,\dots,p\}$ and $i,j \in \{1,\dots,p\}$ the element in the ith row and $s(j)$th column is $a_{is(j)}$ and all other elements are 0. If s is the identity permutation, we get the matrix

$$M_{I,S}(\mathbb{Q}^+) = \begin{pmatrix} a_{11} & \dots & 0 \\ \vdots & \ddots & \vdots \\ 0 & \dots & a_{pp} \end{pmatrix}.$$

This kind of production is said to be *production determined by the input side.*

(3) To determine the input–output matrix belonging to the input–output graph (iii) is left as an exercise to the reader.

(We want to mention here and also for all similar earlier situations that the valuation is 0 if there is no edge.)

Exercise. Determine the input–output graph belonging to the following input–output matrix:

$$M_{I,S,O}(\mathrm{Val}(G_{I-S-O})) := \begin{pmatrix} -M_{I,S}(\mathrm{Val}(G_{I-S})) \\ M_{S,O}(\mathrm{Val}(G_{S-O})) \end{pmatrix},$$

where $M_{S,O}(\mathrm{Val}(G_{S-O}))$ is an arbitrary matrix of type (S,O) and $M_{I,S}(\mathrm{Val}(G_{I-S}))$ has precisely one row which can contain elements different from 0 and all other rows consist of 0 only.

8.6 Multi-stage techniques

If there are outputs of an activity which are inputs of the following activity, we speak of a *multi-stage technique.* If some outputs are inputs of the same activity, we have a *cyclic technique.* In this case the corresponding input–output graph has cycles. By Definition 8.2.1, this is possible. But at first we consider only acyclic multi-stage techniques.

Definition 8.6.1. Let $I \cup S \cup O$ be the set of vertices of the input–output graph of any activity belonging to technique T. If there is an activity in T such that $I' \cup S' \cup O'$ is the set of vertices of the input–output graph of this activity of T and if $O \cap I' \neq \emptyset$, then T is said to be a multi-stage technique (Figure 8.9).

This means that for all activities of T there is a following activity. Since any concrete technique consists only of finitely many activities, there must be an activity having outputs which are inputs of an earlier activity. We consider the following example:

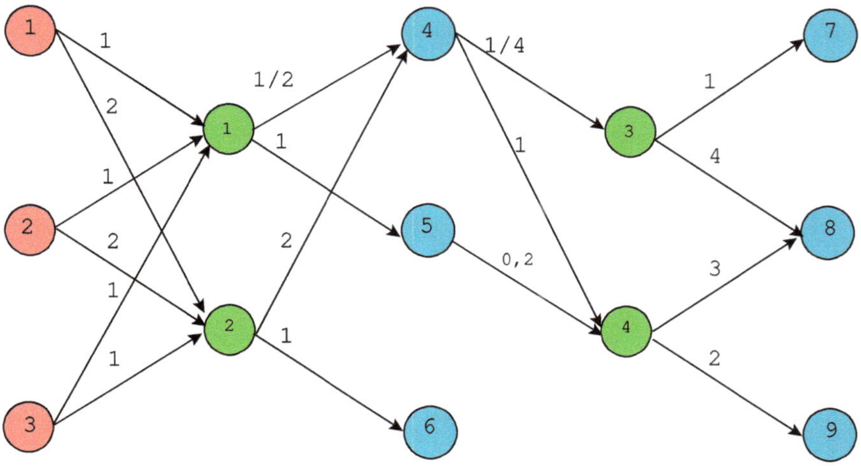

Figure 8.9: Example of a multi-stage technique.

Example 8.6.2. We have $I = \{1,2,3\}$, $S = \{1,2\}$, $O = \{4,5,6\}$, $I' = \{4,5\}$, $S' = \{3,4\}$, and $O' = \{7,8,9\}$. Therefore, $O \cap I' \neq \emptyset$ and T consisting of both activities is a multi-stage technique. The matrices belonging to the input–output graphs G_{I-S-O} and $G_{I'-S'-O'}$ are

$$\begin{pmatrix} \begin{pmatrix} -1 & -2 \\ -1 & -2 \\ -1 & -1 \end{pmatrix} \\ \begin{pmatrix} \frac{1}{2} & 2 \\ 1 & 0 \\ 0 & 1 \end{pmatrix} \end{pmatrix} = \begin{pmatrix} -1 & -2 \\ -1 & -2 \\ -1 & -1 \\ \frac{1}{2} & 2 \\ 1 & 0 \\ 0 & 1 \end{pmatrix}$$

and

$$\begin{pmatrix} \begin{pmatrix} -\frac{1}{4} & -1 \\ 0 & -\frac{1}{5} \end{pmatrix} \\ \begin{pmatrix} 1 & 0 \\ 4 & 3 \\ 0 & 2 \end{pmatrix} \end{pmatrix} = \begin{pmatrix} -\frac{1}{4} & -1 \\ 0 & -\frac{1}{5} \\ 1 & 0 \\ 4 & 3 \\ 0 & 2 \end{pmatrix}, \text{ respectively.}$$

The resulting two-stage technique is given by $I_1 = I \cup I'$, $S_1 = S \cup S'$, $O_1 = O \setminus I' \cup O'$, i. e., $I_1 = \{1,2,3,4,5\}$, $S_1 = \{1,2,3,4\}$, $O_1 = \{6,7,8,9\}$, and the corresponding input–output matrix is

$$\begin{pmatrix} -M_{I,S}(\mathrm{Val}(G_{I-S})) & \mathbf{0} \\ \mathbf{0} & -M_{I_1,S_1}(\mathrm{Val}(G_{I_1-S_1})) \\ M_{O\setminus I',S}(\mathrm{Val}(G_{O\setminus I'-S})) & \mathbf{0} \\ \mathbf{0} & M_{O_1,S_1}(\mathrm{Val}(G_{O_1-S_1})) \end{pmatrix} = \begin{pmatrix} -1 & -2 & 0 & 0 \\ -1 & -2 & 0 & 0 \\ -1 & -1 & 0 & 0 \\ 0 & 0 & -\frac{1}{4} & -1 \\ 0 & 0 & 0 & -\frac{1}{5} \\ 0 & 1 & 0 & 0 \\ 0 & 0 & 1 & 0 \\ 0 & 0 & 4 & 3 \\ 0 & 0 & 0 & 2 \end{pmatrix}.$$

In our second example, not all inputs of the second activity are outputs of the first.

(We notice that input vertices are usually colored red, process vertices are green, and output vertices are blue. In Figure 8.10 vertices 5, 6 are both input and output vertices, but they are colored blue. So, vertices 5 and 6 are an exception, acting as both input and output vertices and colored blue, which is the color for output vertices.)

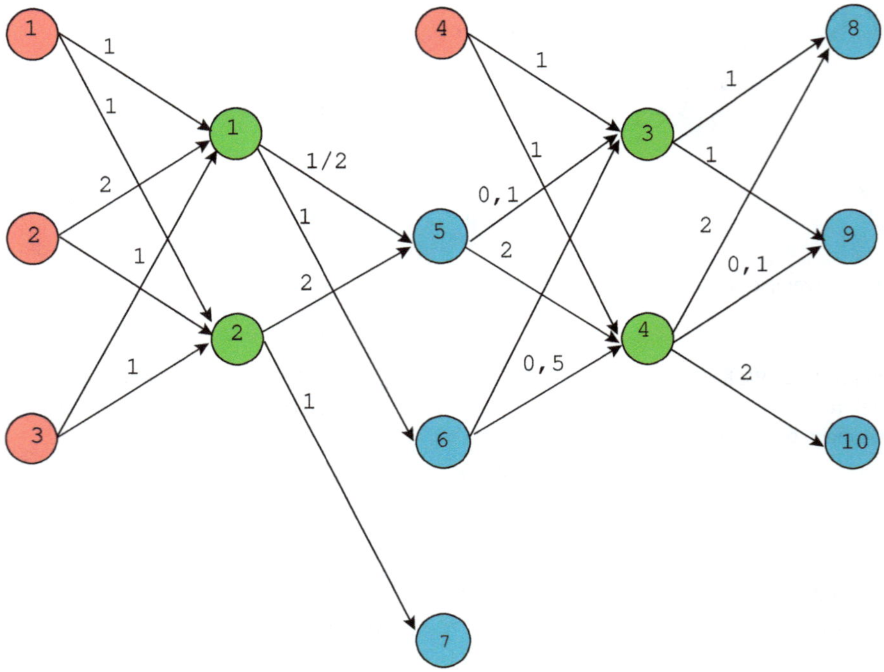

Figure 8.10: Example of a multi-stage technique.

Example 8.6.3. Consider $I_1 = \{1,2,3\}$, $S_1 = \{1,2\}$, $O_1 = \{5,6,7\}$, $I_2 = \{4,5,6\}$, $S_2 = \{3,4\}$, and $O_2 = \{7,8,9\}$. Since

$$O_1 \cap I_2 = \{5,6,7\} \cap \{4,5,6\} = \{5,6\} \neq \emptyset$$

this is a multi-stage (two-stage) technique. The input–output matrices belonging to $G_{I_1-S_1-O_1}$ and $G_{I_2-S_2-O_2}$ are

$$\begin{pmatrix} \begin{pmatrix} -1 & -1 \\ -2 & -1 \\ -1 & -1 \end{pmatrix} \\ \begin{pmatrix} \frac{1}{2} & 2 \\ 1 & 0 \\ 0 & 1 \end{pmatrix} \end{pmatrix} = \begin{pmatrix} -M_{I_1,S_1}(\mathrm{Val}(G_{I_1-S_1})) \\ M_{O_1,S_1}(\mathrm{Val}(G_{O_1-S_1})) \end{pmatrix} = \begin{pmatrix} -1 & -1 \\ -2 & -1 \\ -1 & -1 \\ \frac{1}{2} & 2 \\ 1 & 0 \\ 0 & 1 \end{pmatrix},$$

$$\begin{pmatrix}\begin{pmatrix}-1 & -1\\ -\frac{1}{10} & -2\\ -1 & -\frac{1}{2}\end{pmatrix}\\ \begin{pmatrix}1 & 2\\ 1 & \frac{1}{10}\\ 0 & 2\end{pmatrix}\end{pmatrix} = \begin{pmatrix}-M_{I_2,S_2}(\mathrm{Val}(G_{I_2-S_2}))\\ M_{O_2,S_2}(\mathrm{Val}(G_{O_2-S_2}))\end{pmatrix} = \begin{pmatrix}-1 & -1\\ -\frac{1}{10} & -2\\ -1 & -\frac{1}{2}\\ 1 & 2\\ 1 & \frac{1}{10}\\ 0 & 2\end{pmatrix}.$$

The resulting two-stage technology is given by $I = I_1 \cup I_2$, $S = S_1 \cup S_2$, $O = (O_1 \setminus O_1 \cap I_2) \cup O_2 = \{7, 8, 9, 10\}$, and the corresponding matrix is

$$\begin{pmatrix}-M_{I_1,S_1}(\mathrm{Val}(G_{I_1-S_1})) & \mathbf{0}\\ \mathbf{0} & -M_{I_2,S_2}(\mathrm{Val}(G_{I_2-S_2}))\\ M_{O_1\setminus I_2,S_1}(\mathrm{Val}(G_{O_1\setminus I_2-S_1})) & \mathbf{0}\\ \mathbf{0} & M_{O_2,S_2}(\mathrm{Val}(G_{O_2-S_2}))\end{pmatrix} = \begin{pmatrix}-1 & -1 & 0 & 0\\ -2 & -1 & 0 & 0\\ -1 & -1 & 0 & 0\\ 0 & 0 & -1 & -1\\ 0 & 0 & -\frac{1}{10} & -2\\ 0 & 0 & -1 & -\frac{1}{2}\\ 0 & 1 & 0 & 0\\ 0 & 0 & 1 & 2\\ 0 & 0 & 1 & \frac{1}{10}\\ 0 & 0 & 0 & 2\end{pmatrix},$$

which can be obtained by the following steps:

1. Form the matrix

$$\begin{pmatrix}-M_{I_1,S_1}(\mathrm{Val}(G_{I_1-S_1})) & \mathbf{0}\\ \mathbf{0} & -M_{I_2,S_2}(\mathrm{Val}(G_{I_2-S_2}))\end{pmatrix} = -M_{I,S}\big(\mathrm{Val}(G_{I-S})\big)$$

(the latter matrix is of type (I, S), in the last example $(6, 4)$), with submatrices $-M_{I_1,S_1}(\mathrm{Val}(G_{I_1-S_1}))$, $-M_{I_2,S_2}(\mathrm{Val}(G_{I_2-S_2}))$, $\mathbf{0}$, $\mathbf{0}$ of the corresponding types.

2. Form the matrix

$$\begin{pmatrix}M_{O_1\setminus I_2,S_1}(\mathrm{Val}(G_{I_2-S_1})) & \mathbf{0}\\ \mathbf{0} & M_{O_2,S_2}(\mathrm{Val}(G_{O_2-S_2}))\end{pmatrix}$$

of type $(O = O_1 \setminus I_2 \cup O_2, S)$, in the last example $(4, 4)$ with $M_{O_1\setminus I_2,S_1}(\mathrm{Val}(G_{O_1\setminus I_2-S_1}))$, $M_{O_2,S_2}(\mathrm{Val}(G_{O_2-S_2}))$ as submatrices and with two zero-submatrices of the corresponding types.

3. The matrix corresponding to G_{I-S-O} is

$$\begin{pmatrix}-M_{I,S}(\mathrm{Val}(G_{I-S}))\\ M_{O,S}(\mathrm{Val}(G_{O-S}))\end{pmatrix}.$$

These three steps describe an algorithm to determine the input–output matrix corresponding to the input–output graph G_{I-S-O} for each case with $I = I_1 \cup I_2$, $S = S_1 \cup S_2$, and $O = O_1 \setminus I_2 \cup O_2$ supposing that $O_1 \cap I_2 \neq \emptyset$.

Special cases of $O_1 \cap I_2 \subseteq O_1$ and $O_1 \cap I_2 \subseteq I_2$ are $O_1 \cap I_2 = O_1$ and $O_1 \cap I_2 = I_2$. In this situation all outputs of the first production are used as inputs of the second, $O_1 \subseteq I_2$. In this case there can be more inputs of the second step as come from the outputs of the first step, i. e., $O_1 \subset I_2$.

In the second special case, $O_1 \cap I_2 = I_2$ means $I_2 \subseteq O_1$. In this case there are no inputs of the second step which are not outputs of the first. In general, the first and second steps can be described by the graphs $G_{I_1-S_1-O_1}$ and $G_{I_2-S_2-O_2}$, respectively. By Theorem 8.2.7, the graph $G_{I_1-S_1-O_1}$ uniquely corresponds to the matrix

$$\begin{pmatrix} -M_{I_1,S_1}(\mathrm{Val}(G_{I_1-S_1})) \\ M_{O_1,S_1}(\mathrm{Val}(G_{O_1-S_1})) \end{pmatrix}$$

and $G_{I_2-S_2-O_2}$ uniquely corresponds to the matrix

$$\begin{pmatrix} -M_{I_2,S_2}(\mathrm{Val}(G_{I_2-S_2})) \\ M_{O_2,S_2}(\mathrm{Val}(G_{O_2-S_2})) \end{pmatrix},$$

while G_{I-S-O} uniquely determines the four matrices

$$-M_{I_1,S_1}(\mathrm{Val}(G_{I_1-S_1})), M_{O_1,S_1}(\mathrm{Val}(G_{O_1-S_1})), -M_{I_2,S_2}(\mathrm{Val}(G_{I_2-S_2})), M_{O_1,S_1}(\mathrm{Val}(G_{O_1-S_1})).$$

From $O_1 \cap I_2 \neq \emptyset$ and $O_1 \cap I_2 \subseteq O_1$, we obtain the following four cases:

1. $O_1 \cap I_2 = O_1$ and $O_1 \cap I_2 \subset I_2$. Then $O_1 \setminus I_2 = O_1 \setminus (O_1 \cap I_2) = \emptyset$.
2. $O_1 \cap I_2 \subset O_1$ and $O_1 \cap I_2 = I_2$. Then $O_1 \setminus I_2 = O_1 \setminus (O_1 \cap I_2) \neq \emptyset$.
3. $O_1 \cap I_2 \subset O_1$ and $O_1 \cap I_2 \subset I_2$. Then $O_1 \setminus I_2 = O_1 \setminus (O_1 \cap I_2) \neq \emptyset$.
4. $O_1 \cap I_2 = O_1$ and $O_1 \cap I_2 = I_2$. Then $O_1 = I_2$ and $O_1 \setminus I_2 = O_1 \setminus (O_1 \cap I_2) = \emptyset$.

Then by the following steps we get the matrix corresponding to G_{I-S-O}:

1. Form the matrix

$$-M_{I,S}(\mathrm{Val}(G_{I-S})) = \begin{pmatrix} -M_{I_1,S_1}(\mathrm{Val}(G_{I_1-S_1})) & \mathbf{0} \\ \mathbf{0} & -M_{I_2,S_2}(\mathrm{Val}(G_{I_2-S_2})) \end{pmatrix}$$

   of type (I, S).
2. If $O_1 \setminus I_2 \neq \emptyset$, then form the matrix

$$M_{O,S}(\mathrm{Val}(G_{O-S})) = \begin{pmatrix} M_{O_1\setminus I_2,S_1}(\mathrm{Val}(G_{O_1\setminus I_2-S_1})) & \mathbf{0} \\ \mathbf{0} & M_{O_2,S_2}(\mathrm{Val}(G_{O_2-S_2})) \end{pmatrix},$$

   otherwise, i. e., if $O_1 \setminus I_2 = \emptyset$,
3. form the matrix

$$M_{O,S}(\mathrm{Val}(G_{O-S})) = \begin{pmatrix} \mathbf{0} & \mathbf{0} \\ \mathbf{0} & M_{O_2,S_2}(\mathrm{Val}(G_{O_2-S_2})) \end{pmatrix}$$

   of the type (O, S) with $O = (O_1 \setminus I_2) \cup O_2$.
4. By Theorem 8.2.7, the uniquely determined input–output matrix belonging to G_{I-S-O} is

$$\begin{pmatrix} -M_{I,S}(\mathrm{Val}(G_{I-S})) \\ M_{O,S}(\mathrm{Val}(G_{O-S})) \end{pmatrix}.$$

This proves the following result:

Corollary 8.6.4. *The uniquely determined input–output matrix belonging to the input–output graph of a two-stage technology is the matrix*

$$M_{I-S-O} = \begin{pmatrix} -M_{I,S}(\mathrm{Val}(G_{I-S})) \\ M_{O,S}(\mathrm{Val}(G_{O-S})) \end{pmatrix}.$$

Let $\mathcal{TM}_{I,S,O}(\mathrm{Val}(G_{I-S-O}))$ be the set of all input–output matrices of type $I \times S \times O$, i. e., for finite sets I, S, O the set of all finite matrices of the form

$$\begin{pmatrix} -M_{I,S}(\mathrm{Val}(G_{I-S})) \\ M_{O,S}(\mathrm{Val}(G_{O-S})) \end{pmatrix}.$$

For short, we will write $\mathcal{TM}_{I,S,O}$ or $\mathcal{TM}$ if I, S, O are given. On this set, we define a binary partial operation $\oplus$,

$$\oplus : \mathcal{TM}_{I,S,O} \times \mathcal{TM}_{I,S,O} \multimap \mathcal{TM}_{I,S,O},$$

by

$$\begin{pmatrix} -M_{I_1,S_1}(\mathrm{Val}(G_{I_1-S_1})) \\ M_{O_1,S_1}(\mathrm{Val}(G_{O_1-S_1})) \end{pmatrix} \oplus \begin{pmatrix} -M_{I_2,S_2}(\mathrm{Val}(G_{I_2-S_2})) \\ M_{O_2,S_2}(\mathrm{Val}(G_{O_2-S_2})) \end{pmatrix}$$

$$:= \begin{cases} \begin{pmatrix} -M_{I_1,S_1}(\mathrm{Val}(G_{I_1-S_1})) & \mathbf{0} \\ \mathbf{0} & -M_{I_2,S_2}(\mathrm{Val}(G_{I_2-S_2})) \\ M_{O_1\setminus I_2,S_1}(\mathrm{Val}(G_{O_1\setminus I_2-S_1})) & \mathbf{0} \\ \mathbf{0} & M_{O_2,S_2}(\mathrm{Val}(G_{O_2-S_2})) \end{pmatrix} \\ \text{if } O_1 \setminus I_2 \neq \emptyset \text{ and } I_2 \cap O_1 \neq \emptyset, \\ \begin{pmatrix} -M_{I_1,S_1}(\mathrm{Val}(G_{I_1-S_1})) & \mathbf{0} \\ \mathbf{0} & -M_{I_2,S_2}(\mathrm{Val}(G_{I_2-S_2})) \\ \mathbf{0} & M_{O_2,S_2}(\mathrm{Val}(G_{O_2-S_2})) \end{pmatrix} \\ \text{if } O_1 \setminus I_2 = \emptyset \text{ (then also } I_2 \cap O_1 \neq \emptyset), \\ \text{not defined, if } I_2 \cap O_1 = \emptyset. \end{cases}$$

Then $(\mathcal{TM}_{I,S,O}; \oplus)$ is a partial algebra of type (2) since, if the resulting matrix exists, then it is an input–output matrix in the sense of Definition 8.2.6.

In parallel, we consider the set $\mathcal{G}_{I-S-O}$ of all input–output graphs and define a binary partial operation

$$\bar{\cup} : \mathcal{G}_{I-S-O} \times \mathcal{G}_{I-S-O} \multimap \mathcal{G}_{I-S-O}$$

by

$$G_{I_1-S_1-O_1} \bar{\cup} G_{I_2-S_2-O_2} := \begin{cases} G_{I_1-S_1-O_1} \cup G_{I_2-S_2-O_2} & \text{if } I_2 \cap O_1 \neq \emptyset, \\ \text{not defined} & \text{otherwise,} \end{cases}$$

thus obtaining a partial algebra $(\mathcal{G}_{I-S-O}; \bar{\cup})$ of type (2).

To show that $\bar{\cup}$ is a partial operation on $\mathcal{G}_{I-S-O}$, we have to prove that the result of applying $\bar{\cup}$ on two input–output graphs gives, in the case of existence, an input–output graph. But this becomes clear if we consider $I = I_1 \cup I_2$, $S = S_1 \cup S_2$, $O = (O_1 \setminus (O_1 \cap I_2)) \cup O_2$. This guarantees that the elements of $O_1 \cap I_2$ are considered as inputs of $G_{I_2-S_2-O_2}$ and not as outputs of $G_{I_1-S_1-O_1}$. So, Definition 8.2.1 is satisfied and the result is an input–output graph.

We recall the concepts of weak isomorphisms and weak identities for partial algebras which we have introduced in Chapter 7, Section 7.11. If $\mathcal{A}, \mathcal{B}$ are partial algebras of the same type with indexed sets $\{f_i^{\mathcal{A}} \mid i \in I\}$, $\{f_i^{\mathcal{B}} \mid i \in I\}$ of partial fundamental operations on A and on B, respectively, then a mapping $h : A \to B$ is said to be a *weak homomorphism* if for all fundamental operations the following is satisfied:

if $(a_1, \ldots, a_{n_i}) \in \mathrm{dom} f_i^{\mathcal{A}}$, then $(h(a_1), \ldots, h(a_{n_i})) \in \mathrm{dom} f_i^{\mathcal{B}}$ and then

$$h(f_i^{\mathcal{A}}(a_1, \ldots, a_{n_i})) = f_i^{\mathcal{B}}(h(a_1), \ldots, h(a_{n_i})), i \in I,$$

where $\mathrm{dom} f_i^{\mathcal{A}}$ is the domain of the operation $f_i^{\mathcal{A}}$. Weak endomorphisms are weak homomorphisms with $h : A \to A$, and a weak homomorphism is said to be a weak isomorphism if h is a bijection.

The equation $s \approx t$ is said to be a weak identity of the partial algebra $\mathcal{A}$ if after evaluation there holds: if the right-hand side is defined then the left-hand side is defined, and both sides are equal, or if the left-hand side is defined then the right-hand side is defined, and both sides are equal. For more information on partial algebras, see [13]. A partial algebra of type (2) is said to be a *partial semigroup* if the associative law is satisfied as a weak identity.

Theorem 8.6.5. *$(\mathcal{G}_{I-S-O}; \bar{\cup})$ is a partial semigroup.*

Proof. We have to prove that the associative law

$$F(X_1, F(X_2, X_3)) \approx F(F(X_1, X_2), X_3) \tag{*}$$

is a weak identity in $(\mathcal{G}_{I-S-O}; \bar{\cup})$. Here F is a binary operation symbol for the binary operation $\bar{\cup}$ and X_1, X_2, X_3 are variables for input–output graphs G_{I-S-O}. Assume that the left-hand side of (*) is defined, i. e., $O_1 \cap I_2 \neq \emptyset$, and then

$$G_{I_1-S_1-O_1} \bar{\cup} G_{I_2-S_2-O_2} = G_{I_1-S_1-O_1} \cup G_{I_2-S_2-O_2}$$

and $O_2 \cap I_3 \neq \emptyset$, i. e.,

$$(G_{I_1-S_1-O_1} \bar{\cup} G_{I_2-S_2-O_2}) \bar{\cup} G_{I_3-S_3-O_3}$$

is defined and equal to

$$(G_{I_1-S_1-O_1} \cup G_{I_2-S_2-O_2}) \cup G_{I_3-S_3-O_3}.$$

Since the total operation $\cup$ is associative, the partial operation $\bar{\cup}$ is also associative. (Here we used that the set-theoretical union of two graphs $G_1 = (V_1, E_1)$ and $G_2 = (V_2, E_2)$ is given as $G_1 \cup G_2 = (V_1 \cup V_2, E_1 \cup E_2)$.) □

Theorem 8.6.5 is important for applications for the following reason. For production procedures with more than two steps, after the first step, the result can be combined with one new technique and the new result can be connected with the next step, etc. Because the associative law is satisfied as a weak identity, the two sequences

$$(\cdots((X_1 \bar{\cup} X_2) \bar{\cup} X_3) \bar{\cup} X_4) \cdots \bar{\cup} X_n)$$

and

$$(X_1 \bar{\cup} (\cdots (X_{n-1} \bar{\cup} X_n) \cdots)$$

give the same final result. This means that in this sense each technique can be described by two-step techniques.

Some properties of $(\mathcal{G}_{I-S-O}; \bar{\cup})$ are reflected as properties of $(\mathcal{TM}; \oplus)$ as the following corollary shows.

Corollary 8.6.6. *$(\mathcal{TM}; \oplus)$ is weakly isomorphic to $(\mathcal{G}_{I-S-O}; \bar{\cup})$ and, therefore, it is a partial semigroup.*

Proof. Assume that $(G_{I_1-S_1-O_1}, G_{I_2-S_2-O_2}) \in \operatorname{dom} \bar{\cup}$. Then $O_1 \cap I_2 \neq \emptyset$ and

$$G_{I_1-S_1-O_1} \bar{\cup} G_{I_2-S_2-O_2} = G_{I_1-S_1-O_1} \cup G_{I_2-S_2-O_2}.$$

Let φ be the one-to-one mapping from Theorem 8.2.7 which maps the input–output graph G_{I-S-O} to the input–output matrix

$$M_{I,S,O}(\operatorname{Val}(G_{I-S-O})) = \begin{pmatrix} -M_{I,S}(\operatorname{Val}(G_{I-S})) \\ M_{O,S}(\operatorname{Val}(G_{O-S})) \end{pmatrix}.$$

Then

$$\begin{aligned} &(\varphi(G_{I_1-S_1-O_1}), \varphi(G_{I_2-S_2-O_2})) \\ &= \left(\begin{pmatrix} -M_{I_1,S_1}(\operatorname{Val}(G_{I_1-S_1})) \\ M_{O_1,S_1}(\operatorname{Val}(G_{O_1-S_1})) \end{pmatrix}, \begin{pmatrix} -M_{I_2,S_2}(\operatorname{Val}(G_{I_2-S_2})) \\ M_{O_2,S_2}(\operatorname{Val}(G_{O_2-S_2})) \end{pmatrix} \right) \in \operatorname{dom} \oplus \end{aligned}$$

since $O_1 \cap I_2 \neq \emptyset$.

Moreover,

$$\varphi(G_{I_1-S_1-O_1} \bar{\cup} G_{I_2-S_2-O_2}) = \varphi(G_{I_1-S_1-O_1} \cup G_{I_2-S_2-O_2}) = \varphi(G_{I-S-O})$$

since $O_1 \cap I_2 \neq \emptyset$ with $I = I_1 \cup I_2$, $S = S_1 \cup S_2$, $O = (O_1 \setminus I_2) \cup O_2$.

By the algorithm described before, we have

$$\varphi(G_{I-S-O}) = M_{I-S-O}(\mathrm{Val}(G_{I-S-O})) = \begin{pmatrix} -M_{I,S}(\mathrm{Val}(G_{I-S})) \\ M_{O,S}(\mathrm{Val}(G_{O-S})) \end{pmatrix}$$

$$= \begin{cases} \begin{pmatrix} -M_{I_1,S_1}(\mathrm{Val}(G_{I_1-S_1})) & \mathbf{0} \\ \mathbf{0} & -M_{I_2,S_2}(\mathrm{Val}(G_{I_2-S_2})) \\ M_{O_1\setminus I_2,S_1}(\mathrm{Val}(G_{O_1\setminus I_2-S_1})) & \mathbf{0} \\ \mathbf{0} & M_{O_2,S_2}(\mathrm{Val}(G_{O_2-S_2})) \end{pmatrix} \\ \text{if } O_1 \setminus I_2 \neq \emptyset \text{ and } O_1 \cap I_2 \neq \emptyset, \\ \begin{pmatrix} -M_{I_1,S_1}(\mathrm{Val}(G_{I_1-S_1})) & \mathbf{0} \\ \mathbf{0} & -M_{I_2,S_2}(\mathrm{Val}(G_{I_2-S_2})) \\ \mathbf{0} & \mathbf{0} \\ \mathbf{0} & M_{O_2,S_2}(\mathrm{Val}(G_{O_2-S_2})) \end{pmatrix} \\ \text{if } O_1 \setminus I_2 = \emptyset \ (O_1 \cap I_2 \neq \emptyset) \end{cases}$$

$$= \begin{pmatrix} -M_{I_1,S_1}(\mathrm{Val}(G_{I_1-S_1})) \\ M_{O_1,S_1}(\mathrm{Val}(G_{O_1-S_1})) \end{pmatrix} \oplus \begin{pmatrix} -M_{I_2,S_2}(\mathrm{Val}(G_{I_2-S_2})) \\ M_{O_2,S_2}(\mathrm{Val}(G_{O_2-S_2})) \end{pmatrix}$$

$$= \varphi(G_{I_1-S_1-O_1}) \oplus \varphi(G_{I_2-S_2-O_2}).$$

This shows that $\varphi : (\mathcal{G}_{I-S-O}; \bar{\cup}) \to (\mathcal{TM}_{I,S,O}; \oplus)$ is a weak isomorphism.

Since the associative law is a weak identity in $(\mathcal{G}_{I-S-O}; \bar{\cup})$, we have: if $(G_{I_1-S_1-O_1} \bar{\cup} G_{I_2-S_2-O_2}) \bar{\cup} G_{I_3-S_3-O_3}$ is defined, i. e., if $O_1 \cap I_2 \neq \emptyset$ and $O_2 \cap I_3 \neq \emptyset$, and therefore, $(G_{I_1-S_1-O_1}, G_{I_2-S_2-O_2}) \in \operatorname{dom} \bar{\cup}$ and $(G_{I_1 \cup I_2-S_1 \cup S_2-O_1 \cup O_2}, G_{I_3-S_3-O_3}) \in \operatorname{dom} \bar{\cup}$, and since φ is a weak isomorphism, also $(\varphi(G_{I_1-S_1-O_1}), \varphi(G_{I_2-S_2-O_2})) \in \operatorname{dom} \oplus$ and $(\varphi(G_{I_1 \cup I_2-S_1 \cup S_2-O_1 \cup O_2}), \varphi(G_{I_3-S_3-O_3})) \in \operatorname{dom} \oplus$, then

$$(\varphi(G_{I_1-S_1-O_1}) \oplus \varphi(G_{I_2-S_2-O_2})) \oplus \varphi(G_{I_3-S_3-O_3})$$

is defined. But, if $(G_{I_1-S_1-O_1} \bar{\cup} G_{I_2-S_2-O_2}) \bar{\cup} G_{I_3-S_3-O_3}$ is defined, then also $G_{I_1-S_1-O_1} \bar{\cup} (G_{I_2-S_2-O_2} \bar{\cup} G_{I_3-S_3-O_3})$ is defined, and this means

$$(G_{I_2-S_2-O_2}, G_{I_3-S_3-O_3}) \in \operatorname{dom} \bar{\cup}$$

and

$$(G_{I_1-S_1-O_1}, G_{I_2\cup I_3-S_2\cup S_3-O_2\cup O_3}) \in \operatorname{dom} \bar{\cup}.$$

Since φ is a weak isomorphism, we have

$$(\varphi(G_{I_1-S_1-O_1}), \varphi(G_{I_2\cup I_3-S_2\cup S_3-O_2\cup O_3})) \in \operatorname{dom} \oplus$$

and

$$(\varphi(G_{I_2-S_2-O_2}), \varphi(G_{I_3-S_3-O_3})) \in \operatorname{dom} \oplus .$$

This means that

$$\varphi(G_{I_1-S_1-O_1}) \oplus (\varphi(G_{I_2-S_2-O_2}) \oplus \varphi(G_{I_3-S_3-O_3}))$$

is defined. So we have that if

$$(\varphi(G_{I_1-S_1-O_1}) \oplus \varphi(G_{I_2-S_2-O_2})) \oplus \varphi(G_{I_3-S_3-O_3})$$

is defined, then

$$\varphi(G_{I_1-S_1-O_1}) \oplus (\varphi(G_{I_2-S_2-O_2}) \oplus \varphi(G_{I_3-S_3-O_3}))$$

is defined and both sides are equal. By definition of φ, this means that if

$$\left(\begin{pmatrix} -M_{I_1,S_1}(\mathrm{Val}(G_{I_1-S_1})) \\ M_{S_1,O_1}(\mathrm{Val}(G_{S_1-O_1})) \end{pmatrix} \oplus \begin{pmatrix} -M_{I_2,S_2}(\mathrm{Val}(G_{I_2-S_2})) \\ M_{S_2,O_2}(\mathrm{Val}(G_{S_2-O_2})) \end{pmatrix}\right) \oplus \begin{pmatrix} -M_{I_3,S_3}(\mathrm{Val}(G_{I_3-S_3})) \\ M_{S_3,O_3}(\mathrm{Val}(G_{I_3-S_3})) \end{pmatrix}$$

exists, then

$$\begin{pmatrix} -M_{I_1,S_1}(\mathrm{Val}(G_{I_1-S_1})) \\ M_{S_1,O_1}(\mathrm{Val}(G_{S_1-O_1})) \end{pmatrix} \oplus \left(\begin{pmatrix} -M_{I_2,S_2}(\mathrm{Val}(G_{I_2-S_2})) \\ M_{S_2,O_2}(\mathrm{Val}(G_{S_2-O_2})) \end{pmatrix} \oplus \begin{pmatrix} -M_{I_3,S_3}(\mathrm{Val}(G_{I_3-S_3})) \\ M_{S_3,O_3}(\mathrm{Val}(G_{S_3-O_3})) \end{pmatrix}\right)$$

exists and both sides are equal. Altogether, this shows that $(\mathcal{TM}; \oplus)$ is a partial semi-group. □

8.7 Cyclic techniques and their matrices

Up to now we considered input–output graphs without cycles. Definition 8.2.1 allows cycles. It can be seen that also in this case for two input–output graphs the result $G_{I_1-S_1-O_1} \bar{\cup} G_{I_2-S_2-O_2}$ is an input–output graph if we assume that all outputs which are inputs are considered as inputs. But the corresponding matrices must not be input–output matrices in the sense of Definition 8.2.6. We give some examples for this situation and calculate the matrices of the form $\varphi(G_{I_1-S_1-O_1} \bar{\cup} G_{I_2-S_2-O_2})$.

Definition 8.7.1. Let $0 < k \leq n \leq l$ be natural numbers and let $I = I_1 \cup \cdots \cup I_k \cup \cdots \cup I_n \cup I_{n+1} \cup \cdots \cup I_l$, $S = S_1 \cup \cdots \cup S_l$, $O = O_1 \cup \cdots \cup O_l$. If G_{I-S-O} is the input–output graph of a multi-stage technique consisting of the techniques $G_{I_i-S_i-O_i}$, $0 < i \leq l$, and if there exist $k, n, k \leq n$, such that $O_n \cap I_k \neq \emptyset$, then the technique is said to be cyclic.

Remark 8.7.2. If

$$X_1 \bar{\cup} X_2 \cdots \bar{\cup} X_k \bar{\cup} X_{k+1} \bar{\cup} \cdots \bar{\cup} X_n \bar{\cup} \cdots \bar{\cup} X_l$$

is the sequence of input–output graphs of a multi-stage cyclic technique, then there are natural numbers $k \leq n$ with $O_n \cap I_k \neq \emptyset$ and, using the associative law, we can write

$$(X_1 \bar{\cup} X_2 \cdots \bar{\cup} X_k) \bar{\cup} (X_{k+1} \bar{\cup} \cdots \bar{\cup} X_{n-1} \bar{\cup} X_n) = X_{k'} \bar{\cup} X_{n'}$$

with $I_{k'} = I_1 \cup \cdots \cup I_k$ and $O_{n'} = O_{k+1} \cup \cdots \cup O_n$. Because

$$\begin{aligned}(O_{n'} \cap I_{k'}) &= (I_1 \cup \cdots \cup I_k) \cap (O_{k+1} \cup \cdots \cup O_n) \\ &= (I_1 \cap O_{k+1}) \cup \cdots \cup (I_k \cap O_{k+1}) \cup \cdots \cup (I_1 \cap O_n) \cup \cdots \cup (I_k \cap O_n) \\ &\neq \emptyset,\end{aligned}$$

there are $k' \leq n'$ such that $(O_{n'} \cap I_{k'}) \neq \emptyset$ and the two-stage technique $X_{k'} \bar{\cup} X_{n'}$ is cyclic.

The latter remark proves the following corollary:

Corollary 8.7.3. *If G_{I-S-O} is the input–output graph of a multi-stage technique, then there exist $k', n', k' \leq n'$ such that $G_{I_{k'} \cup I_{n'}, S_{k'} \cup S_{n'}, O_{k'} \cup O_{n'}}$ is the input–output graph of a two-stage cyclic technique.*

This shows once more the importance of the associative law for production systems. Now we consider some examples of two-stage cyclic techniques, determine the corresponding input–output matrices, and try to describe some general properties of these matrices.

Example 8.7.4. Consider $I_1 = \{1,2,3,4,7,10\}$, $I_1{}^* = \{1,2,3,4\}$, $S_1 = \{1,2\}$, $O_1 = \{5,6\}$, $I_2 = \{5,6\}$, $S_2 = \{3,4,5\}$, $O_2 = \{7,8,9,10\}$. (Here I_1^* is the set of immediate inputs of $G_{I_1-S_1-O_1}$). Because $O_1 \cap I_2 = \{5,6\} \neq \emptyset$, $G_{I_1 \cup I_2, S_1 \cup S_2, O_1 \cup O_2}$ is the input–output graph of a two-stage technique and, since $O_2 \cap I_1 = \{7,10\} \neq \emptyset$, it is cyclic.

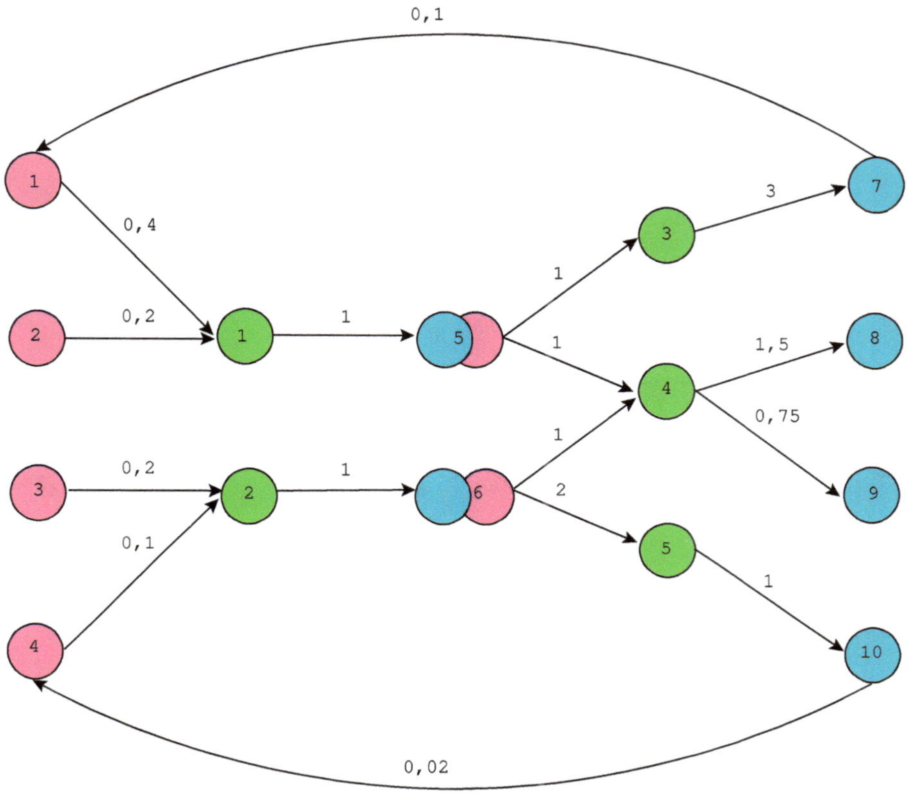

Figure 8.11: Example 1 of a cyclic technique.

The corresponding input–output matrix is given by

$$\begin{pmatrix} -M_{I_1^*,S_1}(\mathrm{Val}(G_{I_1^*-S_1})) & \mathbf{0} \\ M_{O_1,S_1}(\mathrm{Val}(G_{O_1-S_1})) & -M_{I_2,S_2}(\mathrm{Val}(G_{I_2-S_2})) \\ M_{O_2,S_1}(\mathrm{Val}(G_{O_2-S_2})) & M_{O_2,S_2}(\mathrm{Val}(G_{O_2-S_2})) \end{pmatrix} = \begin{pmatrix} -0.4 & 0 & 0 & 0 & 0 \\ -0.2 & 0 & 0 & 0 & 0 \\ 0 & -0.2 & 0 & 0 & 0 \\ 0 & -0.1 & 0 & 0 & 0 \\ 1 & 0 & -1 & -1 & 0 \\ 0 & 1 & 0 & -1 & -2 \\ -0.1 & 0 & 3 & 0 & 0 \\ 0 & 0 & 0 & 1.5 & 0 \\ 0 & 0 & 0 & 0.75 & 0 \\ 0 & -0.02 & 0 & 0 & 1 \end{pmatrix}$$

with

$$M_{I_1,S_1}(\mathrm{Val}(G_{I_1,S_1})) = \begin{pmatrix} M_{I_1^*,S_1}(\mathrm{Val}(G_{I_1^*-S_1})) \\ M_{I_1\cap O_2,S_1}(\mathrm{Val}(G_{O_2-S_1})) \end{pmatrix} = \begin{pmatrix} 0.4 & 0 \\ 0.2 & 0 \\ 0 & 0.2 \\ 0.1 & 0 \\ 0 & 0.02 \end{pmatrix}.$$

A one-stage technique with $I_1 \cap O_1 \neq \emptyset$ is said to be a *one-stage cyclic technique*. In this case we have $k = n = l = 1$. We consider the following example:

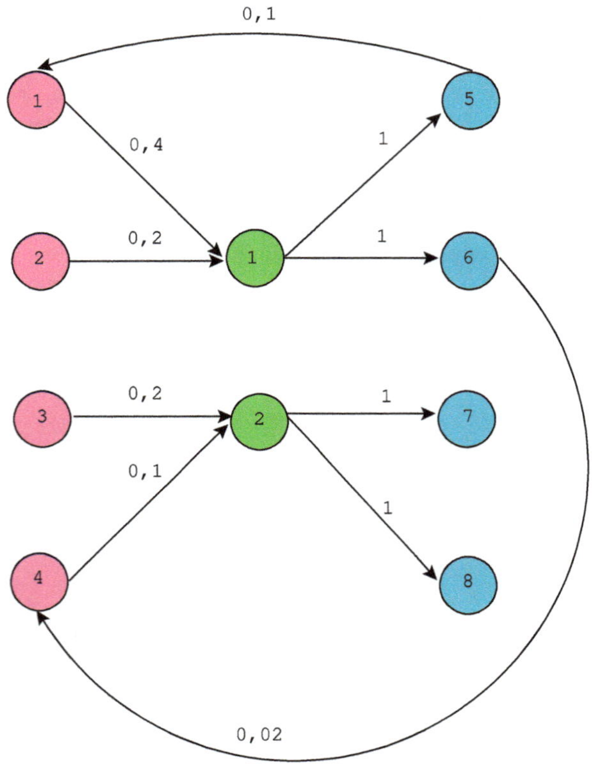

Figure 8.12: Example 2 of a one-stage cyclic technique.

Example 8.7.5. Consider $I_1 = \{1,2,3,4,5,6\}$, $I_1^* = \{1,2,3,4\}$, $S_1 = \{1,2\}$, $O_1 = \{5,6,7,8\}$. Then

$$-M_{I_1,S_1}(\mathrm{Val}(G_{I_1-S_1})) = \begin{pmatrix} -0.4 & 0 \\ -0.2 & 0 \\ 0 & -0.2 \\ 0 & -0.1 \\ -0.1 & 0 \\ 0 & -0.02 \end{pmatrix},$$

$$M_{O_1,S_1}(\mathrm{Val}(G_{O_1-S_1})) = \begin{pmatrix} 1 & 0 \\ 1 & 0 \\ 0 & 1 \\ 0 & 1 \end{pmatrix}, \quad M_{O_1\cap I_1,S_1} = \begin{pmatrix} 0.1 & 0 \\ 0 & 0.02 \end{pmatrix},$$

and we have

$$M_{I_1,S_1,O_1}(\mathrm{Val}(G_{I_1-S_1-O_1})) = \begin{pmatrix} -0.4 & 0 \\ -0.2 & 0 \\ 0 & -0.2 \\ 0 & -0.1 \\ -0.1 & 0 \\ 0 & -0.02 \\ 1 & 0 \\ 1 & 0 \\ 0 & 1 \\ 0 & 1 \end{pmatrix} = \begin{pmatrix} M_{I_1,S_1}(\mathrm{Val}(G_{I_1-S_1})) \\ M_{O_1,S_1}(\mathrm{Val}(G_{O_1-S_1})) \end{pmatrix}.$$

In the general case, for $k = n = l = 1$ and $I_1 \cap O_1 \neq \emptyset$, the solution can be given similarly.

Example 8.7.6. Now we consider a more general example of a two-stage cyclic technique. Let $I_1 = \{1,2,3,8\}$, $S_1 = \{1,2\}$, $O_1 = \{5,6,7\}$, $I_2 = \{4,5,6,10\}$, $S_2 = \{3,4\}$, $O_2 = \{8,9,10\}$, $I = I_1 \cup I_2$, $S = S_1 \cup S_2$, and $O = O_1 \cup O_2$.

Because $O_1 \cap I_2 = \{5,6\} \neq \emptyset$, it is a two-stage technique with $I_2 \setminus O_1 = \{4,10\}$ and, since $O_2 \cap I_1 = \{8\} \neq \emptyset$, $O_2 \cap I_2 = \{10\} \neq \emptyset$, it is a cyclic two-stage technique, while $G_{I_2-S_2-O_2}$ is the input–output graph of a one-stage technique. We form a matrix which corresponds to the graph of Figure 8.13, which is an input–output graph with $I = \{1,2,3,4,5,6,10\}$, $S = S_1 \cup S_2$, and $O = \{7,8,9\}$, namely

$$\begin{pmatrix} -1 & -1 & 0 & 0 \\ -2 & -1 & 0 & 0 \\ -1 & -1 & 0 & 0 \\ 0 & 0 & -1 & -2 \\ 3 & 0 & -1 & 0 \\ 1 & 0 & -1 & -1 \\ 0 & 2 & 0 & 0 \\ -0.2 & 0 & 1 & 0 \\ 0 & 0 & 0 & 1 \\ 0 & 0 & 0.5 & -0.1 \end{pmatrix}.$$

Problem. Define the binary matrix operation for cyclic techniques which gives, for the input–output matrices corresponding to $G_{I_1-S_1-O_1}$ and $G_{I_2-S_2-O_2}$, respectively, the result $\varphi(G_{I_1-S_1-O_1} \bar{\cup} G_{I_2-S_2-O_2})$ where φ is the bijection from the previous section.

8.8 Historical remarks

The history of production systems began when the period of craft production ended. In America, craft production and the importance of the all-round skilled worker were

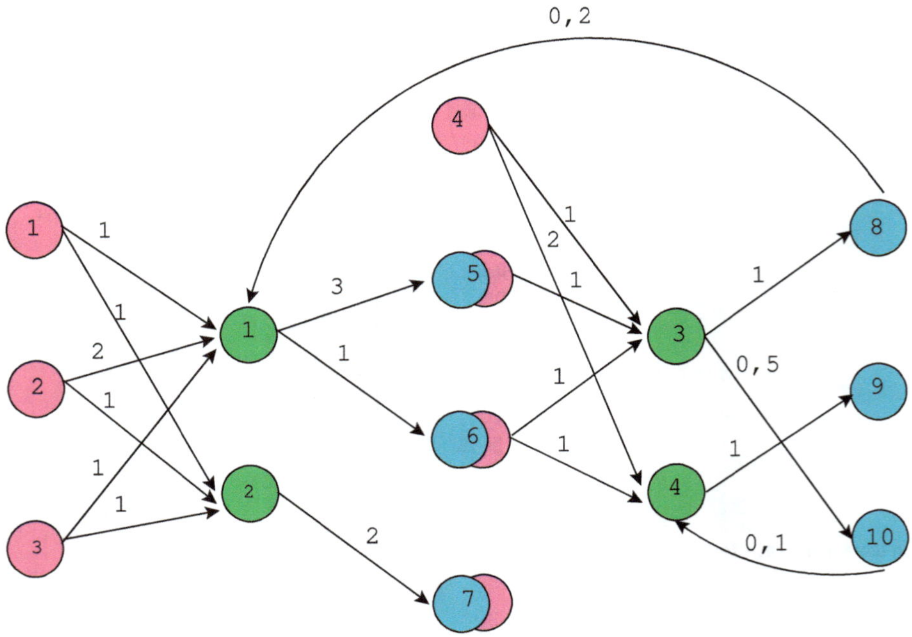

Figure 8.13: More general example.

no longer sufficient to fulfil the standardization demands posed on the arms producers. But also in the automotive industry, the standardization of parts and the requirement to produce such parts made the use of machines inevitable. After *Benz* and *Daimler* used in 1885 for the first time a petrol engine to drive a four-wheel car, the automobile manufacturers developed also the corresponding processes and organizations. Standardization of parts played a key role in this process as *C. Clarke* describes in [17]. One can say that the standardization of parts initiated the standardization of processes and work. The result of the transition from craft production to the American production system was Taylorism and Ford's system of mass production. Taylor's ideas date back to *A. Smith* (1723–1790), especially to his book [62], and to *Ch. Babbage* (1772–1871) [3]. From these sources, *Taylor* deduced two ideas: the fragmentation of labor into small components and the separation of mental from physical work. *Ford* combined these ideas with standards in mass production of cars. In 1909, he decided to produce only one model of cars, the "Model T" and that the chassis would be the same for all cars. As a joke, he remarked "Any customer can have a car painted any color that he wants, so long as it is black." The automobile became a mass product and was available for many people since the price decreased during the period from 1908 to 1924 from 850 US$ to 260 US$. This "success" became possible because the production followed the following five core elements: process orientation based on the principles of mass production, highly fragmented tasks and little responsibility of the direct workers, only a few experts were responsible for production and innovations, human resource management was based on the "learning by doing" – and the "hire and fire"– systems, and there were bad labor relations (distrust, antagonism, and “the boss is always right”). Clearly, in this situation any creativity of the workers was suppressed and identification with the work and its results was impossible. Everybody could clearly recognize that labor under the capitalist mode of production is just a commodity. Earning money to cover the life expenses of their families was the only motivation for the workers.

In our days, the management goes different ways. The Toyota production system started around 1990 and emphasizes waste reduction, continuous improvement of the processes, and involves the employees. Mass customization leads to the efficiency of mass production with the production of personalized commodities. In the smart factories of the *Industry 4.0*, machines and systems are connected, sometimes world-wide by the “Internet-of-Things”, use artificial intelligence, and can communicate with each other. The use of computer science requires mathematical methods, especially those from algebra and discrete mathematics.

One of the researchers which introduced methods from discrete mathematics into production systems was *Tjalling Ch. Koopmanns* (1910–1985). He was born in 1910 in the south of Amsterdam (Netherlands). His parents were teachers. He studied mathematics and statistics at the University of Utrecht under the theoretical physicist *Hans Kramers*. His interests shifted to economics which he studied under *Jan Tinbergen*, one of the leading scientists on this field, at the University of Amsterdam. His PhD dissertation on linear regression was supervised by Kramers in consultation with Tinbergen. He got his PhD in 1936 from the University of Leiden. In 1939, Koopmanns was engaged as a research assistant at Princeton University and taught a course of statistics at New York University. In 1946, he was appointed as an associate professor, and from 1948 to 1955 was a full professor of economics. In 1975, he shared the Nobel Memorial Prize in Economic Sciences with Leonid Kantorovich. He died 1985 in New Haven.

Koopmanns’ contributions deeply influenced the nature of economic research. Between 1939 and 1951, he worked on activity analysis as a novel approach to production theory. As a student, he also published a paper on quantum mechanics which was cited quite often.

9 Graphs and matrices in material management

The goal of material management is to ensure that materials are available at their point of use when needed. A proper implemented material management program can achieve the timely flow of materials and equipment in good quality and in the requested quantity. This can increase the productivity. We are especially interested in the quantitative aspect and will describe some mathematical methods based on the used programs. The main tool is calculation with matrices. We will give some examples (see e. g. [11]). We summarize first what we have learned in Chapter 6 about operations and calculations with matrices.

https://doi.org/10.1515/9783112234433-009

9.1 Operations with matrices

Let $K_{m,n}$ be the set of all matrices of type (m, n) with elements in the commutative field $\mathcal{K}$. Mostly, corresponding to the economical situation, $\mathcal{K}$ is the field of rational numbers. In applications, mainly positive rational numbers are used. We defined already the addition on $K_{m,n}$ as a binary operation,

$$+ : K_{m,n} \times K_{m,n} \to K_{m,n},$$

by $C = A + B$ with $c_{ik} = a_{ik} + b_{ik}$ if $A = (a_{ik})$, $B = (b_{i,k})$, $C = (c_{ik})$ and $i = 1, \dots, m$, $k = 1, \dots, n$. The application of elements of K on elements of $K_{m,n}$ is defined by

$$\circ : K \times K_{m,n} \to K_{m,n}$$

with $\lambda \circ A = (\lambda \circ a_{ik})$ if $A = (a_{ik})$, $\lambda \in K$, $i = 1, \dots, m$, $k = 1, \dots, n$. The addition of matrices is associative, commutative, and invertible. The application of elements from K on elements from $K_{m,n}$ satisfies $\lambda \circ (\mu \circ A) = (\lambda \cdot \mu) \circ A$, $(\lambda + \mu) \circ A = \lambda \circ A + \mu \circ A$, $\lambda \circ (A + B) = \lambda \circ A + \lambda \circ B$, and $e \circ A = A$ if $e \in K$ is the identity element of the field $\mathcal{K}$. Therefore, $(K_{m,n}; +, \circ)$ is a K-vector space and has dimension mn.

For the application of matrices in material management, we need another binary operation, the product of matrices. If $A = (a_{ik})$, $i = 1, \dots, m$, $k = 1, \dots, n$, $B = (b_{jl})$, $j = 1, \dots, p$, $l = 1, \dots q$, are two matrices of type (m, n) and (p, q), respectively, then the multiplication is only defined if $n = p$, i. e., if the number of columns of A is equal to the number of rows of B. The product $C = A \cdot B$ is the matrix $C = (c_{ik})$ with

$$c_{ik} = \sum_{j=1}^{n} a_{ij} b_{jk} = a_{i1} b_{1k} + \dots + a_{in} b_{nk}, \quad i = 1, \dots, m, k = 1, \dots, q.$$

The multiplication is associative, $(A \cdot B) \cdot C = A \cdot (B \cdot C)$, distributive from both sides, $(A + B) \cdot C = A \cdot C + B \cdot C$, $A \cdot (B + C) = A \cdot B + A \cdot C$, and for every $\lambda \in K$ there holds

$$\lambda \circ (A \cdot B) = (\lambda \circ A) \cdot B = A \circ (\lambda \circ B).$$

If we consider the set $K_{n,n}$ of all quadratic matrices, then the multiplication $\cdot : K_{n,n} \times K_{n,n} \to K_{n,n}$ is always defined, i. e., it is a total binary operation on $K_{n,n}$.

9.2 Material requirement planning (MRP)

Material means raw substances, subassemblies, components, parts, other supplies, but also any industrial services required for the manufacturing of products. Material requirement planning helps ensure that the right inventory is available for the production process exactly when it is needed, in quantities which are demanded, and at the lowest possible costs. Material requirement planning improves the efficiency, flexibility, and profitability of production. It increases productivity, improves product quality, and minimizes material and labor costs. It helps manufacturers respond more quickly and increases the demand for their products. So, production delays can be avoided and inventory stockouts can be reduced. There are mainly two possibilities of planning: consumption-based and program-based.

Consumption-based planning procedures use historical data to calculate future requirements. This way of planning is a deterministic approach. The calculation is based on bill of materials, production schedules, and lead times. Then material requirement planning depends essentially on accurate input data, including demand forecasts, lead times, and inventory levels.

Using the program-oriented way, one considers the primary demand which is given by the production program and the secondary or tertiary demand which are determined by individual parts of the production program.

9.3 The gozintograph – an example

A gozintograph is nothing but a directed edge-valued graph and can also be regarded as an input–output graph. The vertices correspond to raw material, semifinished parts, and final products. The edges show the way in which the inputs have to be composed to get the final products. Labeling of edges indicates how many units at a sublevel are needed to produce a unit at a higher level. The labels of the edges are mostly the production coefficients which are the quotients of the quantity of an input divided by the quantity of an output, and say how many units of a kind of input object are needed to produce one unit of a kind of output object. We begin with an example of a two-level production procedure.

Example 9.3.1. A factory produces from the raw materials R_1 and R_2 at the first level of manufacturing the semifinished products Z_1, Z_2, Z_3 and at the second level the final products E_1 and E_2. Calculate the raw material consumption R_1 and R_2 if the factory has to produce 24 units of the main product E_1 and 36 units of the main product E_2. (The corresponding gozintograph is given in Figure 9.1. Since the set of vertices admits a partition into 3 classes such that any edge v has its end in a class different from the class of v and if v and v' belong to the same class of the partition then there are not adjacent, by Chapter 5, Definition 5.4.1, the graph is r-partite for $r = 3$.)

Figure 9.1 shows that for one unit E_1 one needs one unit Z_1, five units Z_2, and four units Z_3, and for one unit E_2 one needs three units Z_1, three units Z_2, and one unit Z_3:

$$1E_1 = 1Z_1 + 5Z_2 + 4Z_3, \quad 1E_2 = 3Z_1 + 3Z_2 + 1Z_3.$$

Similarly, one obtains the equations

$$1Z_1 = 1R_1 + 3R_2, \quad 1Z_2 = 4R_1 + 3R_2, \quad 1Z_3 = 3R_1 + 6R_2.$$

After substitution one has

$$1E_1 = 1(1R_1 + 3R_2) + 5(4R_1 + 3R_2) + 4(3R_1 + 6R_2)$$

and

$$1E_2 = 3(1R_1 + 3R_2) + 3(4R_1 + 3R_2) + 1(3R_1 + 6R_2),$$

and therefore

$$1E_1 = (1 \cdot 1 + 5 \cdot 4 + 4 \cdot 3)R_1 + (1 \cdot 3 + 5 \cdot 3 + 4 \cdot 6)R_2$$

and

$$1E_2 = (3 \cdot 1 + 3 \cdot 4 + 1 \cdot 3)R_1 + (3 \cdot 3 + 3 \cdot 3 + 1 \cdot 6)R_2,$$

i. e.

$$1E_1 = 33R_1 + 42R_2, \quad 1E_2 = 18R_1 + 24R_2.$$

This means that the following quantities of raw materials are needed: for one unit E_1, 33 units R_1 and 42 units R_2; while for one unit E_2, one needs 18 units R_1 and 24 units R_2. For a better understanding, we use the tables

	E_1	E_2
Z_1	1	3
Z_2	5	3
Z_3	4	1

	Z_1	Z_2	Z_3
R_1	1	4	3
R_2	3	3	6

and the corresponding matrices

$$\begin{pmatrix} 1 & 3 \\ 5 & 3 \\ 4 & 1 \end{pmatrix} \text{ of type } (Z, E) = (3, 2), \quad \text{and} \quad \begin{pmatrix} 1 & 4 & 3 \\ 3 & 3 & 6 \end{pmatrix} \text{ of type } (R, Z) = (2, 3).$$

The product

$$\begin{pmatrix} 1 & 4 & 3 \\ 3 & 3 & 6 \end{pmatrix} \cdot \begin{pmatrix} 1 & 3 \\ 5 & 3 \\ 4 & 1 \end{pmatrix}$$

exists since the matrices to be multiplied have the types (2, 3) and (3, 2). The product is a matrix of type (2, 2), namely

$$\begin{pmatrix} 1 \cdot 1 + 4 \cdot 5 + 3 \cdot 4 & 1 \cdot 3 + 4 \cdot 3 + 3 \cdot 1 \\ 3 \cdot 1 + 3 \cdot 5 + 6 \cdot 4 & 3 \cdot 3 + 3 \cdot 3 + 6 \cdot 1 \end{pmatrix} = \begin{pmatrix} 33 & 18 \\ 42 & 24 \end{pmatrix}$$

which corresponds to the table

	E_1	E_2
R_1	33	18
R_2	42	24

Therefore, the result can be easily obtained by matrix multiplication. If $24E_1$ and $36E_2$ are demanded, then the consumption of raw materials can be calculated by

$$\begin{pmatrix} 33 & 18 \\ 42 & 24 \end{pmatrix} \cdot \begin{pmatrix} 24 \\ 36 \end{pmatrix} = \begin{pmatrix} 33 \cdot 24 + 18 \cdot 36 \\ 42 \cdot 24 + 24 \cdot 36 \end{pmatrix} = \begin{pmatrix} 1440 \\ 1872 \end{pmatrix}.$$

To produce 24 units E_1 and 36 units E_2, one needs 1440 units R_1 and 1872 units R_2.

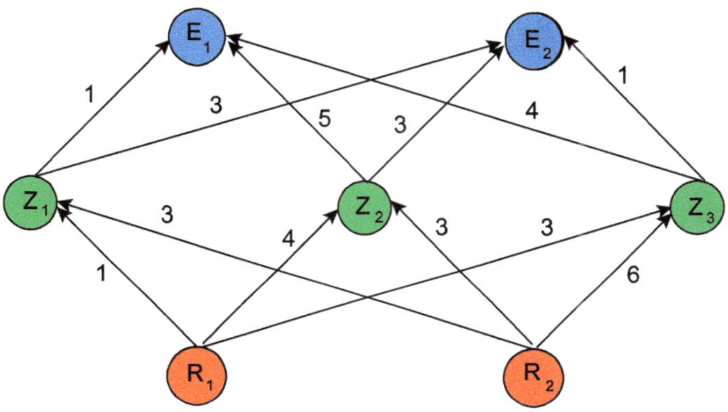

Figure 9.1: Gozintograph of a two-level production.

The example can be generalized. If n final products $1, \ldots, n$ shall be produced from m semifinished products $1, \ldots, m$ and k raw materials $1, \ldots, k$, by a two-level production procedure, then the gozintograph is a 3-partite graph given by Figure 9.2.

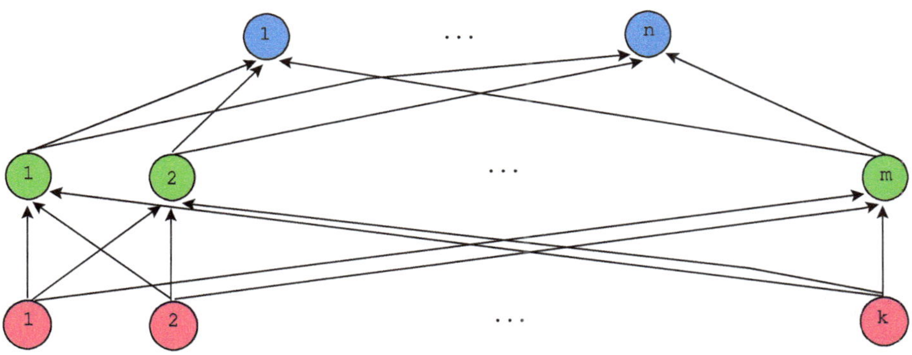

Figure 9.2: Gozintograph of a general two-level production.

If we label the directed edge (of the high level) which connects the semifinished product l ($l = 1, \ldots, m$) with the final product r ($r = 1, \ldots, n$), we can form the matrix

$$\begin{pmatrix} b_{11} & \dots & b_{1n} \\ \vdots & \ddots & \vdots \\ b_{m1} & \dots & b_{mn} \end{pmatrix} = (b_{lr})_{l=1,\dots,m,\, r=1,\dots,n} = B$$

of type $(Z, E) = (m, n)$. The rational numbers b_{lr} have the following meaning. It may happen that not all semifinished products are needed. If, for instance, to produce $E_r, r \in \{1, \dots, n\}$, the semifinished good $z_l, l \in \{1, \dots, m\}$, is not needed, then (l, r) is not a directed edge of the gozintograph. If val is a mapping from the set P of all directed edges of the gozintograph to the set of all nonnegative rational numbers,

$$\text{val} : P \to \mathbb{Q}^+,$$

then we set

$$b_{lr} = \begin{cases} \text{val}(l, r) & \text{if } (l, r) \in Z \times E, \\ 0 & \text{otherwise} \end{cases}$$

and obtain the matrix

$$\begin{pmatrix} b_{11} & \dots & b_{1n} \\ \vdots & \ddots & \vdots \\ b_{m1} & \dots & b_{mn} \end{pmatrix} = (b_{lr})_{l=1,\dots,m,\, r=1,\dots,n} = B$$

of type $(Z, E) = (m, n)$.

For the low level, we proceed in a similar way and define

$$a_{ij} = \begin{cases} \text{val}(i, j) & \text{if } (i, j) \in R \times Z, \\ 0 & \text{otherwise.} \end{cases}$$

This gives a matrix

$$\begin{pmatrix} a_{11} & \dots & a_{1m} \\ \vdots & \ddots & \vdots \\ a_{k1} & \dots & a_{km} \end{pmatrix} = (a_{ij})_{i=1,\dots,k,\, j=1,\dots,m} = A$$

of type $(R, Z) = (k, m)$. The product $A \cdot B$ exists and is a matrix of type $(R, E) = (k, n)$. The matrix $A \cdot B$ is said to be materials requirement matrix. If the market demand is given by a vector

$$X = \begin{pmatrix} x_1 \\ \vdots \\ x_n \end{pmatrix},$$

then the vector

$$Y = \begin{pmatrix} y_1 \\ \vdots \\ y_k \end{pmatrix}$$

of raw materials requirements can be calculated as

$$Y = (A \cdot B) \cdot X.$$

There is an alternative way to calculate the requirements of raw materials. We go back to the starting example and form the input–output table

	R_1	R_2	Z_1	Z_2	Z_3	E_1	E_2
R_1	0	0	0	0	0	0	0
R_2	0	0	0	0	0	0	0
Z_1	1	3	0	0	0	0	0
Z_2	4	3	0	0	0	0	0
Z_3	3	6	0	0	0	0	0
E_1	0	0	1	5	4	0	0
E_2	0	0	3	3	1	0	0

and the corresponding matrix

$$A' = \begin{pmatrix} 0 & 0 & 0 & 0 & 0 & 0 & 0 \\ 0 & 0 & 0 & 0 & 0 & 0 & 0 \\ 1 & 3 & 0 & 0 & 0 & 0 & 0 \\ 4 & 3 & 0 & 0 & 0 & 0 & 0 \\ 3 & 6 & 0 & 0 & 0 & 0 & 0 \\ 0 & 0 & 1 & 5 & 4 & 0 & 0 \\ 0 & 0 & 3 & 3 & 1 & 0 & 0 \end{pmatrix}$$

of type $(R + Z + E, R + Z + E) = (7, 7)$. Let the vector of demand be given by

$$D = \begin{pmatrix} 100 \\ 100 \\ 50 \\ 100 \\ 150 \\ 1000 \\ 1200 \end{pmatrix}.$$

This means that the customer wants to receive 100 units of R_1, 100 units of R_2, 50 units of Z_1, 100 units of Z_2, 150 units of Z_3, 1000 units of E_1, and 1200 units of E_2. Let $X =$

$(x_1\ x_2\ x_3\ x_4\ x_5\ x_6\ x_7)^T$ be the amounts of $R_1, R_2, Z_1, Z_2, Z_3, E_1$, and E_2, respectively. Now we obtain equations for $R_1, R_2, Z_1, Z_2, Z_3, E_1$, and E_2, for instance, $x_3 = 1x_1 + 3x_2$ means that for 1 unit of Z_1 one needs 1 unit of R_1 and 3 units of R_2. This gives the following system of linear equations:

$$\begin{aligned}
x_1 &= 1x_1 + && 100,\\
x_2 &= \quad 1x_2 + && 100,\\
x_3 &= 1x_1 + 3x_2 + && 50,\\
x_4 &= 4x_1 + 3x_2 + && 100,\\
x_5 &= 3x_1 + 6x_2 + && 150,\\
x_6 &= && 1x_3 + 5x_4 + 4x_5 + 1000,\\
x_7 &= && 3x_3 + 3x_4 + 1x_5 + 1200.
\end{aligned}$$

With

$$X' = \begin{pmatrix} x_1 \\ x_2 \\ x_3 \\ x_4 \\ x_5 \\ x_6 \\ x_7 \end{pmatrix},$$

this can be written as

$$X = A'X' + D.$$

We call matrix A' extended matrix of requirements. To solve this system of linear equations, we need the unit matrix E of type $(7,7)$ and obtain

$$E \cdot X' - A'X' = D \Rightarrow (E - A')X' = D.$$

This system of linear equations is uniquely solvable if and only if the determinant $|E - A'|$ is different from 0. The uniquely determined solution is given by $X' = (E - A')^{-1} \cdot D$.

This can be generalized. In the general case, we form the matrix A' which comes from the table with $R_1 \cdots R_k, Z_1 \cdots Z_m, E_1 \cdots E_n$ as horizontal and vertical entries. This matrix has the form

$$\begin{pmatrix} \underline{0}_{(k,m)} & \underline{0}_{(k,k+n)} \\ A^T_{(m,k)} & \underline{0}_{(m,m+n)} \\ \underline{0}_{(n,k+n)} & B^T_{(n,m)} \end{pmatrix}$$

with zero-matrices of the corresponding types (where we wrote the types as indices for checking that the big matrix is quadratic of type $(k+m+n, k+m+n)$).

9.4 More than two levels

We start with an example.

Example 9.4.1. A factory produces from raw materials R_1, R_2, R_3 semifinished goods Z_1, Z_2. From Z_1, Z_2 in the second step, semifinished products X_1, X_2, X_3 will be produced. From X_1, X_2, X_3 in the following step, final products E_1, E_2 will be produced. The production coefficients are given by the following tables.

Production of Z_1, Z_2 from R_1, R_2, R_3:

	Z_1	Z_2
R_1	1	2
R_2	3	5
R_3	4	7

Production of X_1, X_2, X_3 from Z_1, Z_2:

	X_1	X_2	X_3
Z_1	1	3	7
Z_2	2	2	3

Production of E_1, E_2 from X_1, X_2, X_3:

	E_1	E_2
X_1	2	4
X_2	6	8
X_3	5	1

To visualize the material flow, we draw the gozintograph (Figure 9.3). Clearly, the graph is 4-partite. Then we determine the material consumption matrix and calculate the need of raw materials if 100 units of E_1 and 100 units of E_2 shall be produced.

The material consumption matrix will be calculated by

$$M = \begin{pmatrix} 1 & 2 \\ 3 & 5 \\ 4 & 7 \end{pmatrix} \cdot \begin{pmatrix} 1 & 3 & 7 \\ 2 & 2 & 3 \end{pmatrix} \cdot \begin{pmatrix} 2 & 4 \\ 6 & 8 \\ 5 & 1 \end{pmatrix} = \begin{pmatrix} 5 & 7 & 13 \\ 13 & 19 & 36 \\ 18 & 26 & 49 \end{pmatrix} \cdot \begin{pmatrix} 2 & 4 \\ 6 & 8 \\ 5 & 1 \end{pmatrix} = \begin{pmatrix} 117 & 89 \\ 320 & 240 \\ 437 & 329 \end{pmatrix}.$$

From matrix M, we get for a demand of 100 units E_1 and 100 units E_2 that

$$X = \begin{pmatrix} 117 & 89 \\ 320 & 240 \\ 437 & 329 \end{pmatrix} \cdot \begin{pmatrix} 100 \\ 200 \end{pmatrix} = \begin{pmatrix} 29500 \\ 80000 \\ 109500 \end{pmatrix}.$$

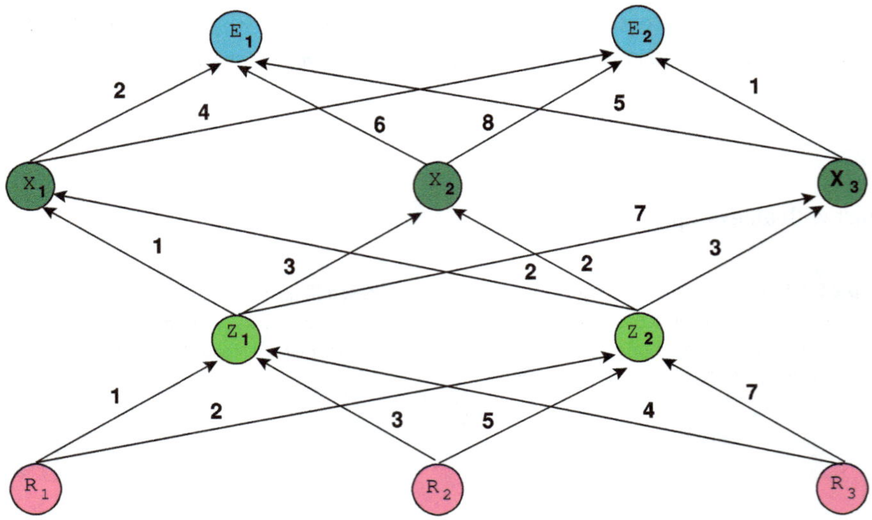

Figure 9.3: Four levels.

Alternatively, from the input–output table with horizontal (input) and vertical (output) entries $R_1, R_2, R_3, X_1, X_2, X_3, Z_1, Z_2$, namely

	R_1	R_2	R_3	X_1	X_2	X_3	Z_1	Z_2	E_1	E_2
R_1	0	0	0	0	0	0	0	0	0	0
R_2	0	0	0	0	0	0	0	0	0	0
R_3	0	0	0	0	0	0	0	0	0	0
X_1	0	0	0	0	0	0	1	2	0	0
X_2	0	0	0	0	0	0	3	2	0	0
X_3	0	0	0	0	0	0	7	3	0	0
Z_1	1	3	4	0	0	0	0	0	0	0
Z_2	2	5	7	0	0	0	0	0	0	0
E_1	0	0	0	2	6	5	0	0	0	0
E_2	0	0	0	4	8	1	0	0	0	0

we obtain the matrix

$$A' = \begin{pmatrix} 0&0&0&0&0&0&0&0&0&0\\ 0&0&0&0&0&0&0&0&0&0\\ 0&0&0&0&0&0&0&0&0&0\\ 0&0&0&0&0&0&1&2&0&0\\ 0&0&0&0&0&0&3&2&0&0\\ 1&3&4&0&0&0&7&3&0&0\\ 2&5&7&0&0&0&0&0&0&0\\ 0&0&0&0&0&0&0&0&0&0\\ 0&0&0&2&6&5&0&0&0&0\\ 0&0&0&4&8&1&0&0&0&0 \end{pmatrix},$$

which contains the transposes of the matrices coming from the first three tables as submatrices. From A' and

$$D = \begin{pmatrix} 0 \\ 0 \\ 0 \\ 0 \\ 0 \\ 0 \\ 0 \\ 0 \\ 100 \\ 100 \end{pmatrix},$$

we obtain

$$X' = (E - A')^{-1} \cdot D = \begin{pmatrix} 29500 \\ 80000 \\ 109500 \\ 0 \\ 0 \\ 0 \\ 0 \end{pmatrix}.$$

This corresponds to the result X calculated by the first method.

9.5 Across levels

Sometimes the intermediate products are not only used in the next level, but in one or more of the following levels. We start with an example.

Example 9.5.1. A factory produces from six raw materials $R_1, R_2, R_3, R_4, R_5, R_6$ four semi-finished products from which final products are manufactured. But raw materials are also directly needed to produce the main products. The material flow can be seen from the following gozintograph (which is a 3-partite graph) and the production coefficients are given by tables. Determine the material consumption matrix. Which quantities of raw materials are needed if 100 units E_1, 200 units E_2, and 200 units E_3 shall be produced? The gozintograph, this time written top-down, looks as follows (see Figure 9.4).

Production of Z_1, Z_2, Z_3, Z_4 from $R_1, R_2, R_3, R_4, R_5, R_6$:

	Z_1	Z_2	Z_3	Z_4
R_1	2	0	0	0
R_2	1	2	0	0
R_3	2	1	3	0
R_4	0	2	1	1
R_5	0	0	0	1
R_6	0	0	0	4

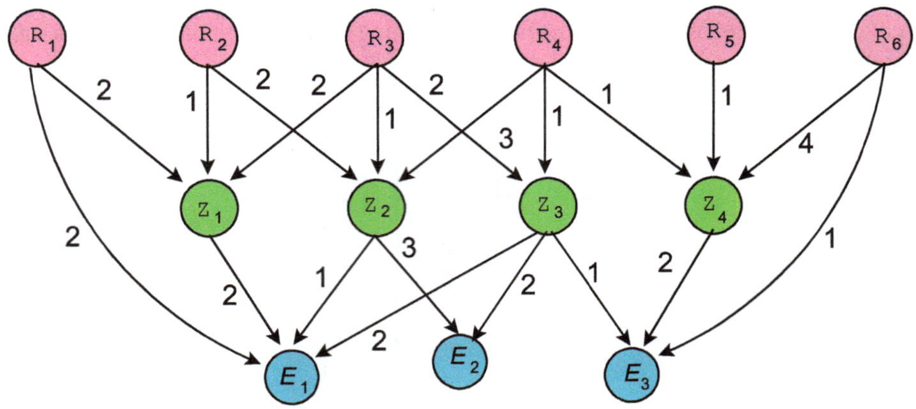

Figure 9.4: Across levels.

Production of E_1, E_2, E_3 from Z_1, Z_2, Z_3, Z_4:

	E_1	E_2	E_3
Z_1	2	0	0
Z_2	1	3	0
Z_3	2	2	1
Z_4	0	0	2

Production of E_1, E_2, E_3 from $R_1, R_2, R_3, R_4, R_5, R_6$:

	E_1	E_2	E_3
R_1	2	0	0
R_2	0	0	0
R_3	0	0	0
R_4	0	0	0
R_5	0	0	0
R_6	0	0	1

Let M_1, M_2, M_3 be the corresponding matrices. Then

$$M_1 \cdot M_2 = \begin{pmatrix} 2 & 0 & 0 & 0 \\ 1 & 2 & 0 & 0 \\ 2 & 1 & 3 & 0 \\ 0 & 2 & 1 & 1 \\ 0 & 0 & 0 & 1 \\ 0 & 0 & 0 & 4 \end{pmatrix} \cdot \begin{pmatrix} 2 & 0 & 0 \\ 1 & 3 & 0 \\ 2 & 2 & 1 \\ 0 & 0 & 2 \end{pmatrix} = \begin{pmatrix} 4 & 0 & 0 \\ 4 & 6 & 0 \\ 11 & 9 & 3 \\ 4 & 8 & 3 \\ 0 & 0 & 2 \\ 0 & 0 & 8 \end{pmatrix}$$

gives the need of raw materials via semifinished products and the matrix

$$\begin{pmatrix} 2 & 0 & 0 \\ 0 & 0 & 0 \\ 0 & 0 & 0 \\ 0 & 0 & 0 \\ 0 & 0 & 0 \\ 0 & 0 & 1 \end{pmatrix}$$

gives the direct need of $R_1, R_2, R_3, R_4, R_5, R_6$ for the production of the final products. Therefore, the raw material consumption matrix can be calculated and gives

$$M = M_1 \cdot M_2 + M_3.$$

The addition is defined since both M_3 and $M_1 \cdot M_2$ have type $(6, 3)$. Then

$$M = \begin{pmatrix} 6 & 0 & 0 \\ 4 & 6 & 0 \\ 11 & 9 & 3 \\ 4 & 8 & 3 \\ 0 & 0 & 2 \\ 0 & 0 & 9 \end{pmatrix}$$

is the raw materials consumption matrix. The total consumption for a demand of 100 units E_1, 200 units E_2, and 200 units E_3 can be calculated as

$$\begin{pmatrix} 6 & 0 & 0 \\ 4 & 6 & 0 \\ 11 & 9 & 3 \\ 4 & 8 & 3 \\ 0 & 0 & 2 \\ 0 & 0 & 9 \end{pmatrix} \cdot \begin{pmatrix} 100 \\ 200 \\ 200 \end{pmatrix} = \begin{pmatrix} 600 \\ 1600 \\ 3500 \\ 2600 \\ 400 \\ 1800 \end{pmatrix}.$$

The factory needs 600 units R_1, 1600 units R_2, 3500 units R_3, 2600 units R_4, 400 units R_5, and 1800 units R_6.

In our example the matrix term $M_1 \cdot M_2 + M_3$ allows calculating the materials consumption matrix. In the general case, the matrix term can be, depending on the complexity of the gozintograph, much more complicated. The second method allows for easier formulation of the general solution. The input–output table has the form

	(x_1) R_1	(x_2) R_2	(x_3) R_3	(x_4) R_4	(x_5) R_5	(x_6) R_6	(x_7) Z_1	(x_8) Z_2	(x_9) Z_3	(x_{10}) Z_4	(x_{11}) E_1	(x_{12}) E_2	(x_{13}) E_3
$(x_1)R_1$	0	0	0	0	0	0	0	0	0	0	2	0	0
$(x_2)R_2$	0	0	0	0	0	0	0	0	0	0	0	0	0
$(x_3)R_3$	0	0	0	0	0	0	0	0	0	0	0	0	0
$(x_4)R_4$	0	0	0	0	0	0	0	0	0	0	0	0	0
$(x_5)R_5$	0	0	0	0	0	0	0	0	0	0	0	0	0
$(x_6)R_6$	0	0	0	0	0	0	0	0	0	0	0	0	1
$(x_7)Z_1$	2	1	2	0	0	0	0	0	0	0	0	0	0
$(x_8)Z_2$	0	2	1	2	0	0	0	0	0	0	0	0	0
$(x_9)Z_3$	0	0	3	1	0	0	0	0	0	0	0	0	0
$(x_{10})Z_4$	0	0	0	1	1	4	0	0	0	0	0	0	0
$(x_{11})E_1$	0	0	0	0	0	0	2	1	2	0	0	0	0
$(x_{12})E_2$	0	0	0	0	0	0	0	3	2	0	0	0	0
$(x_{13})E_3$	0	0	0	0	0	0	0	0	1	2	0	0	0

In the following calculation, we will use $x_1, \ldots, x_{13}$ instead of $R_1, \ldots, E_3$. From the table, we obtain the linear equation system

$$\begin{aligned}
x_1 &= 0x_1 + 0x_2 + 0x_3 + 0x_4 + 0x_5 + 0x_6 + 0x_7 + 0x_8 + 0x_9 + 0x_{10} + 2x_{11} + 0x_{12} + 0x_{13},\\
x_2 &= 0x_1 + 0x_2 + 0x_3 + 0x_4 + 0x_5 + 0x_6 + 0x_7 + 0x_8 + 0x_9 + 0x_{10} + 0x_{11} + 0x_{12} + 0x_{13},\\
x_3 &= 0x_1 + 0x_2 + 0x_3 + 0x_4 + 0x_5 + 0x_6 + 0x_7 + 0x_8 + 0x_9 + 0x_{10} + 0x_{11} + 0x_{12} + 0x_{13},\\
x_4 &= 0x_1 + 0x_2 + 0x_3 + 0x_4 + 0x_5 + 0x_6 + 0x_7 + 0x_8 + 0x_9 + 0x_{10} + 0x_{11} + 0x_{12} + 0x_{13},\\
x_5 &= 0x_1 + 0x_2 + 0x_3 + 0x_4 + 0x_5 + 0x_6 + 0x_7 + 0x_8 + 0x_9 + 0x_{10} + 0x_{11} + 0x_{12} + 0x_{13},\\
x_6 &= 0x_1 + 0x_2 + 0x_3 + 0x_4 + 0x_5 + 0x_6 + 0x_7 + 0x_8 + 0x_9 + 0x_{10} + 0x_{11} + 0x_{12} + 1x_{13},\\
x_7 &= 2x_1 + 1x_2 + 2x_3 + 0x_4 + 0x_5 + 0x_6 + 0x_7 + 0x_8 + 0x_9 + 0x_{10} + 0x_{11} + 0x_{12} + 0x_{13},\\
x_8 &= 0x_1 + 2x_2 + 1x_3 + 2x_4 + 0x_5 + 0x_6 + 0x_7 + 0x_8 + 0x_9 + 0x_{10} + 0x_{11} + x_{12} + 0x_{13},\\
x_9 &= 0x_1 + 0x_2 + 3x_3 + 1x_4 + 0x_5 + 0x_6 + 0x_7 + 0x_8 + 0x_9 + 0x_{10} + 0x_{11} + 0x_{12} + 0x_{13},\\
x_{10} &= 0x_1 + 0x_2 + 0x_3 + 1x_4 + 1x_5 + 4x_6 + 0x_7 + 0x_8 + 0x_9 + 0x_{10} + 0x_{11} + 0x_{12} + 0x_{13},\\
x_{11} &= 0x_1 + 0x_2 + 0x_3 + 0x_4 + 0x_5 + 0x_6 + 2x_7 + 1x_8 + 2x_9 + 0x_{10} + 0x_{11} + 0x_{12} + 0x_{13},\\
x_{12} &= 0x_1 + 0x_2 + 0x_3 + 0x_4 + 0x_5 + 0x_6 + 0x_7 + 3x_8 + 2x_9 + 0x_{10} + 0x_{11} + 0x_{12} + 0x_{13},\\
x_{13} &= 0x_1 + 0x_2 + 0x_3 + 0x_4 + 0x_5 + 0x_6 + 0x_7 + 0x_8 + 1x_9 + 2x_{10} + 0x_{11} + 0x_{12} + 0x_{13},
\end{aligned}$$

more simply written as

$$\begin{aligned}
x_1 &= 2x_{11},\\
x_2 &= x_3 = x_4 = x_5 = 0,\\
x_6 &= x_{13},\\
x_7 &= 2x_1 + x_2 + 2x_3,\\
x_8 &= 2x_2 + x_3 + 2x_4,\\
x_9 &= 3x_3 + x_4,\\
x_{10} &= x_4 + x_5 + 4x_6,\\
x_{11} &= 2x_7 + x_8 + 2x_9,\\
x_{12} &= 3x_8 + 2x_9,\\
x_{13} &= x_9 + 2x_{10}.
\end{aligned}$$

For a given vector D of demand with the unit matrix E of type (13, 13), we obtain $X = (E - A)^{-1} \cdot D$. It is left as an exercise to the reader that in our example with

$$D = (0 \quad 0 \quad 0 \quad 0 \quad 0 \quad 0 \quad 0 \quad 0 \quad 0 \quad 0 \quad 100 \quad 200 \quad 300)^T$$

the result is

$$X = (600 \quad 1600 \quad 3500 \quad 2600 \quad 400 \quad 1800 \quad 0 \quad 0 \quad 0 \quad 0 \quad 0 \quad 0 \quad 0)^T.$$

9.6 Calculation for fixed amounts of raw materials

In some branches of production, for instance, in food industry, the amounts of raw materials are given and all shall be consumed during the production procedure. We will give an example to explain how the problem can be solved.

Example 9.6.1. The raw materials R_1, R_2, R_3 are combined to main products E_1, E_2, E_3, E_4 by the equation

$$\begin{pmatrix} 4 & 2 & 2 & 6 \\ 4 & 1 & 3 & 7 \\ 3 & 3 & 2 & 5 \end{pmatrix} \cdot \begin{pmatrix} E_1 \\ E_2 \\ E_3 \\ E_4 \end{pmatrix} = \begin{pmatrix} R_1 \\ R_2 \\ R_3 \end{pmatrix} \quad \text{with} \quad \begin{pmatrix} R_1 \\ R_2 \\ R_3 \end{pmatrix} = \begin{pmatrix} 6 \\ 6 \\ 6 \end{pmatrix}.$$

This means that the equation $A \cdot E = R$ with

$$A = \begin{pmatrix} 4 & 2 & 2 & 6 \\ 4 & 1 & 3 & 7 \\ 3 & 3 & 5 & 2 \end{pmatrix}, \quad R = \begin{pmatrix} 6 \\ 6 \\ 6 \end{pmatrix}, \quad \text{and} \quad E = \begin{pmatrix} x_1 \\ x_2 \\ x_3 \\ x_4 \end{pmatrix}$$

has to be solved. This is the following system of linear equations consisting of 3 linear equations with 4 unknowns:

$$\begin{aligned} 4x_1 + 2x_2 + 2x_3 + 6x_4 &= 6, \\ 4x_1 + x_2 + 3x_3 + 7x_4 &= 6, \\ 3x_1 + 3x_2 + 2x_3 + 5x_4 &= 6. \end{aligned}$$

The system is solvable if and only if $RgA = Rg(A, R)$, where (A, R) is the matrix

$$(A, R) = \begin{pmatrix} 4 & 2 & 2 & 6 & 6 \\ 4 & 1 & 3 & 7 & 6 \\ 3 & 3 & 5 & 2 & 6 \end{pmatrix}.$$

By substitution it is easy to check that

$$X_0 = \begin{pmatrix} \frac{3}{4} \\ \frac{3}{4} \\ \frac{3}{4} \\ 0 \end{pmatrix}$$

is a special solution of the system and

$$X_H = \begin{pmatrix} -1 \\ 0 \\ -1 \\ 1 \end{pmatrix} \cdot t, \quad t \in \mathbb{R},$$

is the general solution of the homogeneous system. Therefore,

$$\begin{pmatrix} x_1 \\ x_2 \\ x_3 \\ x_4 \end{pmatrix} = \begin{pmatrix} \frac{3}{4} \\ \frac{3}{4} \\ \frac{3}{4} \\ 0 \end{pmatrix} + \begin{pmatrix} -1 \\ 0 \\ -1 \\ 1 \end{pmatrix} \cdot t$$

is the general solution and

$$L = \left\{ \begin{pmatrix} E_1 \\ E_2 \\ E_3 \\ E_4 \end{pmatrix} \mid \begin{pmatrix} E_1 \\ E_2 \\ E_3 \\ E_4 \end{pmatrix} = \begin{pmatrix} \frac{3}{4} \\ \frac{3}{4} \\ \frac{3}{4} \\ 0 \end{pmatrix} + \begin{pmatrix} -1 \\ 0 \\ -1 \\ 1 \end{pmatrix} \cdot t \right\}$$

is the set of solutions. To be a solution of the problem, E_1, E_2, E_3, E_4 have to be nonnegative, i. e., $-t + \frac{3}{4} \geq 0$ and t should be a nonnegative rational number. Therefore, $t \leq \frac{3}{4}, t \in \mathbb{Q}^+$.

9.7 Historical remarks

Materials are all consumables or repetition factors which are used in the form of raw materials, auxiliary products, intermediate products, or in other forms as inputs or outputs of production processes. Graphs describe the relationships between materials, processes, and resources. Matrices are used to manage the quantitative aspects of this relationship for analysis and decision making.

At the beginning of this chapter, we mentioned already that a gozintograph is nothing but a directed edge-valued finite graph. Actually, the name “gozintograph” is the result of a joke. The Hungarian mathematician *A. Vaszonyi* used the word to “honor” the nonexisting Italian mathematician “Zepartzat Gozinto (the part that goes into).” Actually, this notion is not needed, since the notion of an r-partite graph, $r \geq 2$, describes the class of graphs which are used in material management. Note that R-partite graphs, especially complete r-partite graphs which contain all possible edges, were studied by the famous Hungarian mathematician Paul Erdős (1913–1996), who was one of the most important mathematicians of the twentieth century. He solved and posed several mathematical problems in discrete mathematics, especially in graph and number theory, but also in analysis, probability theory, and approximation theory. His life was completely devoted to mathematics. He traveled from one mathematical conference to another, lived “out from the suitcase” and cooperated with more than 500 mathematicians. To express the degree of cooperation inside of the mathematical research community, people defined inductively the *Erdős number*:

Erdős himself has Erdős number 0. *A person has Erdős number* $n > 0$ *if (he) she published a mathematical research article with a person which has Erdős number* $n - 1$.

10 The use of matrices for internal service billings

Cost planning and accounting decide whether a product or a service can be economically implemented on the market. Knowing the costs of a good is necessary to determine its price, i. e., the amount of money a customer pays. Primary costs like costs of raw materials, of semifinished products, of hours of man power, or of energy needed for the production correspond to general ledger accounts. They allow costs to flow between the financial and the cost accounting system. Secondary costs represent internal value flow for allocation or settled purposes. Examples are rents for buildings used for production and administration, costs for water, cleaning, heating, etc. From the service or auxiliary cost centers which have no direct link to production, these indirect costs are transferred to main cost centers, but also to other auxiliary cost centers or to itself by means of assessment and distribution. This is important to determine the overall costs of a product. The primary overhead costs of the auxiliary cost centers and of the main cost centers have also to be distributed internally. (For more details see e. g. [29, 38].)

https://doi.org/10.1515/9783112234433-010

10.1 Methods of internal service billings – an example

We start with an example. In a factory, there exist three auxiliary cost centers: energy production in kilowatt hours (kWh) (auxiliary cost center 1), work of a service station in hours (h) (auxiliary cost center 2), and work of a cleaning department in h (auxiliary cost center 3). In auxiliary center 1, there exist primary overhead costs in an amount of 2000 euros, in auxiliary center 2 arise primary overhead costs in an amount of 9000 euros, and in auxiliary center 3 in an amount of 1000 euros. The following table shows the interdependencies between the cost centers.

↓ to	← from	auxilliary cost center			primary overhead costs
		1	2	3	
auxiliary	1	1000	50	70	2000
cost	2	250	10	50	9000
center	3	300	80	20	1000
main cost center		50000	600	200	-
sum		51550	740	340	-
dimension		in kWh	in h	in h	in euros

We will use the notation $x_i, i = 1, 2, 3$, for the service rates of the auxiliary cost centers. From the table we obtain the following linear equations:

$$\begin{aligned} &\text{I.} & 51550x_1 &= 1000x_1 + 50x_2 + 70x_3 + 2000, \\ &\text{II.} & 740x_2 &= 250x_1 + 10x_2 + 50x_3 + 9000, \\ &\text{III.} & 340x_3 &= 300x_1 + 80x_2 + 20x_3 + 1000. \end{aligned}$$

This is a system of three linear equations with three unknowns which has to be solved. To get smaller numbers, we divide each equation by 10 and obtain

$$\begin{aligned} &\text{I}'. & 5155x_1 &= 100x_1 + 5x_2 + 7x_3 + 200, \\ &\text{II}'. & 74x_2 &= 25x_1 + 1x_2 + 5x_3 + 900, \\ &\text{III}'. & 34x_3 &= 30x_1 + 8x_2 + 2x_3 + 100. \end{aligned}$$

By subtraction on both sides, one gets

$$\begin{aligned} &\text{I}''. & 5055x_1 - 5x_2 - 7x_3 &= 200, \\ &\text{II}''. & -25x_1 + 73x_2 - 5x_3 &= 900, \\ &\text{III}''. & -30x_1 - 8x_2 + 32x_3 &= 100. \end{aligned}$$

We multiply the equations as follows:

$$\begin{array}{lrl} \text{I}''. & 5055x_1 - 5x_2 - 7x_3 = 200, & | \cdot (-5) \mid \cdot 32 \\ \text{II}''. & -25x_1 + 73x_2 - 5x_3 = 900, & | \downarrow \cdot 7 \\ \text{III}''. & -30x_1 - 8x_2 + 32x_3 = 100. & | \downarrow \qquad \cdot 7 \end{array}$$

Calculating $\text{I}'' \cdot (-5) + \text{II}'' \cdot 7$ and $\text{I}'' \cdot 32 + \text{III}'' \cdot 7$, one gets

$$\begin{array}{lr} \text{I}'''. & -25450x_1 + 536x_2 = 5300, \\ \text{II}'''. & 161550x_1 - 216x_2 = 7100. \end{array}$$

Multiplying equation I''' with (–216) and II''' with (–536), and forming the sum of both, gives

$$-81093600x_1 = -4950400,$$

and then

$$x_1 = 0,0610\,\frac{\text{euros}}{\text{kWh}}$$

and, by substitution into I''',

$$536x_2 = 5300 + 25450 \cdot 0.0610,$$

and therefore

$$x_2 = 12.7844\,\frac{\text{euros}}{\text{h}}.$$

Substituting x_1 and x_2 into II'' gives

$$-5x_3 = 900 + 25x_1 - 73x_2$$

and thus

$$-5x_3 = 900 + 25 \cdot 0.0610 - 73 \cdot 12.7844,$$

i. e.,

$$x_3 = 6.3472\,\frac{\text{euros}}{\text{h}}.$$

The found x_1, x_2, x_3 are the service rates of the auxiliary cost centers per unit of the corresponding service.

10.2 General approach

We will use the following notation:

n – number of auxiliary cost centers;
i – index of the auxiliary cost center, $i = 1, \dots, n$;
$K_i^{P'}$ – primary overhead costs of auxiliary cost center i;
a'_{ij} – internal service rate which goes from auxiliary cost center i to j;
q_i – sum of internal service rates of cost center i;
x_i – service rate of auxiliary cost center i.

↓ to	← from	aux. cost center				prim. overh. costs
		1	2	...	n	
aux.	1	a'_{11}	a'_{12}	...	a'_{1n}	$K_1^{P'}$
cost	2	a'_{21}	a'_{22}	...	a'_{2n}	$K_2^{P'}$
centers	⋮	⋮	⋮	⋱	⋮	⋮
	n	a'_{n1}	a'_{n2}	...	a'_{nn}	$K_n^{P'}$
sum		q_1	q_2	...	q_n	
dimension		in kWh	in h	...	in kWh	in euros

For the auxiliary cost centers, one has the following equations:

$$\begin{aligned} q_1x_1 &= a'_{11}x_1 + a'_{12}x_2 + \dots + a'_{1n}x_n + K_1^{P'}, \\ q_2x_2 &= a'_{21}x_1 + a'_{22}x_2 + \dots + a'_{2n}x_n + K_2^{P'}, \\ &\vdots \\ q_nx_n &= a'_{n1}x_1 + a'_{n2}x_2 + \dots + a'_{nn}x_n + K_n^{P'}. \end{aligned}$$

The quotient $\frac{a'_{ij}}{q_i}$ is that part of the sum of all service costs of auxiliary cost center i which goes to j. The quotient $\frac{K_i^{P'}}{q_i}$ is that part of the sum which corresponds to primary overhead costs of auxiliary cost center i. We set $a_{ij} := \frac{a'_{ij}}{q_i}$ and $K_i^P := \frac{K_i^{P'}}{q_i}$, $i, j = 1, \dots, n$.

If we divide the first equation by q_1, the second one by q_2, etc., and the nth equation by q_n, then we get the system

$$\begin{aligned} x_1 &= a_{11}x_1 + a_{12}x_2 + \dots + a_{1n}x_n + K_1^P, \\ x_2 &= a_{21}x_1 + a_{22}x_2 + \dots + a_{2n}x_n + K_2^P, \\ &\vdots \\ x_n &= a_{n1}x_1 + a_{n2}x_2 + \dots + a_{nn}x_n + K_n^P. \end{aligned}$$

We notice that the a_{ij} and the K_i^P have no dimension.

With

$$X = \begin{pmatrix} x_1 \\ x_2 \\ \vdots \\ x_n \end{pmatrix}, \quad A = \begin{pmatrix} a_{11} & a_{12} & \dots & a_{1n} \\ \vdots & & & \vdots \\ a_{n1} & a_{n2} & \dots & a_{nn} \end{pmatrix}, \quad E = \begin{pmatrix} 1 & 0 & \dots & 0 \\ \vdots & & & \vdots \\ 0 & 0 & \dots & 1 \end{pmatrix}, \quad K^P = \begin{pmatrix} K_1^P \\ \vdots \\ K_n^P \end{pmatrix},$$

we obtain the matrix equation

$$(E - A)X = K^p.$$

If $|E - A| \neq 0$, i. e., if $(E - A)^{-1}$ exists, then this equation is solvable in the field of real numbers. Conversely, if the equation is solvable in the field of real numbers, then $|E - A| \neq 0$. In this case, the solution

$$X = (E - A)^{-1} \cdot K^P$$

is uniquely determined. But the solution of this equation must not be the solution of the problem, since X solves the economic problem only if $X \in (\mathbb{Q}^+)^n$. Later we will also solve the economic problem.

10.3 A special case

If interdependencies between auxiliary cost centers are not considered, then we get

$$x_i = \frac{K_i^{P'}}{q_i - \sum_{j=1}^n a'_{ij}} \quad \text{for } i = 1, \dots, n. \tag{*}$$

The sum $\sum_{j=1}^n a'_{ij} = a'_{i1} + \dots + a'_{in}$, $i = 1, \dots n$, consists of a'_{ii}, i. e., the internal consumption of auxiliary cost center i and all costs which go from i to auxiliary cost centers $j, j = 1, \dots, i-1, i+1, \dots, n$. Since this sum will be subtracted from the sum of all internal service rates of cost center i to all cost centers, the difference $q_i - \sum_{j=1}^n a'_{ij}$ is the service rate from auxiliary cost center i to the main cost centers.

The solution $X = (x_1 \ \dots \ x_n)^T$ exists and is a solution of the economic problem, i. e., $x_1, \dots, x_n$ are nonnegative rational numbers, if and only if for all $i \in \{1, \dots, n\}$ the condition $q_i - \sum_{j=1}^n a'_{ij} \neq 0$ is satisfied. This can be formulated as follows:

Proposition 10.3.1. *The solution of* (*) *is uniquely determined and solves the economic problem if and only if*

$$\left(1 - \sum_{j=1}^n a_{1j}\right) \cdot \left(1 - \sum_{j=1}^n a_{2j}\right) \cdots \left(1 - \sum_{j=1}^n a_{nj}\right) \neq 0.$$

Proof. Here $K_i^{P'}$, q_i, $\sum_{j=1}^n a'_{ij}$, $q_i - \sum_{i=1}^n a'_{ij}$, $\frac{K_i^{P'}}{q_i - \sum_{i=1}^n a'_{ij}}$, $i = 1, \ldots, n$ are elements of $\mathbb{Q}^+$. Since the semifield $(\mathbb{Q}^+; +, \cdot)$ is zero-divisor-free, the product is different from 0 if and only if each factor is different from 0. Because $a_{ij} = \frac{a'_{ij}}{q_i}$, for the ith factor we have

$$1 - \sum_{j=1}^n a_{ij} \neq 0 \Leftrightarrow 1 - \sum_{j=1}^n \frac{a'_{ij}}{q_i} \neq 0.$$

We used already that $q_i \neq 0$. Otherwise, the quotients $a_{ij} = \frac{a'_{ij}}{q_i}$ would not exist. Actually, this is not a restriction of the generality since, if $q_i = 0$, there would be no service costs going from i to itself, to other auxiliary cost centers, or to main cost centers, and we would not need to consider the auxiliary cost center i. For all $i \in \{1, \ldots, n\}$, we have

$$1 - \sum_{j=1}^n \frac{a'_{ij}}{q_i} \neq 0 \Leftrightarrow q_i\left(1 - \sum_{j=1}^n \frac{a'_{ij}}{q_i}\right) = q_i - \sum_{j=1}^n a'_{ij} \neq 0.$$

Therefore, by (*) every x_i, $i = 1, \ldots, n$, is defined and is an element of $\mathbb{Q}^+$. □

If we divide each of the equations of (*) by q_i, $i = 1, \ldots, n$, we obtain the equation system

$$\begin{aligned}
x_1 &= \sum_{j=1}^n a_{1j}x_1 + 0 + 0 + \cdots + 0 + K_1^P,\\
x_2 &= 0 + \sum_{j=1}^n a_{2j}x_2 + 0 + \cdots + 0 + K_2^P,\\
&\vdots\\
x_n &= 0 + 0 + \cdots + 0 + \sum_{j=1}^n a_{nj}x_n + K_n^P.
\end{aligned}$$

With

$$X = \begin{pmatrix} x_1 \\ \vdots \\ x_n \end{pmatrix}, \quad A^{an} = \begin{pmatrix} \sum_{j=1}^n a_{1j} & 0 & \ldots & 0 \\ \vdots & & & \vdots \\ 0 & 0 & \ldots & \sum_{j=1}^n a_{nj} \end{pmatrix}, \quad \text{and} \quad K^P = \begin{pmatrix} K_1^P \\ \vdots \\ K_n^P \end{pmatrix},$$

we obtain the matrix equation

$$(E - A^{an}) \cdot X = K^P.$$

This system of linear equations is solvable if and only if $(E - A^{an})^{-1}$ exists, if and only if $|E - A^{an}| \neq 0$, and the solution is $X = (E - A^{an})^{-1} \cdot K^P$.

Because

$$(E - A^{an}) = \begin{pmatrix} 1 - \sum_{j=1}^n a_{1j} & 0 & \dots & 0 \\ \vdots & & & \vdots \\ 0 & 0 & \dots & 1 - \sum_{j=1}^n a_{nj} \end{pmatrix},$$

for the determinant we get

$$|E - A^{an}| = \left(1 - \sum_{j=1}^{n} a_{1j}\right) \cdots \left(1 - \sum_{j=1}^{n} a_{1j}\right) \quad \text{(see Chapter 6).}$$

By Proposition 10.3.1, the system $(E - A^{an}) \cdot X = K^p$ of linear equations is uniquely solvable and the solution solves the economic problem if and only if $|E - A^{an}| \neq 0$.

Therefore, with the particular matrix A^{an}, this case is contained in the general case.

Remark 10.3.2. If (*) is solvable for every $i \in \{1, \dots, n\}$, then $1 - \sum_{j=1}^n a_{ij} \neq 0$. If, in particular, $\sum_{j=1}^n a_{ij} = 0$ for all $i \in \{1, \dots, n\}$, then

$$E - A = E = \begin{pmatrix} 1 & 0 & \dots & 0 \\ \vdots & & & \vdots \\ 0 & \dots & 0 & 1 \end{pmatrix}, \quad |E - A| = 1, \quad (E - A)^{-1} = E,$$

and

$$X = \begin{pmatrix} 1 & 0 & \dots & 0 \\ \vdots & & & \vdots \\ 0 & \dots & 0 & 1 \end{pmatrix} \cdot \begin{pmatrix} K_1^P \\ \vdots \\ K_n^P \end{pmatrix} = \begin{pmatrix} K_1^P \\ \vdots \\ K_n^P \end{pmatrix}.$$

Economically, this means that for each auxiliary cost center i, there are only primary overhead costs which are assigned to the main cost centers.

One more particular case is the *step-by-step* procedure for internal service billings. In this case the internal exchange between auxiliary cost centers will be partly considered.

Problem. How can the *step-by-step procedure* be also written as a special case of the general approach discussed in Section 10.2?

10.4 Historical remarks

Before 1900, if one department of a factory offered some service to another department, for instance, if some equipment was repaired, then this happened without formal billings. After 1900, the industrialization made factories and institutions grow. Dedicated service centers for printing, transportation, cleaning, etc., were established. These centers started to bill others for services. In our days, internal service billing is a standard practice to get transparency and accountability. We showed that the main mathematical tool for internal service billings is the solution of a system of linear equations with nonnegative inputs or, equivalently, the solution of a matrix equation with nonnegative matrices which describes a state of equilibrium.

The Hungarian–American mathematician John von Neumann (1903–1957) formulated at a Princeton seminar in 1932 a restricted model of general economical equilibrium where he also used nonnegative matrices. Von Neumann's model of an expanding economy uses linear matrix pencils $L(\lambda) = A - \lambda B$ for nonnegative matrices A, B. For probability vectors p, q, a positive λ has to be found which solves the equation

$$p^T L(\lambda) q = 0$$

along with two inequality systems expressing economic efficiency. He could solve this problem using his generalization of Brower's fixed-point theorem which states that any continuous function f, mapping a compact convex set to itself, has a fixed point x_0, i. e., satisfying $f(x_0) = x_0$. The probability vector p represents the prices of the commodities to be produced and q represents the "intensity" of the production process. The uniquely determined positive solution λ represents the growth of the economy which is equal to the interest rate.

John von Neumann studied chemistry, was also a physicist, a biologist, an engineer, and one of the fathers of computer science. As a mathematician, he worked in functional analysis, game theory, and other areas. His path-breaking ideas have enriched various disciplines. During the World War II, J. R. Oppenheimer invited von Neumann to join the Manhattan project to develop the first nuclear weapons. He was involved in the selection of a target for the first nuclear bombs in Japan where more than 210 000 people, mostly civilian, died. From 1954 until 1956, von Neumann was a member of the U.S. Atomic Energy Commission and supported D. D. Eisenhower's policy of nuclear deterrence.

11 Matrices in economic systems

Up to now we studied interdependencies between raw materials, intermediate products, and final products (Chapter 9) or between inputs, processes, and outputs of a production system (Chapter 8). The same idea can be adapted in different production areas of an individual enterprise. It can be also used to analyze the connection between various producing and consuming sectors of a national economy and even to analyze and to measure international economic relationships. We showed already that for all considered cases there is a more general approach. The basic idea behind all cases was first developed by W. Leontief [45]. In all considered cases, there is a system of linear equations which can be written as a matrix equation and which has to be solved, if this is possible. The coefficients of these systems result from empirical methods fixed in a statistical input–output table. The data base consists of nonnegative rational numbers and the solutions are vectors (which are special matrices) with nonnegative rational numbers as entries. As Leontief is doing in [45], we start with a simple example consisting of 3 sectors: agriculture, manufacturing, and energy production.

from / into	**agricult.**	**manufact.**	**energy prod.**	**market**	**total output**
agricult.	4	1	5	4	14
manufact.	4	5	5	6	20
energy prod.	10	15	10	45	80

The table entries show the interrelationships between the 3 sectors: 4 million tons of rape seed oil produced by the agricultural sector are used by the agricultural sector itself, 1 million tons by the manufacturing sector, 5 million tons by the energy producing sector, and 4 million tons are demanded by the market. Lines 2 and 3 have a similar interpretation. A real world input–output table of the national economy of a country has thousands of inputs and outputs and a world-wide table can be even much bigger. A problem with the input–output table is that different entries can be given in different physical units and therefore they are not comparable. The price, given in a fixed currency, can serve as a general equivalent. Then the entries of the lines can be meaningfully added. Another way to solve this problem is to use technical constants, so-called production coefficients.

https://doi.org/10.1515/9783112234433-011

11.1 Leontief's theorem

For a general approach, we consider an economic system with n producing sectors. The output of sector i ($i = 1, \ldots, n$) is denoted by x_i. Let x_{ij} be the quantity of that part of the output of sector i which is given to sector j. Then $a_{ij} := \frac{x_{ij}}{x_j}$ is said to be a *production coefficient* and this is the output of sector i used by sector j per unit of sector i; a_{ij} has no physical dimension. Therefore, the entries of line i, $i = 1, \ldots, n$, can be added,

$$a_{i1} + a_{i2} + \cdots + a_{in}.$$

This sum is the total production of sector i.

from / into	agricult.	manufact.	energy prod.	market
agricult.	$\frac{4}{14} = 0.29$	$\frac{1}{20} = 0.05$	$\frac{5}{80} = 0.06$	$\frac{4}{14} = 0.29$
manufact.	$\frac{4}{14} = 0.29$	$\frac{5}{20} = 0.25$	$\frac{5}{80} = 0.06$	$\frac{6}{20} = 0.30$
energy prod.	$\frac{10}{14} = 0.71$	$\frac{15}{20} = 0.75$	$\frac{10}{80} = 0.12$	$\frac{45}{80} = 0.56$

Let y_i be the quantity of the output of sector i given to the demand vector (the requirements of the market) and let

$$M = \begin{pmatrix} y_1 \\ \vdots \\ y_n \end{pmatrix}$$

be the demand vector. Then we get the following system of linear equations:

$$\begin{aligned} x_1 &= a_{11}x_1 + a_{12}x_2 + \cdots + a_{1n}x_n + y_1, \\ &\vdots \\ x_n &= a_{n1}x_1 + a_{n2}x_2 + \cdots + a_{nn}x_n + y_n. \end{aligned}$$

By isolation of y_i in each equation, we obtain

$$\begin{aligned} (1 - a_{11})x_1 - \quad a_{12}x_2 - \cdots - \quad a_{1n}x_n &= y_1, \\ -a_{21}x_1 + (1 - a_{22})x_2 - \cdots - \quad a_{2n}x_n &= y_2, \\ &\vdots \\ -a_{n1}x_1 - \quad a_{n2}x_2 - \cdots + (1 - a_{nn})x_n &= y_n. \end{aligned}$$

With the matrices

$$A := \begin{pmatrix} a_{11} & \dots & a_{1n} \\ \vdots & \ddots & \vdots \\ a_{n1} & \dots & a_{nn} \end{pmatrix}, \quad E := \begin{pmatrix} 1 & \dots & 0 \\ \vdots & \ddots & \vdots \\ 0 & \dots & 1 \end{pmatrix}, \quad X := \begin{pmatrix} x_1 \\ \vdots \\ x_n \end{pmatrix},$$

the system can be written as

$$(E - A) \cdot X = M \tag{*}$$

and the solution is

$$X = (E - A)^{-1} \cdot M,$$

where $(E - A)^{-1}$ is the inverse of $(E - A)$. The inverse $(E - A)^{-1}$ exists if and only if $|E - A| \neq 0$, i. e., if the determinant of $(E - A)$ is different from 0. By Cramér's rule, the uniquely determined solution $X = \begin{pmatrix} x_1 \\ \vdots \\ x_n \end{pmatrix}$ is given by $x_1 = \frac{D_1}{D}, \dots, x_n = \frac{D_n}{D}$, where $D = |E - A|$ and $D_k, k = 1, \dots, n$, is the determinant of the matrix arising from A by replacing the kth row by M.

Let $|E - A|_{ij}$ be the *minor* of $(E - A)$ arising from $(E - A)$ by omitting the ith row and the jth column and calculating the determinant of this matrix, and let

$$(E - A)_{ij} := (-1)^{i+j} |E - A|_{ij}.$$

Then

$$(E - A)^{-1} = \frac{1}{|E - A|} \cdot \begin{pmatrix} (E - A)_{11} & \dots & (E - A)_{1n} \\ \vdots & \ddots & \vdots \\ (E - A)_{n1} & \dots & (E - A)_{nn} \end{pmatrix}^T = \begin{pmatrix} \frac{(E-A)_{11}}{|E-A|} & \dots & \frac{(E-A)_{1n}}{|E-A|} \\ \vdots & \ddots & \vdots \\ \frac{(E-A)_{n1}}{|E-A|} & \dots & \frac{(E-A)_{nn}}{|E-A|} \end{pmatrix}^T$$

is the inverse of $(E-A)$. Moreover, we recall that the determinant of the quadratic matrix

$$A = \begin{pmatrix} a_{11} & \dots & a_{1n} \\ \vdots & \ddots & \vdots \\ a_{n1} & \dots & a_{nn} \end{pmatrix}$$

can also be calculated by *cofactor expansion*.

Let D_{ij} be the minor of A arising from A as the determinant of the matrix which one obtains by omitting the ith row and the jth column and let $A_{ij} = (-1)^{i+j} D_{ij}$. Then

$$|A| = \begin{vmatrix} a_{11} & \dots & a_{1n} \\ \vdots & \ddots & \vdots \\ a_{n1} & \dots & a_{nn} \end{vmatrix} = \sum_{i=1}^{n} a_{ij}A_{ij} = \sum_{j=1}^{n} a_{ij}A_{ij}$$

for $i = 1, \dots, n, j = 1, \dots, n$.

Equation (*) can be economically interpreted as follows. The left-hand side of the system (*) of linear equations gives the total production of the system, which is offered to the market, the total supply, while the right-hand side gives the demand of the market. W. Leontiev [45] formulated the equilibrium condition "supply = demand" as an adaptation of the neoclassical theory of general equilibrium in the sense of L. W. McKenzie, L. Walras, C. Menger, and others [52].

Although solvability and solution of the system (*) of linear equations is known from the main theorem for systems of linear equations in the case $m = n$, the economic problem is not completely trivial. All entries of the matrices A and M are nonnegative rational numbers, and we ask for a solution $X \in (\mathbb{Q}^+)^n$ and are not interested in irrational or negative entries of X. But also in this case, the solvability and solution are not difficult and are known as Leontief's theorem.

Theorem 11.1.1 (Leontief's theorem). *Let $n \geq 2$ be a natural number and let A be a matrix of type (n, n) consisting of production coefficients (production matrix), $A = (a_{ij})_{i,j=1,\dots,n}$, where $a_{ij} \in \mathbb{Q}^+$ are nonnegative rational numbers and let $M = (y_i)_{i=1,\dots,n}$, $y_i \in \mathbb{Q}^+$ be a market vector (vector of demand). Assume that the following condition is satisfied:*

$$\frac{(E - A)_{ij}}{|E - A|} \in \mathbb{Q}^+ \quad \text{for all } i,j \in \{1, \dots, n\}.$$

Then for every vector $M = (y_i)_{i=1,\dots,n}$, $y_i \in \mathbb{Q}^+$, there exists a uniquely determined vector $X = (x_1 \ \dots \ x_n)^T$, $x_i \in \mathbb{Q}^+$ for all $i = 1, \dots, n$, which solves the system $(E - A)X = M$ of linear equations in such a way that the economic system given by the matrices A, M, and X is in equilibrium.

Proof. By the condition, $|E - A| \neq 0$ and, using the main theorem for linear equation systems with $m = n$, there follows that the equation system $(E - A) \cdot X = M$ is solvable, and the solution X is uniquely determined and satisfies $X = (E - A)^{-1} \cdot M$. It is left to prove that all $x_j, j = 1, \dots, n$, are nonnegative rational numbers. From

$$\begin{pmatrix} x_1 \\ \vdots \\ x_n \end{pmatrix} = \begin{pmatrix} \frac{(E-A)_{11}}{|E-A|} & \dots & \frac{(E-A)_{1n}}{|E-A|} \\ \vdots & \ddots & \vdots \\ \frac{(E-A)_{n1}}{|E-A|} & \dots & \frac{(E-A)_{nn}}{|E-A|} \end{pmatrix}^T \cdot \begin{pmatrix} y_1 \\ \vdots \\ y_n \end{pmatrix},$$

there follows

$$x_1 = \frac{(E-A)_{11}}{|E-A|}y_1 + \cdots + \frac{(E-A)_{n1}}{|E-A|}y_n,$$
$$\vdots$$
$$x_n = \frac{(E-A)_{1n}}{|E-A|}y_1 + \cdots + \frac{(E-A)_{nn}}{|E-A|}y_n.$$

By the condition of the theorem, all coefficients are nonnegative rational numbers and by assumption $y_1, \dots, y_n \in \mathbb{Q}^+$. Since $\mathbb{Q}^+$ is closed under addition, we get $x_1, \dots x_n \in \mathbb{Q}^+$. □

Condition

$$\frac{(E-A)_{ij}}{|E-A|} \in \mathbb{Q}^+, \quad 1 \le i,j \le n$$

is not only sufficient but also necessary.

Theorem 11.1.2 (Main theorem for input–output economics). *Let $n \geq 2$ be a natural number, let $A = (a_{ij})_{i,j\in\{1,\dots,n\}}$ be a matrix of type (n,n) with nonnegative rational numbers as entries, $a_{ij} \in \mathbb{Q}^+$ for all $1 \le i,j \le n$, let $M = (y_1, y_2, \dots, y_n)^T, y_i \in \mathbb{Q}^+$ for all $1 \le i,j \le n$ be a vector of demand.*

Then for every $M \in (\mathbb{Q}^+)^n$, there exists precisely one vector $X = (x_1\ x_2\ \dots\ x_n)^T$, $x_i \in \mathbb{Q}^+$ for all $i,j \in \{1,\dots,n\}$ such that $(E-A)\cdot X = M$ if and only if

$$\frac{(E-A)_{ij}}{|E-A|} \in \mathbb{Q}^+$$

for all $i,j \in \{1,\dots,n\}$.

Proof. Economically, the main theorem means that every demand can be satisfied if the condition

$$\frac{(E-A)_{ij}}{|E-A|} \in \mathbb{Q}^+$$

for all $i,j \in \{1,\dots,n\}$ is fulfilled. Since the $n \times n$ nonnegative rational numbers of the condition are the entries of the inverse $(E-A)^{-1}$, the condition means that

$$(E-A)^{-1} \in \mathcal{M}_{n,n}(\mathbb{Q}^+).$$

As $X = (E-A)^{-1}M$, i. e.,

$$\begin{pmatrix} x_1 \\ \vdots \\ x_n \end{pmatrix} = \begin{pmatrix} \frac{(E-A)_{11}}{|E-A|} & \cdots & \frac{(E-A)_{1n}}{|E-A|} \\ \vdots & \ddots & \vdots \\ \frac{(E-A)_{n1}}{|E-A|} & \cdots & \frac{(E-A)_{nn}}{|E-A|} \end{pmatrix}^T \cdot \begin{pmatrix} y_1 \\ \vdots \\ y_n \end{pmatrix},$$

there follows

$$x_1 = \frac{(E-A)_{11}}{|E-A|}y_1 + \cdots + \frac{(E-A)_{n1}}{|E-A|}y_n,$$
$$\vdots$$
$$x_n = \frac{(E-A)_{1n}}{|E-A|}y_1 + \cdots + \frac{(E-A)_{nn}}{|E-A|}y_n.$$

Since all $x_i, i = 1, \dots, n$, are elements of $\mathbb{Q}^+$ and, by assumption, $y_1, \dots, y_n \in \mathbb{Q}^+$, for the quotients we have

$$\frac{(E-A)_{ij}}{|E-A|} \in \mathbb{Q}^+$$

for all $i, j \in \{1, \dots, n\}$. This shows the converse of Leontief's theorem. Altogether, the main theorem is proved. □

The necessary and sufficient condition of the main theorem has the disadvantage that one needs the inverse $(E-A)^{-1}$ to the given matrix A.

Problem. Is there a necessary and sufficient condition for a matrix A which leads for every demand $M = (y_1 \ \dots \ y_n)^T$ to economic equilibrium?

Leontief gives a sufficient condition consisting of (i) and (ii).

Assume that for a matrix $A \in \mathcal{M}_{(nn)}(\mathbb{Q}^+)$ the conditions

(i) $|E-A| \neq 0$;

(ii) $\sum_{j=1}^{n} a_{1j} \leq 1, \dots, \sum_{j=1}^{n} a_{nj} \leq 1$, and $\sum_{j=1}^{n} a_{kj} < 1$ for at least one $1 \leq k \leq n$ (or $\sum_{i=1}^{n} a_{i1} \leq 1, \dots, \sum_{i=1}^{n} a_{in} \leq 1$, and $\sum_{i=1}^{n} a_{il} < 1$ for at least one $1 \leq l \leq k$) are satisfied.

Then $(E-A) \cdot X = M$ has for every $M \in (\mathbb{Q}^+)^n$ exactly one solution $X \in (\mathbb{Q}^+)^n$.

Example 11.1.3. As an exercise, the reader should check that the starting example satisfies the condition from Theorem 11.1.1 (or Theorem 11.1.2). Then the corresponding equation $|E-A| \cdot X = M$ has for every $M \in (\mathbb{Q}^+)^n$ exactly one solution $X \in (\mathbb{Q}^+)^n$. Because

$$\begin{aligned} E - A &= \begin{vmatrix} 0.71 & -0.05 & -0.06 \\ -0.29 & 0.75 & -0.06 \\ -0.71 & -0.75 & 0.88 \end{vmatrix} \\ &= 0.71 \cdot 0.75 \cdot 0.88 - 0.05 \cdot 0.06 \cdot 0.71 - 0.06 \cdot 0.29 \cdot 0.75 \\ &\quad - 0.88 \cdot 0.29 \cdot 0.05 - 0.75 \cdot 0.06 \cdot 0.71 - 0.71 \cdot 0.75 \cdot 0.06 \\ &= 0.38 > 0, \end{aligned}$$

condition (i) is satisfied, but (ii) is not satisfied since $0.71 + 0.75 + 0.12 > 1$ and $0.29 + 0.29 + 0.71 > 1$. This shows that satisfaction of (i) and (ii) is not necessary.

With

$$
\begin{aligned}
(E-A)_{11} &= 0.75 \cdot 0.88 - 0.75 \cdot 0.06 = 0.615;\\
(E-A)_{21} &= (-0.05 \cdot 0.88 - 0.75 \cdot 0.06) \cdot (-1) = 0.089;\\
(E-A)_{31} &= 0.05 \cdot 0.06 + 0.75 \cdot 0.06 = 0.048;\\
(E-A)_{12} &= (-0.29 \cdot 0.88 - 0.71 \cdot 0.06) \cdot (-1) = 0.298;\\
(E-A)_{22} &= 0.71 \cdot 0.88 - 0.71 \cdot 0.06 = 0.58;\\
(E-A)_{32} &= (-0.71 \cdot 0.06 - 0.29 \cdot 0.06) \cdot (-1) = 0.06;\\
(E-A)_{13} &= 0.29 \cdot 0.75 + 0.71 \cdot 0.75 = 0.75;\\
(E-A)_{23} &= (-0.71 \cdot 0.75 - 0.71 \cdot 0.05) \cdot (-1) = 0.57;\\
(E-A)_{33} &= 0.71 \cdot 0.75 - 0.29 \cdot 0.05 = 0.518,
\end{aligned}
$$

and

$$
A = \begin{pmatrix} 0.29 & 0.05 & 0.06 \\ 0.29 & 0.25 & 0.06 \\ 0.71 & 0.75 & 0.12 \end{pmatrix}, \quad M = \begin{pmatrix} 0.29 \\ 0.30 \\ 0.56 \end{pmatrix},
$$

one has

$$
(E-A)^{-1} = \begin{pmatrix} 1.6184 & 0.2342 & 0.1263 \\ 0.7842 & 1.5263 & 0.1579 \\ 1.9737 & 1.4947 & 1.3632 \end{pmatrix}.
$$

Since all entries of $(E-A)^{-1}$ are nonnegative rational numbers, for every $M \in (\mathbb{Q}^+)^3$ there is a vector $X = (x_1\, x_2\, x_3)^T$ such that the economic system is in equilibrium.

For $M = (0.29\; 0.3\; 0.56)^T$, we obtain

$$
\begin{pmatrix} x_1 \\ x_2 \\ x_3 \end{pmatrix} = \begin{pmatrix} 1.6184 & 0.2342 & 0.1263 \\ 0.7842 & 1.5263 & 0.1579 \\ 1.9737 & 1.4947 & 1.3632 \end{pmatrix} \cdot \begin{pmatrix} 0.29 \\ 0.3 \\ 0.56 \end{pmatrix} = \begin{pmatrix} 0.6103 \\ 1.5695 \\ 1.7842 \end{pmatrix}.
$$

Without proof, we mention another condition which is even sufficient and necessary. The so-called *Hawkin–Simon* condition states that the following two propositions are equivalent:

(i) Let $A > \mathbf{0}$ be a matrix with positive rational numbers as entries. Then for every $M = (y_1 \ldots y_n)^T$ with positive rational numbers $y_1, \ldots, y_n$, there exists a uniquely determined $X = (x_1 \ldots x_n)^T, x_1, \ldots, x_n \in \mathbb{Q}^+$ which solves the equation $(E-A) \cdot X = M$.

(ii) All successive leading principal minors of $(E-A)$ are positive, i. e.,

$$
1 - a_{11} > 0, \quad \begin{vmatrix} 1 - a_{11} & -a_{12} \\ -a_{21} & 1 - a_{22} \end{vmatrix} > 0, \quad \ldots,
$$

$$
\begin{vmatrix} 1 - a_{11} & -a_{12} & \ldots & -a_{1n-1} \\ \vdots & \vdots & & \vdots \\ -a_{n-11} & -a_{n-12} & \ldots & 1 - a_{n-1n-1} \end{vmatrix} > 0, \quad |E - A| > 0.
$$

The theorem was proved by Morishima [53], Nikaido [55], and Murata [54], but even earlier by D. Kotelyanski [41] as Gantmacher in his well-known textbook on matrices mentions [31].

Instead of a proof, we consider the case $n = 2$. We suppose that $1 - a_{11} > 0$ and

$$|E - A| = \begin{vmatrix} 1 - a_{11} & -a_{12} \\ -a_{21} & 1 - a_{22} \end{vmatrix} > 0.$$

Since $A > \mathbf{0}$, where $\mathbf{0}$ is the zero-matrix, all entries $a_{11}, a_{12}, a_{21}, a_{22}$ are positive rational numbers. Let $M = \binom{y_1}{y_2}, y_1, y_2 \in \mathbb{Q}^+$ be arbitrary. We want to show that the solution X of $(E - A) \cdot X = M$ is an element of $(\mathbb{Q}^+)^2$, i. e., that $x_1, x_2 \in \mathbb{Q}^+$. We calculate the cofactors of the elements $A_{ij}, i, j \in \{1, 2\}$:

$$\begin{aligned}
(E - A)_{12} &= -(-a_{12}) = a_{12} > 0, \\
(E - A)_{21} &= -(-a_{21}) = a_{21} > 0, \\
(E - A)_{22} &= 1 - a_{11} > 0, \\
(E - A)_{11} &= 1 - a_{22} > \frac{a_{12}a_{21}}{1 - a_{11}} > 0,
\end{aligned}$$

since

$$\begin{vmatrix} 1 - a_{11} & a_{12} \\ a_{21} & 1 - a_{22} \end{vmatrix} = (1 - a_{11})(1 - a_{22}) - a_{12}a_{21} > 0.$$

There follows

$$(E - A)^{-1} = \begin{pmatrix} \frac{(E-A)_{11}}{|E-A|} & \frac{(E-A)_{21}}{|E-A|} \\ \frac{(E-A)_{12}}{|E-A|} & \frac{(E-A)_{22}}{|E-A|} \end{pmatrix} > \mathbf{0},$$

i. e., all entries of $(E - A)^{-1}$ are nonnegative rational numbers. By Leontief's theorem, the uniquely determined solution X of $(E - A) \cdot X = M$ is an element of $(\mathbb{Q}^+)^2$. This shows the implication (ii) $\Rightarrow$ (i).

If, conversely, for $A > \mathbf{0}$ and for every $M > \mathbf{0}$, the uniquely determined solution of $(E - A) \cdot X = M$ is an element of $(\mathbb{Q}^+)^2$ then by the main theorem

$$(E - A)^{-1} = \begin{pmatrix} \frac{(E-A)_{11}}{|E-A|} & \frac{(E-A)_{21}}{|E-A|} \\ \frac{(E-A)_{12}}{|E-A|} & \frac{(E-A)_{22}}{||E-A|} \end{pmatrix} \geq \mathbf{0},$$

i. e., all entries of $(E - A)^{-1}$ are nonnegative rational numbers.

Since $(E - A)_{12} = a_{12} > 0$, we get $|E - A| > 0$. Moreover, $\frac{(E-A)_{22}}{|E-A|} \geq 0$ and $|E - A| > 0$ implies $(E - A)_{22} = 1 - a_{11} \geq 0$. The equation $1 - a_{11} = 0$ would imply $|E - A| = -a_{12}a_{21} < 0$ since $A > \mathbf{0}$ means $a_{12}, a_{21} > 0$, a contradiction. There follows $1 - a_{11} > 0$ and therefore (i) $\Rightarrow$ (ii).

11.2 Algebraic properties of $\mathcal{M}_{(n,n)}(\mathbb{Q}^+)$

Matrices occurring in input–output economics are quadratic of type (n, n) with nonnegative rational numbers, i. e., elements of $\mathbb{Q}^+$, as entries. Let $\mathcal{M}_{(n,n)}(\mathbb{Q}^+)$ be the set of all these matrices. In Chapter 6, calculation with matrices on arbitrary fields was introduced. Here we consider matrices with entries from the semifield $\mathbb{Q}^+$ (see Chapter 3) and their operations.

If $A = (a_{ij})_{i,j\in\{1,\dots,n\}}$, $B = (b_{ij})_{i,j\in\{1,\dots,n\}}$ then $A + B := (a_{ij} + b_{ij})_{i,j\in\{1,\dots,n\}} \in \mathcal{M}_{(n,n)}(\mathbb{Q}^+)$, where $+$ on the right-hand side is the addition in $\mathbb{Q}^+$. This means for $+$ on the left-hand side,

$$+ : \mathcal{M}_{(n,n)}(\mathbb{Q}^+) \times \mathcal{M}_{(n,n)}(\mathbb{Q}^+) \to \mathcal{M}_{(n,n)}(\mathbb{Q}^+)$$

is a binary operation on $\mathcal{M}_{(n,n)}(\mathbb{Q}^+)$. Clearly, this addition is associative and commutative, but not invertible since, except for the zero matrix $A = \mathbf{0}$,

$$A \in \mathcal{M}_{(n,n)}(\mathbb{Q}^+) \Rightarrow -A \notin \mathcal{M}_{(n,n)}(\mathbb{Q}^+).$$

Let $k \in \mathbb{Q}^+$. If $A = (a_{ij})_{i,j\in\{1,\dots,n\}} \in \mathcal{M}_{(n,n)}(\mathbb{Q}^+)$, then $k\circ A := (ka_{ij})_{i,j\in\{1,\dots,n\}} \in \mathcal{M}_{(n,n)}(\mathbb{Q}^+)$ and

$$\circ : \mathbb{Q}^+ \times \mathcal{M}_{(n,n)}(\mathbb{Q}^+) \to \mathcal{M}_{(n,n)}(\mathbb{Q}^+)$$

is a binary two-sorted operation. The sorts are $\mathbb{Q}^+$ and $\mathcal{M}_{(n,n)}(\mathbb{Q}^+)$. This scalar multiplication satisfies:

$$\begin{aligned} k \circ (A + B) &= k \circ A + k \circ B, \\ (k + l) \circ A &= k \circ A + l \circ A, \\ (kl) \circ A &= k \circ (l \circ A), \\ 1 \circ A &= A. \end{aligned}$$

Altogether, $(\mathcal{M}_{(n,n)}(\mathbb{Q}^+); +, \circ)$ satisfies all the defining properties of a *left-semimodule over a semifield* (see Chapter 7).

Proposition 11.2.1. *$(\mathcal{M}_{(n,n)}(\mathbb{Q}^+); +, \circ)$ is a left-semimodule over the semifield $(\mathbb{Q}^+; +, \cdot)$.*

Since we consider matrices of type (n, n) and as addition and multiplication of nonnegative rational numbers give nonnegative rational numbers, the set $\mathcal{M}_{(n,n)}(\mathbb{Q}^+)$ is closed under matrix multiplication,

$$\mathcal{M}_{(n,n)}(\mathbb{Q}^+) \times \mathcal{M}_{(n,n)}(\mathbb{Q}^+) \to \mathcal{M}_{(n,n)}(\mathbb{Q}^+)$$

defined by

$$A \cdot B = (a_{ij}) \cdot (b_{ij}) = \sum_{k=1}^{n} a_{ik} b_{kj}, \quad i,j \in \{1, \ldots, n\}$$

is a binary operation on $\mathcal{M}_{(n,n)}(\mathbb{Q}^+)$.

Matrix multiplication is associative, but in general not commutative, and the distributive laws are satisfied:

$$A \cdot (B + C) = A \cdot B + A \cdot C, \quad (B + C) \cdot D = B \cdot D + C \cdot D.$$

Moreover, we have

$$k \circ (A \cdot C) = (k \circ A) \cdot C = A \cdot (k \circ C) \quad \text{for all } k \in \mathbb{Q}^+ \text{ and all } A, C \in \mathcal{M}_{(n,n)}(\mathbb{Q}^+).$$

The addition is not invertible, i. e., the opposite matrix $-A$ of A does not belong to $\mathcal{M}_{(n,n)}(\mathbb{Q}^+)$. The zero-matrix $\mathbf{0}$ is a zero-element,

$$A + \mathbf{0} = \mathbf{0} + A = A,$$

for all $A \in \mathcal{M}_{(n,n)}(\mathbb{Q}^+)$. In general, the multiplication in $\mathcal{M}_{(n,n)}(\mathbb{Q}^+)$ is not invertible since from $A \in \mathcal{M}_{(n,n)}(\mathbb{Q}^+)$ does not follow $A^{-1} \in \mathcal{M}_{(n,n)}(\mathbb{Q}^+)$ as the following simple example shows:

$$A = \begin{pmatrix} 2 & 2 \\ 2 & 2 \end{pmatrix} \in \mathcal{M}_{(n,n)}(\mathbb{Q}^+),$$

but A^{-1} does not exist since $|A| = 0$. For

$$A = \begin{pmatrix} 1 & 2 \\ 2 & 1 \end{pmatrix} \in \mathcal{M}_{(n,n)}(\mathbb{Q}^+),$$

the determinant $|A|$ exists and $|A| = -3$, but

$$A^{-1} = \begin{pmatrix} -\frac{1}{3} & \frac{2}{3} \\ \frac{2}{3} & -\frac{1}{3} \end{pmatrix} \notin \mathcal{M}_{(n,n)}(\mathbb{Q}^+).$$

Altogether, we have

Proposition 11.2.2. *$(\mathcal{M}_{(n,n)}(\mathbb{Q}^+); +, \cdot)$ is a semiring.*

If we consider the subset $\mathcal{L}_{(n,n)}(\mathbb{Q}^+)$ of $\mathcal{M}_{(n,n)}(\mathbb{Q}^+)$ consisting of all matrices A with $A^{-1} \in \mathcal{M}_{(n,n)}(\mathbb{Q}^+)$, then, because $(A \cdot B)^{-1} = B^{-1} \cdot A^{-1}$ and $(A^{-1})^{-1} = A$, this set is closed under multiplication and formation of inverses, i. e., it is a subgroup of the multiplicative semigroup of $\mathcal{M}_{(n,n)}(\mathbb{Q}^+)$.

Proposition 11.2.3. *$(\mathcal{L}_{(n,n)}(\mathbb{Q}^+); \cdot)$ is a group and $(\mathcal{L}_{(n,n)}(\mathbb{Q}^+); +, \cdot)$ is a semifield.*

The last both propositions show that semirings and semifields are important algebraic structures not only for number systems or for applications in computer science but also in economics.

11.3 Historical remarks

Wassily W. Leontief (1905–1999) was one of the most famous economists of the twentieth century. He has been called the father of input–output analysis in econometrics and won the Nobel prize for Economics in 1973. In his own words, a main problem of economics is that, on the one hand, there exist many economical facts and data and, on the other hand, there are economic theories independent of facts. His idea was to connect both sides and to work out a model which combines the various economic branches of a country, but also the economics of different countries in the world.

His personal background was a good preparation in doing this work. He was Russian (the pronunciation of his family name is probably "Leontjev") born in Munich and grew up in Petrograd, the later Leningrad, studied philosophy, sociology, and economics in Leningrad, but wrote his dissertation thesis in Berlin. From 1931 on, he worked at Harvard University on a general equilibrium theory. Up to that time, economists studied only individual parts of the economy. His equilibrium theory examined processes in the entire economy in the US and in the whole world. With large input–output tables, he described the relationships between different industries and developed the theory of input–output economics, using well-known facts from linear algebra.

A funny detail – although he was used to work with mountains of data, he was not sure about his birthday and assumed to be born in 1906. His relatives found later his birth certificate from 1905, however, the Nobel foundation still uses the old data.

12 Formal concept analysis in economics

In the philosophy of Aristotle, concepts are essential to reflect reality, their conditions and relationships, and human being's thinking, with the aim of understanding the world. A concept consists of its extension, i. e., of the collection of all objects belonging to the concept, and its intension, which is the collection of all attributes or properties applicable to all the objects of the extension. Concepts exist related with other concepts, where the subconcept–superconcept relation plays an essential role. Clearly, a concept X is a subconcept of a concept Y if each object belonging to X also belongs to Y. A subconcept has more attributes than each of its superconcepts. The set of attributes of a superconcept is a subset of the set of attributes of each of its subconcepts. Concerning the properties of objects, the subconcept–superconcept relation leads to implications. For a mathematical foundation of formal concept analysis, we need lattices, especially the concept of a complete lattice and the concept of a Galois connection.

In some sense, this chapter continues the mathematical modeling of the epistomological ideas leading to concepts which we started in Chapter 2 where we used equivalence relations and the diagram lemma. In this chapter we will apply complete lattices and Galois connections.

https://doi.org/10.1515/9783112234433-012

12.1 Complete lattices and Galois connections

A lattice $\mathcal{V} = (V; \wedge, \vee)$ was defined in Chapter 7, Section 7.2, as an algebra of type (2, 2) where the binary operations $\wedge$ and $\vee$ satisfy the commutative, associative, and idempotent laws, as well as the absorption laws:

$$\forall x, y \in V\ (x \vee (x \wedge y) = x),$$
$$\forall x, y \in V\ (x \wedge (x \vee y) = x).$$

Equivalently, a lattice is a partially ordered set where for any two-element subset $\{x, y\}$ the infimum $\bigwedge\{x, y\}$ and supremum $\bigvee\{x, y\}$ exist (Chapter 7, Theorem 7.2.2). Extending this for arbitrary subsets $X \subseteq V$ of a lattice, we have

Definition 12.1.1. A lattice $\mathcal{V} = (V; \wedge, \vee)$ is said to be complete if for arbitrary subsets $X \subseteq V$, the infimum $\bigwedge X$ and supremum $\bigvee X$ exist.

By Chapter 7, Theorem 7.3.7, for every algebra $\mathcal{A}$ the set of all its subalgebras $\mathrm{Sub}(\mathcal{A})$ is a lattice $(\mathrm{Sub}(\mathcal{A}); \wedge, \vee)$. It is not difficult to see that $\mathcal{S}ub(\mathcal{A})$ is even a complete lattice since for any subset $X \subseteq \mathrm{Sub}(\mathcal{A})$ of subalgebras of $\mathcal{A}$, the infimum and supremum exist. The congruence lattice $\mathcal{C}on(\mathcal{A})$ of an algebra $\mathcal{A}$ is also a complete lattice. In Chapter 7, Theorem 7.7.5, we proved that the collections of all equational classes of type τ and of all equational theories of type τ are lattices. Both lattices are complete. It is clear that any finite lattice is complete since from the existence of the infimum and supremum for two-element subsets, using the associativity of the supremum and infimum, there follows the existence of the supremum and infimum for any finite subset of V. Since economic data are always finite sets, the reader could ask why we defined complete lattices. The reason is that for the characterization theorem of concept lattices, we need complete lattices.

For any subset $\Sigma \subseteq W_\tau(X) \times W_\tau(X)$ of equations of type τ and any subclass $K \subseteq \mathrm{Alg}(\tau)$ of algebras of type τ, we defined

$$\mathrm{Mod}\,\Sigma := \{\mathcal{A} \in \mathrm{Alg}(\tau) \mid \forall s \approx t \in \Sigma\ (\mathcal{A} \models s \approx t)\} \quad \text{and}$$
$$\mathrm{Id}\,K := \{s \approx t \in W_\tau(X)^2 \mid \forall \mathcal{A} \in K\ (\mathcal{A} \models s \approx t)\}.$$

Let τ be a fixed type. The pair (Mod, Id) has the following properties (Chapter 7, Theorem 7.7.3):

(i) For all subsets Σ and Σ' of $W_\tau(X) \times W_\tau(X)$ and all subclasses K and K' of $\mathrm{Alg}(\tau)$, we have $\Sigma \subseteq \Sigma' \Rightarrow \mathrm{Mod}\,\Sigma \supseteq \mathrm{Mod}\,\Sigma'$ and $K \subseteq K' \Rightarrow \mathrm{Id}\,K \supseteq \mathrm{Id}\,K'$;

(ii) For all subsets Σ of $W_\tau(X) \times W_\tau(X)$ and all subclasses K of $\mathrm{Alg}(\tau)$, we have $\Sigma \subseteq \mathrm{Id}\,\mathrm{Mod}\,\Sigma$ and $K \subseteq \mathrm{Mod}\,\mathrm{Id}\,K$.

The properties (i) and (ii) define the concept of a Galois connection as a pair (ι, μ) of mappings, $\iota : \mathcal{P}(A) \to \mathcal{P}(B)$ and $\mu : \mathcal{P}(B) \to \mathcal{P}(A)$, between the power sets $\mathcal{P}(A)$, $\mathcal{P}(B)$ of two sets A, B which satisfy properties corresponding to (i) and (ii) (see, e. g., [23]).

Formal concept analysis considers the concepts and attributes in a given *context*.

Definition 12.1.2. A context is a triple (G,M,I) where G and M are sets, the set of objects and the set of attributes, and where $I \subseteq G\times M$ is a binary relation which is defined by

$$I = \{(g,m) \mid g \in G, m \in M, \text{ and } g \text{ has property } m\}.$$

For $X \subseteq G$ and $Y \subseteq M$, we define two mappings between the power sets of M and G by

$$\sigma : \mathcal{P}(G) \to \mathcal{P}(M), \quad \tau : \mathcal{P}(M) \to \mathcal{P}(G),$$

with

$$\sigma(X) := \{m \in M \mid \forall g \in X\ ((g,m) \in I)\}$$

and

$$\tau(Y) := \{g \in G \mid \forall m \in Y\ ((g,m) \in I)\}.$$

Then the pair (σ,τ) is a Galois connection between G and M, i. e., it satisfies for all $X, X' \subseteq G$ and all $Y, Y' \subseteq M$ the conditions

(i) $X \subseteq X' \Rightarrow \sigma(X) \supseteq \sigma(X')$, $Y \subseteq Y' \Rightarrow \tau(Y) \supseteq \tau(Y')$, and

(ii) $X \subseteq \tau\sigma(X)$, $Y \subseteq \sigma\tau(Y)$.

12.2 Concept lattices

We start with the definition of a concept.

Definition 12.2.1. A concept of a context (G,M,I) is a pair $(X,Y), X \subseteq G, Y \subseteq M$, with $\sigma(X) = Y$ and $\tau(Y) = X$. Here X is said to be the extent of the context, and Y is said to be the intent of the concept.

By this definition, concepts are precisely those pairs $(X,Y), X \subseteq G, Y \subseteq M$, which satisfy $\tau\sigma(X) = X, \sigma\tau(Y) = Y$. Let $\mathcal{B}(G,M,I)$ be the set of all concepts which belong to the context (G,M,I).

As a Galois connection, the pair (σ,τ) satisfies conditions (i) and (ii), but also the following one.

Lemma 12.2.2. *Let J be a set of indices and $j \in J$. Then for $X_j \subseteq G$, we have $\sigma(\bigcup_{j\in J} X_j) = \bigcap_{j\in J}\sigma(X_j)$. For $Y_j \subseteq M$, there holds $\tau(\bigcup_{j\in J} Y_j) = \bigcap_{j\in J}\tau(Y_j)$.*

Proof. From $X_j \subseteq \bigcup_{j\in J} X_j$ for all $j \in J$, there follows $\sigma(X_j) \supseteq \sigma(\bigcup_{j\in J} X_j)$, and thus $\sigma(\bigcup_{j\in J} X_j) \subseteq \bigcap_{j\in J}\sigma(X_j)$. If, conversely, $m \in \bigcap_{j\in J}\sigma(X_j)$, then $(g,m) \in I$ for all $g \in \bigcup_{j\in J} X_j$, i. e., $m \in \sigma(\bigcup_{j\in J} X_j)$ and then $\bigcap_{j\in J}\sigma(X_j) \subseteq \sigma(\bigcup_{j\in J} X_j)$. Altogether, we have equality. The second equation can be proved in a similar way. □

On the set $B(G,M,I)$, we define a partial order relation $\le$ by

$$(X_1,Y_1) \le (X_2,Y_2) :\Leftrightarrow X_1 \subseteq X_2 (\Leftrightarrow Y_1 \supseteq Y_2).$$

In this case, (X_1, Y_1) is said to be a subconcept of (X_2, Y_2) and (X_2, Y_2) is a superconcept of (X_1, Y_1).

The following theorem on the binary operations of $(B(G, M, I); \leq)$ corresponds to the definition of the binary operations of the complete lattices $\mathcal{S}ub(\mathcal{A})$ and $\mathcal{C}on(\mathcal{A})$. Its proof can be found in [32].

Theorem 12.2.3 ([32]). *Let (G, M, I) be a context. Then $(B(G, M, I); \leq)$ is a complete lattice and the lattice operations are given by*

$$\bigvee_{j\in J}(X_j, Y_j) = \left(\tau\sigma\left(\bigcup_{j\in J} X_j\right), \bigcap_{j\in J} Y_j\right),$$

$$\bigwedge_{j\in J}(X_j, Y_J) = \left(\bigcap_{j\in J} X_j, \sigma\tau\left(\bigcup_{j\in J} Y_j\right)\right).$$

Now we ask the question which complete lattices are isomorphic to concept lattices. The answer will characterize concept lattices. To ask such questions is very typical in algebraic research. Recall that in Chapter 3 we asked which ordered semirings are o-isomorphic to the ordered semiring of natural numbers. The answer was given in Chapter 3, Theorem 3.1.18. Concept lattices were characterized in [32] using the following properties of a partially ordered set Q: a subset P of Q is said to be *supremum-dense* in Q if for any $s \in Q$ there exists a subset $A \subseteq P$ such that $s = \bigvee A$. An *infimum-dense* subset of Q is defined in the dual way.

Theorem 12.2.4 ([32]). *A complete lattice $\mathcal{L}$ is isomorphic to a concept lattice $\mathcal{B}(G, M, I)$ if and only if the following conditions are satisfied:*

(i) *There are mappings $\gamma : G \to L$ and $\mu : M \to L$ such that $\gamma(G)$ is supremum-dense in L and $\mu(M)$ is infimum-dense in L.*
(ii) *$(g, m) \in I$ is equivalent to $\gamma(g) \leq \mu(m)$ for any $g \in G$ and any $m \in M$.*

12.3 The concept lattice of the main legal forms of a company

An important problem for the founders of a business is to choose the right legal form of that business. The legal form should suit the business idea. To make the right decision, it is helpful to form a table where the entries are all possible legal forms of a business and all properties which are important for the business plan, and then to form the concept lattice. To show the main ideas and the calculations which are necessary, we consider a very simple context consisting of 4 forms of business and 6 properties. For bigger contexts, the user has several possibilities to choose the appropriate software (FCA4J, Snelting, MindMeister).

The set G of objects is the set of the four main legal forms of a company G = {sole proprietorship (SP), partnership (PS), limited liability cooperation (LLC), cooperation (C)} = {SP, PS, LLC, C}.

The set M of properties is given by $M = \{$foundation with partnership agreement (fa), minimum capital requested (minc), equity injection is possible (ei), equity withdrawal is possible (ew), profit sharing (ps), limit of liability (ll)$\} = \{$fa, minc, ei, ew, ps, ll$\}$. The relation $I \subseteq G \times M$ is given by a cross-table or by a dyadic matrix (with entries from $\{1, 0\}$, 1 for "yes" and 0 for "no").

	SP	PS	LLC	C
fa		X	X	X
minc		X		
ei	X	X	X	X
ew	X	X	X	X
ps	X	X	X	X
ll			X	

$$\begin{pmatrix} 0 & 1 & 1 & 1 \\ 0 & 1 & 0 & 0 \\ 1 & 1 & 1 & 1 \\ 1 & 1 & 1 & 1 \\ 1 & 1 & 1 & 1 \\ 0 & 0 & 1 & 0 \end{pmatrix}.$$

Now we determine all concepts of this context. We choose an element of the set G of objects, say SP. Object SP has the properties ei, ew, and ps. Now we are looking for all objects which have properties ei, ew, ps, and obtain {SP, PS, LLC, C}. Therefore, ({SP, PS, LLC, C}, {ei, ew, ps}) is a concept of the given context. Now we choose step by step all subsets of set G, determine all properties of the objects from the subset chosen, ask for all objects which have the same properties, and obtain a concept. The following table shows all the calculations. The first row shows the selected subset of G, the second row gives the attributes satisfied by the selected subset, the third row gives the set of all objects of the context which satisfy these attributes, and the last row gives the concept.

subset	attributes	corresponding objects	concept
{PS}	fa, minc, ei, ew, ps	{PS}	({PS}, {fa, minc, ei, ew, ps})
{SP}	ei, ew, ps	{SP, PS, LLC,C}	({ SP, PS, LLC,C},{ei, ew, ps})
{LLC}	fa, ei, ew, ps, ll	{LLC}	({LLC}, { fa, ei, ew, ps, ll})
{C}	fa, ei, ew, ps	{PS, LLC,C}	({PS, LLC,C}, {fa, ei, ew, ps})
{SP, PS}	ei, ew, ps	{SP, PS, LLC,C}	({ SP, PS, LLC,C},{ei, ew, ps})
{SP, LLC}	ei, ew, ps	{SP, PS, LLC,C}	({ SP, PS, LLC,C},{ei, ew, ps})
{SP,C}	ei, ew, ps	{SP, PS, LLC,C}	({ SP, PS, LLC,C},{ei, ew, ps})
{PS, LLC}	fa, ei, ew, ps	{PS, LLC,C}	({ PS, LLC,C},{fa, ei, ew, ps})
{PS,C}	fa, ei, ew, ps	{PS, LLC,C}	({ PS, LLC,C},{fa, ei, ew, ps})
{LLC,C}	fa, ei, ew, ps	{PS, LLC,C}	({ PS, LLC,C},{fa, ei, ew, ps})
{SP, PS, LLC}	ei, ew, ps	{SP, PS, LLC,C}	({ SP, PS, LLC,C},{ei, ew, ps})
{SP, PS,C}	ei, ew, ps	{SP, PS, LLC,C}	({ SP, PS, LLC,C},{ei, ew, ps})
{SP, LLC,C}	ei, ew, ps	{SP, PS, LLC,C}	({ SP, PS, LLC,C},{ei, ew, ps})
{PS, LLC,C}	fa, ei, ew, ps	{PS, LLC,C}	({ PS, LLC,C},{fa, ei, ew, ps})
{SP, PS, LLC,C}	ei, ew, ps	{SP, PS, LLC,C}	({ SP, PS, LLC,C},{ei, ew, ps})

In the given context there are precisely the following concepts:

$$(\{\text{SP, PS, LLC,C}\}, \{\text{ei, ew, ps}\}), (\{\text{PS, LLC,C}\}, \{\text{fa, ei, ew, ps}\}),$$
$$(\{\text{LLC}\}, \{\text{fa, minc, ei, ew, ps}\}), (\{\text{PS}\}, \{\text{fa, minc, ei, ew, ps}\})$$

and the empty concept $(\emptyset, \emptyset)$. The concepts

$$(\{LLC\}, \{fa, minc, ei, ew, ps\}), (\{PS\}, \{fa, minc, ei, ew, ps\})$$

are incomparable. The corresponding partial order relation is not linear.

The concept lattice including the empty concept is given in Figure 12.1.

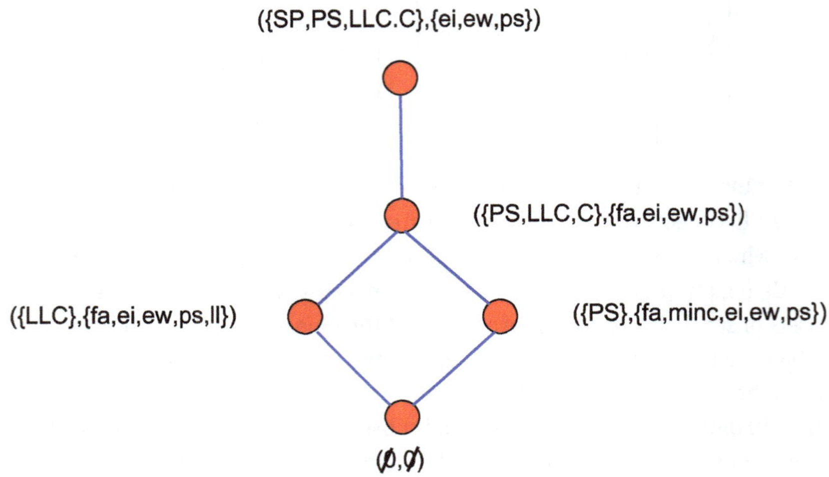

Figure 12.1: The main legal forms of companies.

This concept lattice is very simple. Considering more legal forms of companies or more attributes, i. e., a more complicated context, one obtains a more complicated graph. The calculation inside of the table shows an algorithm which produces the concept lattice starting with a cross-table of data. For more information on those algorithms, see, e. g., [66].

Clearly, this result is strongly dependent on the context.

12.4 Historical remarks

Since 1979, formal concept analysis became one of the main research projects of the group of researchers around *P. Burmeister, B. Ganter,* and *R. Wille* at the Technical University Darmstadt. Since then formal concept analysis found many applications, for instance, in data analysis, knowledge discovery, and software engineering.

In Chapters 1 and 2, we used a different approach to concepts and defined them as equivalence classes of objects. The mathematical background of that approach are equivalence relations, the main theorem for equivalence relations, the diagram lemma, and the kernel of a function. In this chapter we applied the theory of ordered sets, lattice theory, and Galois connections to define concepts and concept lattices.

The origins of the main contributions to lattice theory trace back to the early twentieth century and are connected with the important contributions of *Garett Birkhoff.* He was born in 1911, in Princeton, New Jersey, USA. He entered Harvard University in 1928 to study mathematics and physics following suggestions of his father who was a famous mathematician. Garett Birkhoff himself said about his decision: "I decided to become a mathematician. I liked mathematics, and my father, being a mathematician, was no reason I should not become one, too." In 1932, he graduated from Harvard University and continued his studies at Cambridge University in England. Influenced by *Philip Hall*, the main person of the British school of group theory, his interests changed from mathematical physics to abstract algebra. During a short stay in Munich in 1933, he became familiar with van der Waerden's algebra book [65] and *Andreas Speiser*'s book on group theory [61]. In 1933, he returned to the United States and became an instructor at Harvard University in 1936. His paper "The structure of abstract algebras" [7] from 1935 founded universal algebra as a new branch of mathematics. G. Birkhoff held no PhD and did not obtain MA since the British educational system did not emphasize this qualification at that time. Nevertheless, he spent his academic career from 1936 onward teaching at Harvard. In the following years, he wrote his successful textbooks "Lattice Theory" which appeared in 1940 [8] and together with *Saunders Mac Lane* "Survey of Modern Algebra" in 1941 [9]. In 1970, he wrote the text "Modern Applied Algebra" with *Thomas Bartee* [5]. He received the honorary degrees "Doctor honoris causa" from six universities around the world, in 1986 from the Technical University Munich and in 1991 from the Technical University Darmstadt. During and after World War II, he was interested in engineering mathematics, especially in mathematical questions concerning the development of weapons. G. Birkhoff was a member of several academic societies. He died 1996 in Water Mill, New York.

Bibliography

[1] K. Appel, W. Haken, J. Koch, Every planar graph is four colorable, II Reducibility, *Ill. J. Math.* **21**, 3, 491–567 (1977).

[2] K. Appel, W. Haken, *Every Planar Map is Four Colorable*, American Mathematical Society 1989.

[3] Ch. Babbage, *On the Economy of Machinery and Manufactures*, Cosimo Classics 2012.

[4] C. C. Bachet, *Diophanti Alexandrini Arithmeticorum, Libri Sex, Graece & Latini Editi Atque Absolutissimis Commentariis Illustrati*, Hieronymus Drovard, Paris 1621.

[5] Th. Bartee, G. Birkhoff, *Modern Applied Algebra*, McGraw-Hill 1970.

[6] M. Berliant, T. T. Raa, A foundation of location theory, Consumer preferences and demand, *J. Econ. Theory* **44**, 2, 336 (1988). https://doi.org/10.1016/0022-0531(88)90008-7.

[7] G. Birkhoff, The structure of abstract algebras, *Proc. Camb. Philos. Soc.* **31**, 433–454 (1935).

[8] G. Birkhoff, *Lattice Theory*, American Mathematical Society Colloquium Publication, vol. 25, American Mathematical Society, New York 1940.

[9] G. Birkhoff, S. Mac Lane, *Survey of Modern Algebra*, The Mac Millan Company 1951.

[10] G. Birkhoff, J. D. Lipson, Heterogeneous algebras, *J. Comb. Theory* **8**, 115–133 (1970).

[11] M. Bliefernich, M. Gryck, M. Pfeifer, C-J. Wagner, *Aufgaben zur Matrizenrechnung und linearen Optimierung*, Verlag Die Wirtschaft, Berlin 1971.

[12] S. Brehmer, H. Belkner, *Einführung in die Analytische Geometrie und Lineare Algebra*, VEB Deutscher Verlag der Wissenschaften, Berlin 1968.

[13] P. Burmeister, *A Model-Theoretic Oriented Approach to Partial Algebras. Introduction to Theory and Applications of Partial Algebras-Part I.* Mathematical Research Vol. 32, Akademie-Verlag, Berlin 1986.

[14] A. Cayley, On the theory of the analytic forms called trees, *Philos. Mag.*, Series IV, 172–176 (1857).

[15] P. M. Cohn, *Elements of Linear Algebra*, Chapman & Hall Mathematics 1994.

[16] G. Cantor, Beiträge zur Begründung der transfiniten Mengenlehre, 2. Artikel *Math. Ann.* **49** (1897).

[17] C. Clarke, *Automotive Production, Systems and Standardization, from Ford to the Case of Mercedes-Benz*, Contributions to Management Science, Physica-Verlag, a Springer Company 2005.

[18] C. Debreu, *Theory of Value. An Axiomatic Analysis of Economic Equilibrium*, John Wiley and Sons, New York 1959.

[19] K. Denecke, K. Todorov, *Algebraische Grundlagen der Arithmetik*, Heldermann Verlag, Berlin 1994.

[20] K. Denecke, K. Todorov, *Allgemeine Algebra und Anwendungen*, Aachen 1996.

[21] K. Denecke, S. L. Wismath, *Universal Algebra and Applications in Theoretical Computer Science*, Chapmann & Hall, a CRC Press Company, Boca Raton, London, New York, Washington D.C. 2002.

[22] K. Denecke, *Algebra und Diskrete Mathematik für Informatiker*, Teubner-Verlag, Stuttgart, Leipzig, Wiesbaden 2003.

[23] K. Denecke, M. Erné, S. L. Wismath, (Eds.) *Galois Connections and Applications*, Kluwer Academic Publishers, Dordrecht, Boston, London 2004.

[24] K. Denecke, *Partial Clones of Terms*, Eliva Press 2024.

[25] R. Dedekind, *Was sind und was sollen die Zahlen?*, Braunschweig 1988.

[26] R. Dedekind, *Stetigkeit und irrationale Zahlen*, Braunschweig 1892.

[27] R. Diestel, *Graph Theory*, 5th edition, Springer 2017.

[28] H. Dyckhoff, Th. S. Spengler, *Produktionswirtschaft. Eine Einführung*, 3. Auflage, Springer, Berlin, Heidelberg 2010.

[29] G. Fandel, A. Fey, B. Heuft, Th. Pitz, *Kostenrechnung*, Springer-Verlag, Berlin, Heidelberg 2009.

[30] P. de Fermat, *Bemerkungen zu Diophant*, Ostwalds Klassiker der exakten Wissenschaften, Akademische Verlagsgesellschaft m.b.H., Leipzig 1932.

[31] F. Gantmacher, *The Theory of Matrices*, pp. 71–73, Chelsea, New York 1959.

[32] B. Ganter, R. Wille, *Formale Begriffsanalyse, Mathematische Grundlagen*, Springer-Verlag 1996.

[33] Y. Q. Guo, K. P. Shum, G. T. Shu, *Linear Algebra*, Science Press, Beijing 2007.

[34] J. S. Golan, *Semirings and Their Application*, Springer Science and Business Media, B.V., 1999.

https://doi.org/10.1515/9783112234433-013

[35] J. R. Hicks, *Value and Capital*, 2nd edition, Oxford University Press, Clarendon 1946.
[36] P. J. Higgins, Algebras with a scheme of operators, *Math. Nachr.* **27**, 115–123 (1963).
[37] K. Kieckkäfer, K. M. Gubitz, *Internes Rechnungswesen und Funktionale Steuerung*, Einheit 3: Einführung in die Produktion und Logistik, Fernuniversität Hagen.
[38] R. Kinney, C. A. Raiborn, *Cost Accounting, Foundations and Evolutions*, South-Western Cengage Learning 2010.
[39] D. König, *Theorie der endlichen und unendlichen Graphen*, Leipziger Akademische Verlagsgesellschaft, Leipzig 1936.
[40] T. C. Koopmanns, Analysis of production as an efficient combination of activities, in: T. C. Koopmanns: *Activity Analysis of Production and Allocation*, John Wiley and Sons, New York 1951.
[41] D. M. Kotelyanski, On some properties of matrices with positive elements, *Mat. Sbornik N.S.*, **31**, 3, 497–506 (1952).
[42] E. Landau, *Grundlagen der Analysis, Das Rechnen mit ganzen, rationalen, reellen und komplexen Zahlen*, Akademische Verlagsgesellschaft, Leipzig 1930.
[43] D. Lau, *Algebra und Diskrete Mathematik 1,2*, Springer-Verlag, Berlin Heidelberg New York 2004.
[44] S. Lekkoksung, *Hyperidentities and Hypersubstitutions of many-sorted Algebras*, Dissertation, Potsdam 2010.
[45] W. W. Leontief, *Input-Output Economics*, Oxford University Press, New York 1986.
[46] H. Lugowski, Eine axiomatische Grundlegung des anschaulich-genetischen Aufbaus der Arithmetik in der Schulmathematik, *Wiss. Z. Pädagog. Hochsch. Potsdam*, **7**, 1,2, 197–230 (1962).
[47] H. Lugowski, H.-J. Weinert, *Einführung in die Algebra, Arithmetik und Zahlentheorie*, Lehrbriefe für das Fernstudium der Lehrer, Potsdam 1973.
[48] H. Lugowski, *Der Aufbau der Zahlenrechnung unter Verwendung außerarithmetischer Motivierungen*, Potsdam 1985.
[49] H. Lugowki, *Grundzüge der Universellen Algebra*, B. G. Teubner, Verlagsgesellschaft, Leipzig 1976.
[50] A. I. Mal'cev, *Algebraic Systems*, Akademie-Verlag, Berlin 1973.
[51] K. Marx, *Capital (Das Kapital)*, Grapevine India Publishers, Delhi, Mumbay 2022.
[52] L. W. McKenzie, The classical theorem on existence of competitive equilibrium, *Econometrica*, **49**, 4, 819–841 (1981).
[53] M. Morishima, *Equilibrium, Stability and Growth; a Multisectoral Analysis*, pp. 15–17, Oxford University Press, 1964.
[54] Y. Murata, *Mathematics for Stability and Optimization of Economic Systems, Input-Output*, Academic Press, New York 1977.
[55] H. Nikaido, *Convex Structures and Economic Theory*, pp. 90–92, Academic Press 1968.
[56] H. Peters, *Wirtschaftsmathematik* (especially Chapter 5), Verlag W. Kohlhammer 2022.
[57] N. Robertson, S. Sanders, P. D. Seymour, R. Thomas, The four-color theorem, *J. Comb. Theory, Ser. B*, **70**, (1997).
[58] O. Schulz, *Vorbemerkungen zum Buch"Arithmetik"von Diophantus v.Alexandria*, Schlesingersche Buch-und Musikhandlung, Berlin 1822.
[59] D. Struik, *Abriss der Geschichte der Mathematik*, VEB Deutscher Verlag der Wissenschaften, Berlin 1965.
[60] E. Silberberg, W. Suen, *The Structure of Economics. A Mathematical Analysis*, McGraw-Hill, 2001.
[61] A. Speiser, *Theorie der Gruppen endlicher Ordnung mit Anwendungen auf algebraische Zahlen und Gleichungen sowie auf die Kristallographie*, Die Grundlehren der mathematischen Wissenschaften in Einzeldarstellungen mit besonderer Berücksichtigung der Anwendungsgebiete, Julius Springer, Berlin 1923.
[62] A. Smith, *Wealth of Nations*, Wordsworth Classics of World Literature, Wordsworth Editions Ltd., 2012.
[63] P. Tannery, *Diophantus Alexandrinus, Opera Omnia*, 2 Volumes, Teubner 1893.
[64] H. W. Turnbull, *The Theory of Determinants, Matrices and Invariants*, Dover Publications Inc., London 1960.
[65] B. L. van der Waerden, *Algebra I, II*, Springer-Verlag Berlin, Heidelberg, New York 1966.
[66] S. Yevtushenko, Computing and visualising concept lattices, Dissertation thesis, Technical University Darmstadt, 2004.
[67] M. Zorn, A remark on method in transfinite algebra, Bull. Am. Math. Soc., **41**, 667–670 (1935).

Index

3-partite graph 114
n-ary relation 17
n-factorial 28
r-dimensional vector space 132
r-partite graph 114, 260

absolutely free algebra 181
absorption law 162, 282
absorption laws 14
activity 26, 210
addition of matrices 122
additive technique 214
adjacency matrix 104
adjacent 104
admissible activity 25, 210
algebra 160, 161
– absolutely free 181
– Boolean 163
– directly irreducible 176
– factor 173
– indexed 161
– non-indexed 161
– quotient 173
– relatively free 186
– simple 171, 178
– subdirectly irreducible 177
– term 181
alphabet 179
antisymmetric 20
Appel 118
approximation fraction
– lower 60
– upper 60
Archimedean ordered 60
area of restrictions 221
associate element 71
associative law 4, 122, 161
associative laws 14
asymmetric 21
automorphism 168, 170
– of a graph 104
automorphism group 170
Axiom of comprehension 13

base set 161
basis 131, 217
bijective function 27

binary operation 14, 47
binary relation 17
Birkhoff, G. 178
Birkhoff's Theorem 189
budget 92
budget equation 92

cancellation rule 54
cardinal number 44
cardinality 44
cartesian power 15
cartesian product 15
Ch. Goldbach 73
chromatic number 118
class
– equational 183
clone generated by C 193
closed Eulerian walk 109
closed homomorphism 195
closed interval 22
closed set 184
closed walk
– Hamiltonian 110
closure operator 165, 166, 183, 187
closure properties 172
closure property 165
coefficient 147
colorable graph 118
commutative group 122
commutative law 5, 122
commutative laws 14
commutative monoid 214
comparable elements 20
compatibility 171
complement 14, 163
complement laws 163
complete input-output graph 203
complete lattice 184, 282
complete r-partite graph 260
complex number 3
complexity 180
composite number 72
concept 283
conclusion 7
congruence
– fully invariant 185
– generated by θ 172

https://doi.org/10.1515/9783112234433-014

- relation 171
congruence lattice of an algebra 173
congruence modulo m 79
conjunction 7
connectedness 107
connectedness components 107
consumer problem 92
context 283
contradiction 8
convergent production 226
Cramér's rule 154
cycle 107
- directed 108
cyclic techniques 238

de Morgan rules 15
decimal fraction
- finite 61
- infinite 60
- mixed periodic 64
- periodic 60
- purely periodic 60
- without 9 as period 61
decimal system 52, 60
decreasing marginal utility 92
degree
- of a vertex 106
desired input and output 25
determinant of order n 137
diagonal form 150
difference 68
difference ring 68, 87
difference set 14
dimension 132
direct product 174
directed graph 202
directly irreducible algebra 176
disjunction 7
distributive law 5, 162
distributive laws 14
divergent production 226
division algorithm 61
divisor 71
- proper 72
- trivial 72
divisor relation 21
dodecahedron 117
domain 27, 194, 195

edge-valuation 203
edge-valued graph 109
edges 104
efficient activity 26
elementary form of the value 36
elementary technique 224
embedding 168
empty relation 17
endomorphism 168, 170
equation 182
equational class 183
equational system
- homogeneous 147
- inhomogeneous 147
equational theory 183
- Main Theorem 190
equationally definable class of algebras 183
equinumerous 44
equivalence 7
equivalence class 18
equivalence relation 18
equivalent equation systems 148
Erdös number 260
Euclid 74
Euclidean algorithm 21, 77
Eulerian graph 109
Eulerian walk 109
Euler's φ-function 82
Euler's Formula 115
exchange-value of a commodity 36
expanded form of the value 36
expansion by the ith row 142
expansion by the jth column 142
Extensionality axiom 13
extensivity 166, 172
extent of a concept 283

factor algebra 173
factor set 18
field 87, 162
finite deterministic automaton 226
finite graph 104
finite set 44
finitely based 131
finitely generated 216
finitely generated semimodule 217
fixed point 170, 184
forest 111
formal language 179

formula 7
four-color-problem 118
fully invariant congruence 185
functors 7
fundamental operation 161

Galois connection 183
Gaussian algorithm 150
Gaussian elimination 149
generating set 216
Goldbach hypotheses 73
graph 104
– connected 108
– directed 108
graph of a function 160
greatest common divisor 21
greatest element 22
groupoid 162
– abelian 162

Haken 118
Hamiltonian graph 110
Hasse-diagram 22
Heawood 118
hexahedron 117
Homomorphic Image Theorem 174
homomorphism 50, 168
– of a graph 104

icosahedron 117
idempotency 166, 172
idempotent law 162, 282
idempotent laws 14
identity 182
image 27
implication 7
incident 104
inclusion 16
incomparable elements 20
independence on the representatives 55
indifference relation 97
induced subgraph 105, 202
infimum 21
infimum-dense 284
infinitely based 131
infinitely dimensional vector space 132
Infinity axiom 13
injective 63
injective function 27
input vertices 203
input-output graph 202
input-output matrix 205
integer 3
integral domain 67
intent of a concept 283
intersection 13
intersection of algebras 165
inversion 29
invertibility 122, 161
invertibility of an operation 5
irreflexive 21
irreflexive order relation 87
isomorphism 51, 168
– of graphs 104
isomorphism problem
– for graphs 105

join operation 162

K-free algebra 186
Koch 118

L. Euler 82
lattice 162
– bounded 163
– complete 163
– distributive 162
leaf 112
least common multiple 21
least element 22
left zero-divisor 67
left-SR-semimodule 216
legal forms of companies 285
Leibniz 137
length of a walk 107
linear combination 216
linear congruence 84
linear equational system 147
linear group 146
linear independence 216
linear order 58
linear span 127
linear technique 214
linear term 196
linearly dependent 131
linearly independent system of vectors 131
linearly ordered set 20, 97
lower bound 21

Main theorem for homogeneous linear equation systems 153
Main theorem for inhomogeneous linear equation systems 154
Main theorem of equational theory 190
matrix addition 123
matrix multiplication 123
maximum 21
meet operation 162
minimal naturally ordered ring 70
minimal ordered field 87
minimal upper semifield 54
minimum 22
minor 141
monoid 161
monotone order relation 218
monotonicity 166, 172
- of the multiplication 59
multi-stage technique 228

natural number 3, 44
natural order 50
negation 7
non-desired input and output 26
non-negative rational number 3
non-satiation 92
number system 2

o-homomorphism 51
o-isomorphism 51
octahedron 117
open interval 22
operation
- finitary 160
- fundamental 161
- n-ary 160
- nullary 160
- term 182
operators H, S, P 187
order relation
- irreflexive 57, 70
original 27
output vertices 203

partial algebra 195
partial algebra of type τ 194
partial Menger algebra of rank n 198
partial operation 194
partial order 162
partial order relation 20
partially ordered set 20
path 107
- directed 108
Peano structure 46
permutable relations 175
permutation
- even 29
- odd 29
permutation of order n 28
permutations of order n 28
planar graph 115
Platonian graphs 117
Platonian solids 117
positional system 52
predicate logical equivalences 10
preimage 27
premise 7
prime number 72
prime remainder classes modulo m 81
prime twins 74
process vertices 203
product
- direct 174
- subdirect 177
production space 221
projection 175
projections 193
proof by induction 46
proper subset 16
proposition 7
proposition logical equivalent 8

quantifiers 10
quotient 53
quotient algebra 173
quotient semifield 54, 56

r-regular graph 106
rank 132, 150
rational number 3
real number 3
reflexive 18, 20
reflexivity 26
regrouping production 226
regular graph 106
relation
- trivial 171

relative subalgebra 195
relatively free algebra 186
remainder class 80
remainder class ring modulo *m* 80
restriction of a function 160
right zero-divisor 67
ring 162
- zero-divisor-free 67
root of a tree 180
root order 112
rooted tree 112

satisfaction 182
scalar multiplication 122
semantic tree 180
semifield 53, 87
semifield of non-negative rational numbers 57
semigroup 162
semilattice 163
semimodule 216
semiring 162, 216
Sieve of Eratosthenes 73
signum 29
simple algebra 171, 178
size-proportional technique 214
small Fermat's theorem 83
smooth production 226
socially necessary labour 36
span 216
statement 7
statement form 10
statement variable 6
strong identity 195, 198
strong partial Menger algebra 197
subalgebra 164
Subalgebra criterion 165
subdirect product 177
subdirectly irreducible algebra 177
subdivision 117
subfield 87
subgraph 105, 202
subset 16
subspace 127
Subspace Criterion 127
successor 45
successor function 45
supremum 21
supremum-dense 284
surjective 63
surjective function 27
symmetric 18
symmetry 26

tautology 7
technique 213
technique matrix 217
technique without output 226
term
- inductive definition 179
term algebra 181
term operation 182
ternary relation 17
tetrahedron 117
totally ordered set 97
transitive 18, 20
transitive hull operator 172
transitivity 26
transpose of a matrix 124
transposition 28
tree 111, 180
trivial relation 171
truth function 7
type 179
- of an algebra 161
- of terms 179

unary relation 17
union 13
union of graphs 109
unit 71
unit element 47
unitary Menger algebra of rank *n* 196
unitary partial Menger algebra of rank *n* 198
universe of an algebra 161
unknown 147
upper bound 21
use value of a commodity 36
utility function 92

variable 10, 179
variety 178, 188
vertices 104

walk 107
- closed 107
- directed 108
- Hamiltonian 110

– open 106
weak endomorphism 234
weak homomorphism 195, 234
weak identity 195, 198, 234
weak isomorphism 234
weak subalgebra 194

www.ingramcontent.com/pod-product-compliance
Lightning Source LLC
Chambersburg PA
CBHW061201240926
51846CB00006B/1078
* 9 7 8 3 1 1 2 2 3 4 4 2 6 *